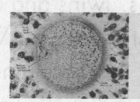

How
New Humans
Are Made

How New Humans Are Made

Cells and Embryos, Twins and Chimeras, Left and Right, Mind/Self\Soul, Sex, and Schizophrenia

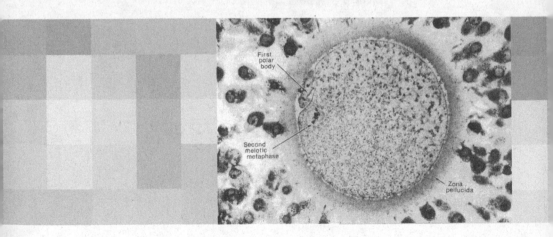

First polar body

Second meiotic metaphase

Zona pellucida

Charles E. Boklage

East Carolina University, USA

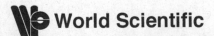

World Scientific

NEW JERSEY · LONDON · SINGAPORE · BEIJING · SHANGHAI · HONG KONG · TAIPEI · CHENNAI

Published by

World Scientific Publishing Co. Pte. Ltd.
5 Toh Tuck Link, Singapore 596224
USA office: 27 Warren Street, Suite 401-402, Hackensack, NJ 07601
UK office: 57 Shelton Street, Covent Garden, London WC2H 9HE

Library of Congress Cataloging-in-Publication Data
Boklage, Charles E.
   How new humans are made : cells and embryos, twins and chimeras, left and
right, mind/self/soul, sex, and schizophrenia / Charles E. Boklage.
      p. ; cm.
   Includes bibliographical references and index.
   ISBN-13: 978-981-283-513-0 (pbk. : alk. paper)
   ISBN-10: 981-283-513-X (pbk. : alk. paper)
 1. Embryology, Human. 2. Twins. 3. Schizophrenia--Genetic aspects. 4. Human genetics. I. Title.
   [DNLM: 1. Reproduction--physiology. 2. Functional Laterality. 3. Schizophrenia.
   WQ 205 B686h 2009]
   QM601.B64 2009
   612.6'4--dc22

                                                                    2009049302

British Library Cataloguing-in-Publication Data
A catalogue record for this book is available from the British Library.

Typeset by Stallion Press
Email: enquiries@stallionpress.com

Printed in Singapore.

For the twins and chimeras and their parents and sibs and offspring
who have asked me to help them know how they became how they are;
for all my teachers and my students and everyone who wanted to help;

to open eyes, to open minds, and may their kind increase.

Foreword

> "When even the brightest mind in our world has been
> trained up from childhood in a superstition of any kind,
> it will never be possible for that mind, in its maturity,
> to examine sincerely, dispassionately, and conscientiously
> any evidence or any circumstance which shall seem
> to cast a doubt upon the validity of that superstition.
> I doubt if I could do it myself."
>
> *Mark Twain*

There are few other examples in the history of Medicine that are such a blend of myths, superstition, and unfounded beliefs as is to be found in the history of twin studies. It might be hard to believe that for many centuries it was "accepted" that twins are a result of some divine punishment for sexual misconduct, either in the form of superfecundation (from different partners) or as a consequence of female movements during coitus at the time of conception.

Only well within the last two centuries has it become clear that twins have a unique and fascinating embryonic development, quite different from that of singletons. Still, there few other examples in Medicine where doctrinaire views turned into what is usually considered "common knowledge" among students and teachers of early embryonic development. Admittedly, our ability to look at these early stages of human life in singletons, as well as in twins, is extremely limited. However, our understanding of this blurred scientific void is becoming clearer owing to two relatively new areas of scholarship. One is examining human gametes and zygotes *in vitro* during the application of assisted reproductive technology, and the second is inferential genetic research. The latter is the background from which Prof. Charles E. Boklage developed into one of the most prominent scholars in twin research.

My ways with Charles crossed in the past owing to our mutual interest in the development of twins. Charles emerged as an expert on twinning during his long and remarkable scientific career until his present position at East Carolina University in Greenville, North Carolina. Prof. Boklage has his primary and secondary specialties in behavioral and developmental genetics and neurosciences and biostatistics, respectively, whereas his keen interest in research is directed to reproduction and embryonic development, where the process of twinning has naturally a central position.

The innovative ideas of Prof. Boklage are in every way *avant garde*, pushing the boundaries of what is considered the "common knowledge" in the presently "accepted" theories of the twinning process. As in decorative art, being avant-garde is the hallmark of open-mindedness. One might agree with Charles' opinions, one might not, but none can avoid his fresh ideas and remain impartial to his scientific perspective of the early stages of human development. This monograph entitled *How New Humans are Made* is further example of Charles' pure, unbiased, open-minded, *avant-garde* scientific endeavor.

Isaac Blickstein, MD
The Hadassah-Hebrew University School of Medicine
Jerusalem, Israel
Chairman of the Working Party on
Multiple Birth of the International Society of Twin Studies
and the Research Committee on Multiple Pregnancies
of the World Association of Perinatal Medicine

Acknowledgments

The author is grateful for support for various parts of this work from the National Institute for Child Health and Human Development, grant number R01-HD-22507, the National Human Genome Research Institute Center for Inherited Disease Research at Johns Hopkins University grant number N01-HG-65403, and the Children's Miracle Network.

There was something undifferentiated and yet complete,
which existed before heaven and earth.
Soundless and formless,
it depends on nothing and does not change.
It operates everywhere and is free from danger.
It may be considered the mother of the universe.
I do not know its name; I call it Tao.

(Tao-te Ching 25)

Tao is its own source, its own root.
Before heaven and earth existed
it was there, firm from ancient times.
It gave spirituality to the spirits and gods;
it gave birth to heaven and to earth.
(Chuang Tzu 6)

Contents

Introduction

A human Self, a human Person, is — first and last — a biological thing. A human person is a living system, a supersystem of many different tissue subsystems, which are themselves supersystems of many different cell systems, and the human self system is a subsystem of numerous levels of organization up to and beyond its species. You may imagine a human self or person to be also, or instead, some sort of psychological entity, or a philosophical entity, or a moral entity, or even (a container for) a sacred and immortal something or other.

None of those ways of seeing a human person can have any demonstrable reality before or after, or independent of in any other dimension, the biological, cellular entity. The business of being or becoming human can never be understood at any of those levels or from any other viewpoint without an understanding of the structures and functions of the cellular substrates of Life and of Mind.

Everything about human life is cellular in origin and in execution, in the same ways with all other kinds of life we have ever been able to know about. It has beginnings and middles and ends that are like those of all other kinds of life. That is sacred enough.

I became interested in the genetics of schizophrenia [a disorder, as far as we know, entirely peculiar to the human] when I was at or near its most usual age of onset and a close family member was given that diagnosis. The diagnosis was in error; it went away when they made him stay for a while where no amphetamines were available. But I did not learn about amphetamine psychosis until a good little while later. I did already know that some smart people believed schizophrenia to be genetic, and that my ways of thinking had never been exactly entirely "normal" (as I erroneously supposed that those of most other people were), nor were those of most of my family — so I wanted to know more about this thing I take some comfort in supposing that I have managed to avoid a lot of trouble in my life by knowing things. That could perhaps be an illusion, or a delusion, but it has worked for me; it is my story, and I am sticking to it.

The twin-study approach was very popular in human genetics at that time, as it has been widely considered a very powerful approach to learning the genetic bases of traits like schizophrenia. Schizophrenia definitely "runs in families," but it does not follow any known simple pattern of genetic transmission. And that is exactly the sort of thing for which the genetic twin study has been considered most useful. More than 40 years later, we still do not know exactly how that works (neither for schizophrenia nor for any of the other disorders that are distributed over the human population more or less the same way schizophrenia is).

About 1973, while I was brushing up to lead a discussion-group kind of seminar for the psychology department at Kansas State University, studying a book about Irv Gottesman's and Jerry Shields's then-recently-published still-definitive genetic twin study of schizophrenia of all time, my wife looked over my shoulder at the case history I was reading and asked: "Why do they tell you that guy is lefthanded?" I didn't know. I went to the index and looked up everything the authors had to say about handedness. Most of it was in one paragraph not quite three inches long, but that paragraph spread out before me an enthralling puzzle that I have worked on until it now seems to have very few if any pieces missing. I am now reasonably certain that I know what the picture is and can defend my belief.

Briefly, at the primitive level of my understanding at that time, I found an important relationship between twinning and the asymmetry of brain function development, about the asymmetries of brain function in handedness and in schizophrenia. I set out to move myself closer to understanding all of that and preferably to bring the world along with me if I could cause it to work out that way. The question soon enough became "Where do left and right come from in human development?" and I thought I saw a way to approach the question, using twins in a completely different way from any way they had ever been studied before.

The story at the time was that twins have an excess of lefthandedness, and that the excess is concentrated among the "~~identical~~" monozygotic twins because each monozygotic pair must arise from the "splitting" of a single embryo. Somehow a group of cells that "should" make a single embryo must be separated into two subsets that will go on to build two embryos. This was expected to disturb the asymmetries of forming their embryos to the extent that the split interrupted the building of asymmetries of structure and function that were already underway.

> When Mangold and Spemann tied newt embryos in half, back in the 1920s, if they did it just so, they sometimes caused a single newt embryo to develop into twins, and sometimes one of the twins grew its internal organs left-right reversed from the normal configuration. Many people decided that splitting, like that, was a good way to explain the cellular origins of monozygotic twinning and the twin excess of lefthandedness all at once, and it quickly became the standard understanding.

I am not certain that I have ever understood "nature abhors a vacuum" except in the light of how uneasy most humans become at the prospect of having to admit to a complete absence of understanding, and how avidly they will grasp at any prospect of relieving that discomfort.

"~~Fraternal~~" dizygotic twins, on the other hand, have been understood (according to the common knowledge, as if it were a matter of fact without need of any further consideration) to have arisen from "double ovulation" — the more-or-less simultaneous release of two egg cells from two separate, independent, parallel ovulations, from two unrelated follicles that need not even be in the same ovary or the same ovulatory cycle. The symmetries and asymmetries of their embryonic development are therefore not disturbed, have no reason ever to be disturbed, and therefore dizygotic twins can be expected to be exactly representative of development among the embryos and fetuses of the single-born remainder of the population.

So, my idea was about as straightforward as anything ever is in the study of human development: By studying the differences between monozygotic twins with their odd embryogenesis, and all the normal, everyday singletons like the rest of us, I should be able to learn something about how the "normal" embryogenic processes work by studying how the people who turn out differently get that way. The "rest of us" could be represented perfectly by dizygotic twins, serving as experimental controls for any possible difference that might be due to some aspect/s of simply physically being twins…, such as sharing the space and resources of a womb.

As it turned out, as things went forward, my results, time after time, every time, from the very beginning, from every angle from which I could find a way to question it, told me that the old stories were all wrong. I also began to learn how severely limited are the powers of fact and truth on the one side, against false but comfortable common knowledge on the other.

Arguably, the first "big" thing I did was to collect handedness data from a large sample of twins and their families. With the very substantial help of our National Organization of Mothers of Twins Clubs, I gathered 773 usable three-generation twin families, with handedness data on over 10,000 people. [I got over 800 responses from 500 questionnaires I sent to their annual meeting that year — I have never seen or heard of any 160% questionnaire efficiency before or since.] It *is* true that twins are more often lefthanded or ambidextrous than are general population singletons. It is *not* true that it has anything to do with the circumstances of twin pregnancy. It has nothing to do with which twin spends more time "on top" in the uterus, or with anything we understand about how a pregnancy turns out to be twins. It is *not* true that monozygotic twins account for all or even most of the difference. Monozygotic twins do not in fact have any higher frequency of lefthandedness than the dizygotic twins have.

Nothing about twinning itself causes nonrighthandedness, and I know this because the single born siblings of the twins in these families also include more lefthanders than the singleton population, statistically no different from the frequency among the twins.

The parents of the twins have a higher frequency of lefthandedness than their own siblings (the aunts and uncles of the twins) who are not themselves parents of twins.

So … twins do not get their excess lefthandedness from anything about the biology of becoming twins or anything about just physically *be*ing twins. They inherit their excess lefthandedness from their parents, and so do their singleborn siblings. Each lefthanded parent (mother or father equally) increases the chance of lefthandedness among their children by a factor of about 1.5 (children of one lefthanded parent and one righthander are about half-again as likely to be lefthanded as the children of two righthanders). Families with both mother and father lefthanded have about 2.25 (=1.5 × 1.5) times as many lefthanded children as two righthanded parents have.

There are also certain malformations that are excessively frequent among twins, again attributed to the monozygotics for the same reason, because the malformations in question are symmetry malformations, "midline" or "fusion" malformations, presumably disturbed by the "splitting." They are anomalies of structures formed in the process of building the embryo by the fusion, in the midline, of not-exactly symmetrical left and right half-structures. In anencephaly and spina bifida, the neural tube does not close properly, leaving the spine or the skull open and the nervous tissue inside poorly formed, if at all. In the orofacial cleft anomalies, the parts of the face are not fully and properly grown together into nose and lips and such. In the congenital heart defects, the cardiac tubes that will form the heart after fusion in the midline and remodeling either skip part/s of the process or do it wrong.

All of these most common major malformations are also in excess among the parents, siblings, and/or offspring of twins, and there is no difference between the families of monozygotic twins and the families of dizygotic twins. All of these malformations are also associated with an excess of lefthandedness in their non-twin victims, and among the first-degree relatives thereof.

I did a good bit of work with some dental diameters data I was given by Professor Rosario Potter, comparing monozygotic and dizygotic twins, and later singletons — more fascinating results. Examining those data with multivariate statistical procedures, the programs can in fact tell the monozygotic and dizygotic twins apart (the individuals, not the pairs; nothing to do with pairwise relationships), but they are much more alike than either is like the singletons. Both groups are far more different from singletons than they are from each other.

The greatest multidimensional distances are between singletons and twins (of both zygosities, equally again). Classification is about as perfect as biology ever is. Not one person identified as a twin in that sample was ever misclassified as a singleton. A very small number of singletons were classified as twins in some of the results. By the time we get to the end of this story, that fact will make all the sense in the world because most people who grow from twin embryos to live birth are born single. [Explaining how I know that deserves a whole chapter, later.]

Those multivariate discriminant function analyses can distinguish the size measurements of teeth in left sides of the mouth from those on the right sides, among the singletons, but not among the twins of either "kind." The twins — both zygosity groups equally — are extraordinarily symmetrical in the sizes and shapes of their teeth — missing the small normal asymmetries of dental development.

Those statistical programs can also tell the sets of tooth size measurements apart by sex with very high accuracy, as long as the subjects are singletons or members of same-sex pairs. Boy-girl twins cannot be sorted by sex with these measurements of craniofacial structures. As it turns out, a number of other variables can be lined up with these. The members of girl-boy twin pairs are not quite exactly like any other people of either sex, twin or single. That, too, deserves and gets a chapter of its own.

The old common knowledge is false.

The observations I have collected are loudly incompatible with the idea that monozygotic and dizygotic twinning are very different processes. Every anomaly that is present in excess among twins has been attributed to the monozygotics because of their "splitting," and dizygotic twins have been presumed to be innocent of all such variations. That is absolutely wrong. Everything that is odd in any way about the embryogenesis of monozygotic twins is more or less exactly as odd among dizygotic twins, except when it is more so among the dizygotics.

In every situation where there is a difference between monozygotic twins and dizygotic twins as groups that is too large to be due to sampling error, the dizygotics are the ones who have it worse.

Dizygotic twins cannot be arising from any kind of embryogenesis except exactly the same system of processes that result in monozygotic twins.

How can they do that, starting with two independent egg cells?!

They cannot.

So, how do they do it if they don't do it the way "everybody knows" they do it?

This is the story of how a classically-trained, philosophy- and theology- and psychology-literate behavioral and developmental geneticist has gone about seeking answers to such questions, set in a frame of what it means to become a human in the first place.

If there is a science a person needs,
then it is the science I teach, that is to show a person his place in the world
and what you should be in order to be a person.

Immanuel Kant

There is no more complex question anywhere than the ones about how humanness in general, or any particular individual human, happens. I cannot make it quite simple and still tell you the truth as we best understand it, but I think I can make it much clearer for you than it probably is now.

… the history of man for the nine months preceding his birth
would probably be far more interesting
and contain events of greater moment,
than all the three score and ten years that follow …

S. T. Coleridge

In almost 200 years since Mr. Coleridge left that thought behind in passing, we have actually made considerable progress on many fronts toward learning much of what it seems he would have liked to see. It may be that the only part he had wrong was the word "probably." Except perhaps for some variation in personal points of view among those considering the question, I believe we have advanced the question well beyond the probability that he conjectured … .

A true scientist is bored by knowledge;
it is the assault on ignorance that motivates him.

Matt Ridley

Any sufficiently advanced technology
is indistinguishable from magic.

Arthur C. Clarke

Yes, so be it, and all living systems are sufficiently complex that they are indistinguishable from miracles. However, to call anything magic or miracle without even trying to understand it is at best no fun, and arguably, from at least any scientific sort of viewpoint, irresponsible.

I have thought for years that it would be wonderful to start some sort of written thing with the sentence, "Each of us was once a single cell," or to use that as a title for a writing effort about human developmental genetics. Of course, the single cell I had in mind was the zygote — the single cell that is often simply, but not quite correctly, called "the fertilized egg cell." That is a fascinating thought, very evocative, and I thought for a while that it would make a great title for a book more or less like this one. Further reflection brought me very near, for a different while, to abandoning that idea no matter how lovely it sounds. You see, that lovely-sounding idea is one of those lovable, comfortable, simple ideas that are just a little too simple to be true. It is, however, a great lead line and a wonderful little lie because it is the theme on which the truth is a fascinating variation. I have decided to use it after all and clean up the part that is not true later. The difference, the reason why and the extent to which it is not quite true, is a major component of why this particular story needs to be told. As I hope and expect to make clear in these pages, an astonishing fraction of us never were entirely contained within a single cell, even in "potential" form or substance. There will be more about that when we talk in greater depth about chimeras, and about "potential".

No question has ever arisen in any human mind that is more complex or compelling than the questions many of us have about who and what we humans are and how we came to be as we are. Those of you who are satisfied with answers to those questions that have been delivered as articles of faith may want to put this down and walk away, now — to just step around the possibility that this will make all of that harder for you to hang on to. If you would be lost without those comforts, it is just as well for all concerned that you stay where you are and do not try to follow along. Much of this material does have strong prospects of increasing your comfort with rational versions of answers like those, but only with the rational versions.

Along one of its several paths, this is all about efforts on the part of human minds to understand human minds as a part of the business of becoming and being human — as individuals and as species. I hope to give you a comfortable understanding that being human has no need of meaning outside of the place it holds in that grand continuum of being that is occupied by all living things, and that all we need of miracle and wonder and transcendence and ineffability is comfortably contained in thoughts just such as those.

Life on Earth may well have begun in any number of times and places, but every scrap of the huge body of evidence that we have been able to discover so far tells us that only one of those beginnings succeeded and endures today. The life sciences of the last century or so have gathered and made available an overwhelming mass of sound evidence that *all* of *Life as We Know It* has come from a single source/"event"/process, has not finished unfolding, and will not finish unfolding in any foreseeable part of the future.

Human life is a part of all that, in no demonstrable way above it or beyond it, and cannot in any way be understood as separate from it. Human life cannot be understood except in the perspective of the place human life holds among all living systems.

I have had occasion to see and hear philosophers (real live professional living-and-working-for-a-living-today philosophers) acting as if they are certain that questions such as those belong exclusively to their domain. However, I can assure you that any answer to those questions which is not biological in substance is empty nonsense and a waste of time. Everything you are, everything you do, everything that happens between your scalp and your toenails, between your ears, and between your legs ... is done by cells, and I can prove it.

There is a glossary at the back of the book to give you definitions for words the meanings of which might not be right away comfortably clear to you. There certainly are other definitions elsewhere for some of those words, but the definitions in this book's glossary will be the ones about what the word means in our present context. There is also an index that should help you look ahead to find more quickly any of the ideas in here that I don't get around to writing about close enough to the front of the book to suit you.

At this time in human cultural evolution, when the Internet puts an enormously generous sample of all the information in the world at our fingertips, all of the background material in here is within reach of the kind of Web search that anyone who can read this can do. We have available the general-purpose search engines like Yahoo® and Google®, and "metasearch" engines like Dogpile®

and WebCrawler*. Then there is PubMed, a very usable route to the primary biomedical literatures of the world, provided by the National Library of Medicine at the US National Institutes of Health. All of them have built-in ways to help you learn how to make them work. The most important variable component of your effectiveness and efficiency as a Web-searcher is the business of guessing the best search keywords for getting what you want. As far as I know, that can only be improved with practice. It will evolve, you see, under the selective pressure of your curiosity and ideas. If the first bunch of answers you get contains anything at all useful, find one or more words that the useful returns have in common, and use those for a new search.

Some profoundly fundamental concepts will show up at issue here, where most of what people have believed about fundamental events or components of human prenatal development has been wrong, some of it for well over a century so far. In all such cases, I hope to explain a plausible alternative and to offer ways to understand whatever might have been plausible about the substance and the origins of those old errors. It is often of serious interest to consider how so many otherwise apparently reasonable and intelligent people could have gotten it so thoroughly wrong for so long.

The story of human prenatal development is nothing short of fascinating and full to bursting with wonders. I am convinced, and I hope that I will leave you convinced, that the cellular truths are even much more wonderful than the simpler superstitious imaginings of the past which I hope and expect to replace. When you know something about how, and how many, and how often, things can and do go wrong in the unfolding of what might eventually become a new human individual from the contents of a single cell or two, the wonder grows still more.

If you will think through that last thought again without restricting your thinking to the "human" situation, you will find that the wonder is no less. When you think of the human organism as a bridge to be designed and built — which it is, for the "purpose" of continuing the species by carrying a particular, unique version of the human genome from one generation to the next — then all concerns you might have about the intelligence behind its design should find their focus dramatically shifted. There is much more and greater intelligence in the Universe than any imagined counterpart or version or extension of Mind in the image of Man.

The development of any complex organism is without room for doubt not an event, but a process — a proceeding, a procession, of events most of which cannot be explained in detail within the confines of what we know of natural law. That, by the way, is as good a definition of "miracle" as there is to be found. It is my duty, my passion and my mission as a scientist to demystify mysteries and to run miracles to ground, but I am under no illusion of expecting ever to take definitive care of all of them.

Chapter One

Orientation

From wonder into wonder existence opens.

Lao Tzu

There are only two mistakes one can make along the road to truth;
not going all the way, and not starting.

Buddha

Give me a little of your time and I will try to stretch your head to make room for one more story, a story I hope and believe you will agree can reasonably be considered very important. This is a real live life-or-death tale of mystery and adventure and myth and of a seeking of the truth. It is also ALL about sex, start to finish, every page!... well, about reproduction, anyway. It is a seven-billion-times-over true story of chance and necessity and rare success against very large odds. In all of this Universe-as-we-know-it, i.e., within the reach of human intellectual efforts to date, there is no bigger story than the one about how each human person comes to be as s/he is.

This is a story of *twins* and *chimeras* and actual monsters, and how small and subtle, and yet so terribly important, the differences may really be between them and "normal" single babies. Some people, before they understand, think of chimeras as two people in one, and of twin pairs as the closest possible thing to being two copies of one person. In both directions the answer is "no such thing!" Even with *monozygotic* twins, there is no question that they are not "identical" but are indeed two, different people. There is no question that they have separate lives to live. They almost always have full, separate sets of everything they need for doing that. Even when that is not the case, when they share some body parts, the reality of two *Selfs*, two "souls," two *different persons* may be inescapable.

The working definition of individual humanity seems to be centered on the face and head. If there are two faces on separate heads, we will give them different names and regard them as two different people, instead of one person with extra parts... as long as one of the extra parts contains a face and a brain. Two faces on one head is called the Janiceps malformation (after Janus, the Roman god of gates and doors, and beginnings and endings, who always faces both ways). That structure appears to be unanimously not compatible with life. A newborn thus afflicted seldom breathes long enough to get a name for either face. Undoubtedly human by virtue of coming from human parents, undoubtedly alive for a while, but not ever going to be considered for membership in the human + life = person club. It takes more than being human and alive to qualify for personhood.

It will be worth your time to learn what it really means when I tell you "there is no such thing as 'identical' twins." I will also be telling you about how there is indeed such a thing as **a** person, *one* person, with two genetically different kinds of cells that "*might* have become" two people (but did not), and that such people are not even rare, let alone incredibly rare. There are in fact rather a great many of these people. They probably account for at least (deliberately and specifically meaning "no less than and probably more than") 10% of us, and it really does not matter how few of you may be prepared to believe that... you will see. You will have to get over your reluctance to accept it, and I will do what I can to help. What we see is seldom all there is, and is often only a tiny fraction. We do have ways to "know" (infer quite soundly) what we are missing.

This is a story about how hugely important some differences as tiny as small molecules really are, a story of all the things that have to be done well in order to unpack and unfold the contents of one cell to build a body that can support a life that may, or may not, turn out to be fully and proudly, or even just barely recognizably, human.

There are a good many things we understand fairly well about human life and about some of the parts of the human organism and how they must be assembled and arranged to form a cellular structure that might come to embody a human life. When we look at how all of that unfolds from the contents of a single cell, we can look back and forth between those early steps and the full-grown body, and each view shows us things we should expect to find in the other. All of it from one end to the other is done by cells, and cells have only the physics and chemistry of atoms and molecules to work with.

The only kind of thinking that ever allows me to make any progress in this neighborhood is mechanical — about what appears to need to be done, and how cells might go about doing it with the resources at their disposal. It is all about how the parts and contents of cells do what they need to do, that makes it possible for cells to do what they need to do, which in turn makes it possible for bodies and minds to do what they need to do. ["need"? Yes. According to my favorite definition (from Jacques Monod, in *Chance and Necessity*), every living thing, every instance of Life, has two properties which together distinguish it from that which is not alive: a Need (to maintain and reproduce an emergent highly ordered structure against a gradient of entropy) and the Means (whereby to meet that Need).]

This story includes some threads about some of the many things that can go wrong and thereby generate a situation such that the normal developmental processes cannot proceed to a satisfactory conclusion.

I will give you a tour of the landscape of what various people have offered in the way of answers to questions about how human lives begin, and I will tell you about how very many of those "human" "lives," in fact by far the majority of them, end before they can plausibly be said to have begun.

This is a story of living cells and of things that are made of living cells. It is a story of things that living cells do and that things that are made of living cells do. These include you and me and all the things we can do, the way we look, the way we act and the way we think and feel.

None of that would be the way it is except for specifically differentiated cells being where they are, and doing what they came to where they are to do. All of that, in turn, came from cells that were there before, which came from cells that were there before them, and, so... on, going farther and farther back to a point beyond our reach that none of us can really fathom.

Some may consider it wrong to say anything that might imply any inkling of purpose on the part of a single cell, as in "...doing what they came to where they are to do" in the previous paragraph. I dare say you knew what I meant when you read that, and it will be worthwhile at some point to spend some time considering what it is about our definitions of behavior that even caused that little uncertain pause there. Besides, there are numerous excellent examples of single cells, which are to all appearances much less complex than the cells in your body and mine, showing unarguable sentience, single celled organisms that are able to respond effectively to changes in their circumstances by behaving in clearly directed ways that fit any reasonable definition of purposive behavior. If you want to continue believing that only humans do any such thing, Get Down! off that high horse so I can quit worrying about you falling and making an even bigger fool of yourself! Being human is a very special thing without having to tell ourselves outrageous obvious lies about things like that.

This will be about issues of the nature of life, and of the nature of mind, and of systems, and complexity, and emergence, all of which are basic considerations that will arise to be dealt with in this story about unpacking, unfolding, and assembling the cellular substrate of the human version of each. It will be "about" those concepts primarily at "practical" levels of specific, concrete, cellular examples.

Living cells are wonderful things, and they can do wonderful things, but they are finite. Everything any cell does comes from what it has to work with, and all it has to work with depends on what molecules it "knows how" to gather or to build, in order to equip and maintain itself and to perform its necessary functions and its conditional functions, and to change its functions appropriately according to changes in circumstances.

Most cells can, and must, change what they do at various times, according to changes they can sense, mechanically (as in touch or kinesthetic feedback) or chemically (as in taste or smell) in their surroundings. Their surroundings are, of course, composed primarily of the products of other genes in the same cell, and of the products of genes in other cells.

To get this story told all the way we will have to involve almost all of everything anybody has ever known (about anything, actually, but especially about humans and the biology thereof). We will have to do some wondering about how we could go about knowing whatever it is that we think we know, and, especially, how we go about knowing *that* we know it. Unless we can keep all of that inside the boundaries of common sense, and build it mostly from simple words, it could turn out to be a huge waste of your time and mine.

The business of becoming alive, of becoming human and being alive in a human way, is something most of us never even consider, never even think about. We just do it. For most people and most purposes, there is of course a lot to be said for that. Some of us, however, will think about it, even if most people we know think we are wasting our time.

One layer of this story is about a career spent studying those questions, about answers I have gotten from watching living things living (sometimes tweaking them this way or that and doing some more watching to see if my tweaks changed anything about what they do or the ways that they do it — but more often from what we call "experiments in nature" — looking at living things at large, in large populational groups, tweaking themselves or each other in certain ways), and from other students with the same questions, and about how I see it after all that.

I will be careful about what I say and how I say it because I do not want to mislead either of us by using words or statements that sound like we know what we are talking about when we really do not. I will not, however, waste any significant amount of time and energy beating around all the politically correct bushes that might pop up along the way. I will especially do my best to avoid getting tangled up in things we know, that some people want very badly to think we do not know, because the answer is not what they have been told to believe it should be.

> What makes you think human beings are sentient and aware? There's no evidence for it. Human beings never think for themselves; they find it too uncomfortable. For the most part, members of our species simply repeat what they are told and become upset if they are exposed to any different view. The characteristic human trait is not awareness but conformity, and the characteristic result is religious warfare. Other animals fight for territory or food; but, uniquely in the animal kingdom, human beings fight for their "beliefs." The reason is that beliefs guide behavior, which has evolutionary importance among human beings. But at a time when our behavior may well lead us to extinction, I see no reason to assume we have any awareness at all. We are stubborn, self-destructive conformists. Any other view of our species is just a self-congratulatory delusion.
>
> *Michael Crichton*

Many people seem to believe that they are not supposed to understand the making of new human Selfs, that human reproduction, the making of new humans, is so sacred and mysterious a thing that they never can understand how it happens and probably really should not try. It may be the most complex example we know about, but it is really not any harder to know about and understand than any other part of appreciating Life and living things, when we can start out with confidence and curiosity and take it one step at a time...

Oh! and by the way, go ahead and plan to tolerate some ambiguity, at least from time to time, because we cannot expect to get all of it right and have it all make sense, on the first try...

The human organism is built from cells, like every other kind of living thing we think we understand at all, and those cells are built from recycled, previously-owned molecules, which are made from other previously-owned molecules... according to a plan, and with tools, that are largely the same as those of all other organisms.

With only a few exceptions — and even those are only partial, if and when they have any concrete reality at all — every animal in the world grows into a bilaterally a/symmetric body structure much like the layout of your body and mine. We are still a long way from knowing everything about that, but we are continuing to make very gratifying progress, and every scrap of what we know will be important before we finish here.

I hope and believe that most of you should have no great difficulty reading this and understanding it. Even if and when I get it all into the simplest reasonable choice of words, not all of the ideas are simple, but there is no point in even waking you up if I would end up leaving you stuck with a bunch of old simple stuff that is simply wrong. I will tell you the straight story to the best of my ability, and I will not talk to you as if I think you could not understand the straight story.

There are some parts that may drag you out of your comfort zone, away from the just-so stories that you have been told about various parts of human life and the origins thereof and the living thereof and the scientific study thereof. I will move you out of your comfort zone as frequently and as far I can, on purpose, for a variety of reasons, but for one reason most especially. That is because, in the end, the (factual, objective, empirical, scientific) truth, as near as we can get to knowing it, is always the least distorted window through which to look at anything.

This is meant to be a work of science, and will therefore be concentrated, to the best of my abilities, on products of our senses and our use of reason — on factual and repeatable observations. Scattered throughout you will also find a number of opportunities to learn about how there may be nothing in this world of which we can be certain, if we can set aside all notions of certainty that are not based on factual observation and held together with reason. Certainty is not, you see, not ever, a product of science, if only because anybody who is paying any attention at all will always know that there is more to learn.

Science does not deal in certainties, and sources which do claim to have certainties to offer never seem able to provide even a credible background for the ideas on which they base their claims, let alone sound and reproducible evidence for them.

> Mediocre spirits demand of science the kind of certainty
> which it cannot give, a sort of religious satisfaction.
> Only the real, rare, true scientific minds can endure doubt,
> which is attached to all our knowledge.
>
> *Sigmund Freud*

> Our knowledge of nature remains provisional because
> we can never know if we have final Truth ...
> What separates science from all other human activity
> is its belief in the provisional nature of all conclusions.
> In science knowledge is fluid and certainty fleeting.
> That is the heart of its limitations, and also its greatest strength.
>
> *Michael Shermer*

> The test of a first-rate intelligence is the ability
> to hold two opposed ideas in the mind at the same time
> and still retain the ability to function.
> One should, for example, be able to see that things are hopeless
> and yet be determined to make them otherwise.
>
> *F. Scott Fitzgerald*

> When one admits that nothing is certain one must, I think, also admit that
> some things are much more nearly certain than others.
>
> *Bertrand Russell*

There are problems here that are nowhere near as important as they seem to some people, primarily dealing with things that people believe they are obligated to believe, independent of, or at the expense of, reason. Too many people fail to appreciate the enormous difference between **believing** something and **believing** *in* something. A great many people sincerely *believe in* things they have been told they MUST *believe in* as an act of faith, but which they could never *believe* as an act of reason.

> Convictions are more dangerous enemies of truth than lies.
>
> *Friedrich Nietzsche*
>
> Reality is that which, when you stop believing in it, doesn't go away.
>
> *Philip K. Dick*

It can be astonishing and appalling the loads we often expect mere words to bear — especially when we must deal with words from one language that other perfectly useful languages do not even have. Augment the complexity of that situation with multiple streams of tradition and multiple translations, filtered through points of view with no counterpart in the reality of when and where and from whom the words first came, and things may become really difficult.

Much of science depends on certain kinds of faith, primarily in the form of assumptions we cannot prove but without making which we cannot proceed. A major component of this story concerns the enormous influence of a particular set of superficially plausible assumptions. That set of assumptions and the results of their applications, over the past century or so, have been cemented together into a system of false beliefs that have hidden the underlying truth even from productive investigation, to say nothing of understanding.

My very best reason for writing this is to share with you the marvels of a body of knowledge that most people seem to consider beyond their reach, if they have thought about it at all. If you have tried at all to learn about the business of becoming human, especially with regard to the building of new humans, you have almost certainly been deceived. This is not so much because anyone has set out to deceive you as because they have passed on to you just the same unquestioned answers with which and about which they have for so long been deceiving themselves and each other. A very large fraction of what we are here to deal with lies somewhere between falsely simplified and simply false.

Scientists are human too, and in general we just do not like to say "I don't know" any more than other mere mortals (quite probably even less), even though Ignorance is the Mother of Science. Ignorance, you see, is the reason we scientists and teachers have jobs, trying to diminish its range and influence, trying to replace it with understanding.

I love magic and miracles and mysteries as much as anybody else does; great fun. The thing I least like about them is that they are so often invoked as excuses for failure to accept a responsibility to challenge ignorance and superstition. The thing I like best about magic and mysteries and miracles is the prospect of finding a credible possibility for how the reality underneath actually works and then finding ways to prove and explain that. The business of making new humans fits right in there. It is such an outrageously complex process, at one and the same time both logically outrageously unlikely and probabilistically quite inevitable, that magic, mystery and miracle are much more easily in reach than human efforts to understand it in detail. Many people just will not entertain the prospect of doing the work, and instead readily take someone else's word for the "answer."

In this, as in any other human endeavor, the easy way of doing things is very often not the right way and hardly ever the best way. The ignorance and superstition surrounding human reproduction probably exceeds that associated with any other biological process, or any other piece of the business of being alive or becoming human.

The defining characteristic of science —
the one that gives science its extraordinary explanatory power —
is the objective use of evidence
to distinguish between alternative guesses.

Paul Ewald

Science is a wonderful thing if one does not have to earn one's living at it.

Albert Einstein

It has been said that the primary function of schools is
to impart enough facts to make children stop asking questions.
Some, with whom the schools do not succeed, become scientists.

Knut Schmidt-Nielsen

I have been learning and teaching human genetics, concentrating on development and behavior, in the company of graduate students and postdoctoral fellows and medical students and college students, and sometimes even elementary and junior high school and high school classes too, for over two thirds of my life. The thing I have come to know the most about is how very much I still don't know and quite possibly never will. For better or for worse, however, near as I can tell from what I can find to read, nobody has done any more thinking than I have about this particular story taken as a whole. If there is such a person out there, why has she or he not written about it for the rest of us to understand? I must acknowledge a tendency to do a great many things the hard way, but, if this story had already been worked out and this book or one very much like it had already been written somewhere, I could have spent all these years doing some other things — the hard way if it came to that.

To understand human developmental genetics well requires a better-than-basic understanding of all aspects of human biology and behavior, and there is in turn no way to have that without understanding genetics and development. It is mutual and reciprocal and circular and self-referential, and often paradoxical, and I believe that amounts to what Hofstadter has called a "strange loop." This may well be as it is because all of that is very intimately about life and the innermost workings thereof, and life itself is probably the prototype of strange loops, most certainly with respect to passing it from one generation to another — which is exactly our focus here.

All of my favorite teachers have been strong on perspective. Putting things in terms of relationships among the parts of knowledge makes learning work better, and makes all of the parts more meaningful when each can be seen in context of all the others. In teaching human genetics, I have always tried to put the bits and pieces in the perspective of "the business of becoming human" — as individuals and as species. The business of becoming human, as individuals and as species, with diversity and development as fundamental themes, is probably the most complex thing you and my university students and I will ever try to understand.

The Universe — with the upper case U, considered as a whole to the extent that any such kind of thinking is actually possible — clearly must be acknowledged as being more complex than the human species, simply because it includes the human species as well as huge numbers of parts other than those involved with the definition of and membership in the human species.

The Universe is in fact statistically certain beyond a reasonable doubt to contain forms of life more complex than the human organism. However unlikely any such outcome or component thereof, however unlikely any step toward such outcome, the functionally infinite number of trial-and-error opportunities provided by the size, complexity and duration of the Universe make it virtually certain that each unlikely branch of the outcome has in fact occurred at least once and quite possibly on an infinite number of occasions. However, within the current reach of the efforts of earthbound human minds to understand anything, we can be reasonably satisfied that we will not encounter anything more complex than the human organism itself, at least before we can get finished with this version of this story.

If anyone really wants to consider the human organism as a product of supernatural conscious design, it is easy enough to uncover any number of ways in which it clearly might have been designed and executed more functionally and/or more simply — especially if the capabilities and resources of the imagined designer are supposed to be infinite.

> We are a bit of stellar matter gone wrong. We are physical machinery — puppets that strut and talk and laugh and die as the hand of time pulls the strings beneath. But there is one elementary inescapable answer. *We are that which asks the question.*
>
> Arthur Eddington

A major turning point in this story, that might even be fairly called the beginning of this adventure, came from a part of the graduation honors protocol at my undergraduate college (now known as Bellarmine University, in Louisville, Kentucky). Most colleges these days give graduation honors based on grade point average only. If you get all A's or very nearly so, you may graduate summa cum laude (with highest honor). Mix in enough B's so that your average can confidently be considered measurably less than perfect, and you have come down to magna cum laude (with great honor), etc. How absolutely! boring! especially now that all of our children are above average and all of the students are so much smarter than we were back then, that B's are now average grades in many schools. When I graduated from Bellarmine College, so very long ago, we did not base graduation honors on grades. We had a series of essay and oral exams. As I understood it, the object was to assess the extent to which we had integrated all they had given us the opportunities and tools to learn, many very important parts of which might not be readily palpated by exam scores and course grades.

The first step in our graduation honors examination process consisted of two essay questions. I have long since forgotten almost all of the rest of the examination process and the Dean's question for the whole graduating class on that first step, but I will never forget the first question from the Biology Department faculty for their own seniors. It was, in more or less so many words:

> "How may we know that our senses tell us the truth?"
> [This question and all efforts to answer it form the core of epistemology
> (the study of knowledge and knowing): "What does it mean 'to know'?
> What is required? How do I know? How is knowing to be done? How do
> I know what I know? How do I know that I know what I know?"]

The question of what or how we know is more or less obviously a vital launch point for most excursions into philosophy. On that particular occasion, our College's Philosophy Department faculty joined hands, jumped up and down and collectively soiled themselves: "No *biologist* can be expected to answer that question!" Our department chairman calmly answered "*Only* a biologist can be expected to answer that question." Of course, I thought that was just incredibly cool! Doc Sames was even smarter than I already knew he was, especially as I was destined to spend the rest of my life so far trying to get closer to understanding what he thought that question has to mean.

I got my consummate honors beginning with an argument from natural selection to answer that question: That any species whose members could not reliably distinguish sustenance from sudden death or sex from slaughter would not long have survived. I still find myself considering occasional prospects for improving upon that answer, but in general I believe it will suffice. The part of this that has to do with building human brains and with the functioning of human minds is descended from that experience.

I went on to spend most of my time in graduate school learning about how genes regulate themselves and each other to guide cell differentiation and development. Most of what I learned in that time came from bacteria and bacterial viruses as model systems, with the prospect of understanding more complex living systems set as a goal for "some (other) day." Meanwhile, I annexed human behavioral and developmental genetics as a back-burner sort of study hobby. It may not be immediately obvious to all of you (perhaps not for any of you ...), but there are certain profound and pervasive principles in common between the building and filling of the "head" of a bacterial virus particle and the head of a human. More about that later, maybe, but until then ... it's about how neither kind of head can be built or can function properly once built without certain molecules getting certain other molecules to where they need to be to do what they need to do next...

From within that kind of thinking, I began to discover how fundamentally important it is to the structure and function of human minds that the left and right sides of human brains have different structures and functions. While looking for the basis for a grown-up research career of my own, I thought perhaps that the left-brain, right-brain business might be a respectable chunk that I could bite off and have some hope of productively chewing. That is what I thought, and for reasons that I can hope will become clear it did seem like a good idea at the time...

Where in development does left and right come from, and how does it get to where it can make so much difference in human brain function? How hard can that be ...?

I could not find anyone who knew anything much about that, and I did like the-road-less-traveled aspect of the situation better than the prospect of joining in among or behind a group of people who were all working on various aspects of some one same thing. So here we are. I have a great deal left to learn, even after I have been at it for over 40 years. For reasons that I hope will be clear before we finish here, I think the story — although somewhat short of having all the details screwed down — is now about ripe for telling and there are some things that might be seen as evidence that the hour grows late, so let's get on with it and see how far we can get.

What I have learned about making new humans, about how human embryos build themselves, how they sort the pieces and parts into the right places with an acceptable level of reliability, and how twins and chimeras do those things in out-of-the-ordinary ways that might just tell us a great deal about how the ordinary ways work, has kept me fascinated for decades. Give me a few hours and I will bring you along a fair part of the way.

Science has promised us truth.
It has never promised us either peace or happiness.

Gustave Le Bon

Scientific theories are judged by the coherence they lend
to our natural experience and the simplicity with which they do so.
The grand principle of the heavens balances on the razor's edge of truth.

Pravin Lal

In science and in any other kind or use of logic, simplicity is a virtue. The 14th century logician William of Ockham has gotten rather more credit than I believe he really deserved for his particular often-cited "Ockham's Razor" statement of the simplicity rule-of-thumb, as follows:

"entia non sunt multiplicanda praeter necessitatem"

("entities are not to be multiplied beyond necessity") That is, in English and in context: the bits and pieces and parts you include in any logical operation — particularly in any hypothesis about how this happens or how that works — should not be made any more numerous than necessary.

The idea behind that helpful guide to learning and logic was old long before Ockham. The idea had, in ironic fact, been previously stated by others in several rather simpler, but perhaps less well known, ways.

I suppose my favorite version is "simplex veri sigillum." The simple is the seal (the shield, the symbol, the hallmark) of the true. A good clear translation into English is "simplicity is a characteristic of truth."

Albert Einstein has been quoted as saying
"Everything should be made as simple as possible, but no simpler."

The aim of science is to seek the simplest explanations of complex facts.
Seek simplicity — and distrust it.

Alfred North Whitehead

NASA engineering culture of the 1960s seems to be the source of
the K.I.S.S. Principle ("keep it simple, stupid"):

Leonardo DaVinci is reported to have said
"simplicity is the ultimate sophistication"

And of course, there is: "less is more," a popular motto in any number of applications since arising from an unknown source in the 19th century — The fewer the parts, the easier the thing will be to build and use, the harder to wear out or break ...

Some say there can be no such thing as an excess of virtue. Others say an excess of any particular virtue is a correspondingly particular vice. If simplicity is a virtue in scientific reasoning, then I believe an excessive clinging to simplicity in scientific reasoning may generate a state of error and vice indistinguishable from stupidity. To be sure of distinguishing it well from simple ignorance — i.e., merely not having had occasion for knowing — we will define stupidity here today as the inability, from whatever cause, to overcome ignorance and superstition by learning and reason. Yes, prejudice and denial both provide quite adequate pathways.

Believing anything you have accepted without having examined it against the rest of what you believe you know, is always a giant step in the wrong direction for anyone who really wants to learn what can rationally be believed. Unquestioned answers are no help at all when there are unanswered questions at issue.

Simplicity may be a hallmark of truth, but it is by no means the one and only sign, and it is certainly no guarantee. I am certain, and I believe that you will see and understand before we finish here, that not all simple things are true, and that not all true things are simple. Hold these notions close. Perhaps you could make a plaque and hang it on the wall, or put it in a flashy screen saver:

> Not All True Things Are Simple
> Not All Simple Things Are True

What is most to be desired and sought by science is not the simplest of all possible hypotheses, but the simplest hypothesis that fits all the data, the simplest idea that can explain all of the confirmed observations (including the inconsistencies) and generate valid predictions of observations that might yet be made.

No matter how simple, any hypothesis that does not fit all available confirmed observations is at best problematic, and more often than not it really needs to be remodeled or replaced. It is a hardship, though; it really, really is tough, to find that one needs to put a favorite old theory to sleep and move on to build a new one.

The common knowledge version of twin biology is as simple as generations of researchers and writers have been able to make it. As it turns out, it is easy enough, with a reasonable level of attention, to see that the standard version of the twin biology story is much too simple to fit all the data. The truth about the biology of human twinning is not as simple as dogma would have it. I will do my best to make what follows here simple enough for many of you who are not embryologists or developmental geneticists to understand it, **and** true enough to be worth the effort.

I will not be able to make it simple enough to seem easy
without making it too simple to be true.

Much of what I have to tell you here will involve differences between all of what is real and true on the one hand, and, on the other hand, those parts of what is real and true that we can readily observe and measure. Much of what is real and true is beyond the reach of our unaided senses, perhaps not forever, but here and now and so far. With instruments like microscopes and telescopes and fluorescence-activated cell sorters and confocal microscopes and mass spectrometers and photomultipliers and many wonderful new forms of digital imaging, we extend the reach of our senses. Beyond even that point, logic and mathematical modeling, deduction and induction, extrapolation and interpolation can be used to give us reliable ways to understand much that we still cannot see even with miraculously powerful instruments.

In extrapolation, we project beyond the range of observations already collected. In interpolation, we estimate values in the spaces between observed values. Interpolation is generally considered more reliable than extrapolation. That makes sense. Going out there beyond observed values is generally more likely than going in there between observed values to involve moving into parts of the event-space where the behavior of the system being observed may be more likely to change importantly.

There is one huge, at least nearly universal, fundamental systematic assumption that we all make, and of which we need to keep ourselves mindful. That is the very common pattern of thought whereby we assume that what we are measuring bears some important and consistent and explanatory relationship with what we think we should be coming to understand, and that the information we are gathering is accurate and complete. You may have already been told that things are not always, not even often, that good. In truth, it may well be that things are never that good. However, if we could not believe at least some approximation of that, at least provisionally, we would have to quit all of this.

Sometimes we have no choice but to make simplifying assumptions and depend on them…but we should keep those situations temporary and provisional. Assumptions are by definition unproven. That is not usually, and in general should not be, the same thing as being totally unsupported by evidence. There are situations in most mathematical and scientific endeavors, for example, in which we have no choice but to suppose, somewhere well short of certainty, that some part of the system we are investigating does indeed work in the way it seems to work from observations collected to date.

It is not fair therefore to demand of ourselves or others that we should never depend on assumptions. I will hold my position with the advisory plea that you should *never depend on any assumption that you are not aware of making.*

> …preconceived ideas…unconsciously held
> are the most dangerous of all.
>
> *Henri Poincaré*

At first glance, that may seem rather circular, and you may wonder "How can one depend on an assumption without being aware of making that assumption?" Why, but, of course, I am very much afraid that it happens all the time, all the time, all over the world, all the time, and one major instance of it is a very important thread through the weaving of this story.

One of the pillars of the Orthodox Biology of Twinning for the past century is The Weinberg Difference Method. The Weinberg Difference Method is an algorithm for estimating the fraction of any population of twin pairs that must be monozygotic, in the absence of means to determine that number accurately by sound and direct analysis. We will discuss that algorithm and its consequences at length later. Most of those who have used it are aware of the simplest few of its underlying assumptions, and some of its users are even aware of the fact that those are absolutely required assumptions, but almost none of its users have shown a useful level of awareness of *all* of its fundamental and necessary assumptions. As I will hope to show you, the difference that can make is huge and fundamental.

In order that you might minimize your use of unconscious assumptions, always — always — always — in any new or ongoing effort you make in search of new understanding — search diligently for any assumptions that underlie any th(ink)ing you do. As far as is humanly possible, always keep your assumptions in front of you — or else they will, they surely will, perhaps not today or tomorrow but sooner or later I confidently promise you that they will... bite you from behind. That may not destroy you, but it will hurt at least quite enough to interrupt your train of thought!

What we understand about development has been derived in good part from, and must be illustrated by, some understanding of some known deviations from the plan that cause it not to work well or not to work at all. Twins and the biology of how twins develop can bring us crucial insights into those stories, because twins do important parts of development in ways that are different from the ways the rest of us do it.

> Whenever any living thing does any part of what it needs to do in a different way from the way that others of its kind usually do that same thing, then we have an opportunity to learn, from the unusual, something about how the usual works.

To help get you off to a good start toward understanding the major components of the substance of this story, a few definitions:

TWINS, as the term is generally understood, are people who were delivered as members of a double pregnancy. That simple, visible part of reality is, however, only a small part of what is worth knowing, because twins born alive as twins are no more than about 2% of all naturally-conceived pregnancies that began as twin embryos! The fun and phenomenology of their two-ness quite apart, all that is really worthy of our biological attention to twins is rooted in their embryology and embryogenesis, not in the delivery.

We will spend some time with this later: Most twins (at conception) are not twins (at delivery). The world has had clear explanation of that system of observations available for a little less than 20 years as of this writing. That is not a very large fraction of the time we have spent trying to put a scientific frame under and around our understanding of the biology of twinning.

Most twins by our definition here (products of twin embryogenesis) are not twins by that generally understood definition (born in pairs). To drag that truth out into better light, we are obliged to use mathematical inference to look into parts of development that we cannot see. We have to extrapolate patterns from the visible parts of the distribution of prenatal survival into the parts we can see only through logic and math and inference. I have been doing things like that for a long time and have become quite accustomed to it. I find it impossible to function in biological science without those mathematical approaches. I expect that most of you are not accommodated to that way of seeing things, so I will try to remember to take time and take care to explain whenever that is the way something has to be done.

What matters most for what we need to do here is the understanding that the core of twin biology, the source of almost everything that makes the development of twins interesting and special, is in the cellular and molecular events of the first few cell divisions, in the first three or four days after fertilization. Nothing that happens in the course of clinical pregnancy, delivery or development after birth as twins is as important for understanding the important differences between twins and singletons.

The most important things are all happening weeks before the existence of the pregnancy is normally likely to become known, with the result that we will never see it and must investigate it with indirect, inferential methods. Much of that is a matter of drawing lines along the paths through time of the parts we can see, and following those lines into the parts of developmental time that we cannot see — extrapolation and interpolation.

In biology, a CHIMERA (ky-mare'-uh) is an organism whose body contains cells of two or more different genetic types. The first use of the term in biological science seems to have been about a hundred years ago, when the German botanist Hans Winkler (the same who later coined the term "genome") grafted tomato and nightshade plants together and called the result a chimaera, after the mythical fire-breathing monster. The best-known classical representation seems to be the statue from Arezzo in Tuscany showing a creature with heads of goat and lion on its shoulders and the head of a snake at the end of its tail.

Tissue transplants — kidney, heart, liver, or any other living tissue, including blood transfusions, can make an adult artificially chimeric. The transplanted tissue came from somebody else, from some other body. No two individual human bodies have exactly, entirely the same set of genes, and only monozygotic twins even come close. Therefore, if the transplant succeeds, such that the donated tissue lives on in the recipient's body, then there are cells in the recipient's body which are of two different genotypes (genetic constitutions), and this makes the recipient now by definition chimeric.

It is common for cells from a fetus to enter its mother's blood stream during pregnancy. Sometimes, some of these cells can establish permanent colonies in the mother, causing her to become chimeric. We routinely make good use of this knowledge for a noninvasive approach to prenatal diagnosis. A sample of the mother's blood containing some cells from the fetus is much easier and safer to take than any sample we can get directly from the fetus.

Fetal cells have been found in maternal circulation more than 50 years after the pregnancy from which they entered her body.

There are those who believe that many, maybe even most, of us may carry a few cells that moved the other way, from mother to fetus, but I have seen no evidence. (DNA profiles would make it quite straightforward to recognize such cells, if and when they might be found. A parent-child relationship is hard to miss in a competent DNA profile. I have never seen or heard of a parent-child relationship between the cell lines in any chimera of the sort most central to this story. If any such individual has been found, there seems to have been no report in the literature.)

Spontaneous embryonic chimeras will be of the most interest here. These are not the result of pregnancy or of any kind of tissue transplant or other medical artifice. They have happened entirely without external intervention, with no contribution from outside the embryo in which they occur.

We have known of the existence of such chimeras among humans for less than 60 years (the earliest report I have found in the literature is from 1953). Throughout that time they have consistently been thought of as an outrageously, freakishly rare occurrence.

The very idea carries a significant "yuck factor." Many people, maybe most, have great difficulty in understanding that there are people among us, the great majority of whom we will never recognize, whose bodies are built of cells of two different genetic types. [These are NOT cells "from two different people" — neither set of cells ever constituted a separate person.] The idea might be acceptable as myth, perhaps, or a reasonable subject for science fiction or television story plots… but to suppose that it is in fact happening many times every day, in real live people like you and me, is simply asking too much, for many people, maybe even for most people. It smacks of alien invasion and demonic possession, does it not?

Many people will see conclusive evidence of such an odd little truth, find no place in all of their knowledge or experience to put such an idea, no way to make sense of it, and then walk away as if it never happened.

Human spontaneous chimeras are not mythical, not magical, not even fictional, and most certainly not even rare, but they have indeed been only rarely discovered and demonstrated.

There is an association in many minds between chimerism and severe anomalies such as hermaphroditism (the state of being a hermaphrodite, a single person with both male and female reproductive tissues). The basis for this association is that many of the human chimeras that have been discovered before the routine use of DNA genotyping in the last few years, have been discovered because of some major abnormality that caused the closer investigation that resulted in the discovery of their chimerism.

Now, with the current and still-rapidly-increasing availability of much more precise and much less expensive genetic profiling, we have begun to discover other chimeric individuals "by accident," for reasons that have nothing whatever to do with how they look or how they function or anything else about how their lives are unfolding. They are perfectly "normal" people, and they give us every reason to believe that the great majority of chimeras have two normal cell lines of the same sex and will never directly give us reason to investigate closely enough to discover their chimerism. For these reasons, it is easy enough to understand that the number of chimeras so far discovered represents only a very small fraction of all the chimeras among us.

Spontaneous embryonic chimeras are dizygotic twins one or both of whom carry some cells "belonging to" the other twin. There is in fact good reason to suppose that chimeras are more or less exactly as frequent as dizygotic twins (in conceptions, not live twin births). Most, and probably all, *naturally conceived* dizygotic twins are at least temporarily chimeric (but I won't spoil the plot here; that comes farther along, much closer to the culmination of the story).

Twins and chimeras are both outcomes of what appear to be unusual paths through embryogenesis (the building of the embryo). I say "appear to be unusual" because it is not in fact possible to state with any precision the frequencies at which either phenomenon occurs, so as to know how usual or unusual they may be. They all arise, and most of them fail and disappear, in a part of pregnancy we cannot see, much of which we also cannot penetrate with instrumentation. The present best available estimate is that about one conception in eight (~12.5%) is a twin conception, which is in fact a conservative estimate.

Defining the "Human Embryo"

Meanwhile, it is surely not trivial, and perhaps not even possible, to state with any grace and skill and confidence exactly what constitutes a human *embryo*. We routinely use that word as if we know, and expect everyone hearing it to know, what it means. It is, however, just not that simple, for several different, all very important, reasons.

It seems that any realistic definition of a "human embryo" must include being an aggregate of living cells of human origin, with the potential to develop into a fully realized and functional human individual like you and me. In fact, it would be hard to improve on that definition. Suppose we keep it, at least for a while?

> A human embryo is an aggregation of living cells of human origin, with the potential to develop into a fully realized and functional human individual.

For the application of this definition in arguments about clinical abortion and the use of embryonic stem cells, the "potential" in there is often argued as if it were already fully actualized.

It should be fairly clear to all open minds that anything that can be called a human embryo by any reasonable definition is not in fact a fully realized and functional human individual.

All things, including all such aggregations of cells, are actual — they are what they are when observed, while any potential (the noun) the thing is imagined to possess is always potential (the adjective) and never actual.

Potential is never a physical or a biological entity. Potential can never be an observable fact. Potential is abstract, hypothetical, imaginary. It is always the result of an extrapolation from things we can see and understand to things we cannot see or understand.

We can never know whether any particular thing has any particular potential until that thing has proven the potential in question by bringing that potential into actuality, at which point it is no longer potential. We attribute to certain things the properties of other things we have seen before, that looked like what this thing looks now, and which did go on from looking like this to becoming some other thing, some other kind of thing, or a more fully realized thing of the same kind.

We also do know quite confidently that the great majority of the things most of us might be likely to call "human embryos" *will* die before birth. They do not in any part of reality have any such potential, even in the abstract, even in any imagined possible future actualization. Most of them will die before birth, and never become real human persons.

To the extent that the definition depends on any given cell or group of cells having the potential to become the cellular substrate for a fully realized human life, the great majority of all such entities cannot meet that requirement for definition as human embryos. The great majority of all such entities will fail before the birth of even a beginner human, even the newborn larval form of the species.

That particularly and especially includes everything artificially made of human cells by any means other than by natural development from a zygote produced by fusion of human sperm and egg — because no such artificially-constructed entity — not one — has ever been known to proceed to the birth of a human individual. There is therefore no rational basis for imagining and assigning to such an entity any such potential, because of which fact all such things fail any reasonably competent definition of an embryo.

EMBRYOGENESIS is the process of forming, or building, the embryo. In order for a sperm cell to penetrate the egg cell, a great many things have to be built right and many processes must be performed properly for even that "first step" to happen. Then, after the sperm penetrates the egg cell, there are absolute requirements for many more things to be built right and events and processes to be performed correctly, to get from there to the proper fusion of the two half-genomes to form a proper zygote genome, and to install the very complex protein machinery necessary for accurate DNA replication and proper cell division. Not before all that happens, not before that is all completed properly, can embryogenesis get underway. Fertilization is not complete until all of that is done, and it will not be successful until and unless all of that has been done properly. That is something we cannot know until it has finished actualizing what we are now giving it imaginary credit for being able to do ("potential").

When fertilization has succeeded, resulting in the formation of a viable zygote, then the work gets harder yet. The cell divisions of embryogenesis begin with the division of that single zygote cell, and continue, through a marvelously complex system of processes, to unpack the contents of the zygote. The business of building the embryo proceeds according to a plan built into the egg cell during its formation, activated by machinery from the sperm, eventually to establish the whole structure of an organism that might some day grow into a fully creditable human.

Even LEFTHANDERS are also only a fractional representation of the group they represent, of the biological realm they occupy.

Nonrighthandedness is the simplest and currently still the best sign we have by which to recognize the members of a minority group of people whose brains are built differently from those of the majority. Ambidexterity (the ability to use either hand with a reasonable level of grace and skill) is genetically equivalent to left-handedness, so in here you will find the whole group of left-handers plus ambidexters being called "nonrighthanders." Nonrighthanders are different in many ways from righthanders (those whose right hands are so much more adroit that their left hands have almost nothing to do besides holding things for their right hands to work on).

The asymmetric functional differences are subtle, but clearly can only have arisen from structural asymmetries the development of which is set in motion within the first few cell divisions of embryogenesis, as we will discuss in more depth later. There is good reason to believe that the "normal" (usual or majority) developmental process specifies the asymmetries associated with righthandedness. In the minority situation, in the absence of that normal specification, the outcome is random, L:R::50:50. It is therefore probable that the nonrighthanders among us are a random half of that special subpopulation whose development has proceeded without full proper embryonic specification of the normal asymmetry.

Twins do embryogenesis differently. Substantially more often than singletons, they are members of that nonrighthanded minority with the unusual embryogenic asymmetries. We might therefore expect to learn from twins and chimeras some important lessons about when and how body (including brain) symmetries are established in the early embryo.

This writing is meant to be a way of bringing together and interpreting what can be seen in relationships between twinning and developmental asymmetries and between twinning and chimerism. We have every reason to believe that this will result in an improved understanding of the events of the first few cell divisions in embryogenesis. There are processes behind these phenomena that are sufficiently complex that they might as well be magic and miracles. The nature of the connection between these facts is what we are here to drag out into better light.

Maybe I Should Write a Book …

There must be at least a dozen different sets of files on one or another of these computers around here, in which I have at one time or another started "writing a book" about the business of becoming human. It should be about human development as individuals and as a species. It should especially be about what I have learned from twins and chimeras and left-handedness and schizophrenia (to be defined and discussed at some length in later chapters), about how that all gets done.

I want this to be enjoyable reading for people who want to know something more about how Humans have grown into the place we occupy in the framework of Life in the Universe, who may or may not be fluent in the jargons of modern biology. The tale might best be told by focusing on how each of us has somehow built a human body (including mind, which is somehow a function of brain tissues) by unpacking and unfolding the contents of a single cell, and what we know about the way that has to be done so that the very complex product will work, and preferably work well.

Chapter Two

Human by [Cellular & Developmental] Definition:

Emergence: of Mind, of Self, of Soul, of Person

There is an art
of which every man should be a master; the art of reflection.
If you are not a thinking man, to what purpose are you a man at all?

William Hart Coleridge

The longest journey of any person is the journey inward.

Dag Hammerskjvld

… the ultimate and most creative miracle of mind
is its awareness of itself,
a phenomenon not found in other creatures.

Morton Hunt

Human life is unique in being the result
of three coevolving information inheritance systems:
genes, minds and technology.

Robert Aunger

The only form of intelligence that really matters is
the capacity to predict.

Colin Blakemore

There is no basis, not even a roughly rational basis, let alone a fine scientific basis, for supposing that Life exists only on Earth, or that the way Life works on Earth is The Only Way There Can Be Life.

Even considering how oddly stringent the precise conditions of our physical, cosmic, planetary environment are in favor of the way Life-As-We-Know-It works, the prospect that Earth is the only planet in the Universe with this system of conditions is vanishingly remote. Some call this the Anthropic Principle, and others the Goldilocks Phenomenon, about how of course everything is just right because if everything were not just right we would not be here to see that it is just right. Prolonged reflection indicates that this is a far more profound notion than it may seem at first glance, and that it grows more so with any increase in attention. There is, further, excellent reason to consider the possibility that all that is is as it is entirely and only because we see it as it is. Neither Universe nor Mind is contained within the other; they are so purely congruent that to see and to say they are the same is a pale and inadequate vision of the intimacy of their union. One useful way to think of it is to suppose that Life Creates Reality and not the other way round. Study the concept of biocentrism for further opportunities for insight.

For that matter, before we can go even that far, there is no reason to suppose that we know anywhere near everything there is to know about The Way Life Works On Earth. Actually, that is demonstrably not the case, because we keep finding new forms of life in microenvironments that we would have sworn could not possibly harbor "life as we know it." We keep on finding living things, which keep on turning out to be further examples of Life As We Have Always Known It, in corrosively alkaline pools in the desert or in superheated acidic brines among Yellowstone's geysers, in supercooled brine channels in polar sea ice, in the wildly corrosive effluents of volcanic vents in the sea floor, even making a living digesting rocks thousands of feet under the sea floor, or miles deep in a mine shaft, operating exclusively on energy captured from radioactive decay in the surrounding rock...

> Sometimes I think we're alone in the Universe
> And sometimes I think we're not.
> In each case, the idea is quite staggering
>
> *Arthur C. Clarke*

Some of those organisms do some fantastic variations on the basic themes, but each and all of the life forms we have found and properly examined so far gather or make their daily bread using variations on the same carbon-based structures and biochemical mechanisms that the rest of us previously known organisms use. They use DNA, several kinds of RNAs, proteins, lipid bilayer membranes with their hydrophobic cores and hydrophilic exteriors, etc., etc., etc.

So far every one of them of which we have asked the right questions use ribosomes and transfer RNAs, to read messenger RNAs, to make proteins for structures and for enzymatic catalysis using a code of three nucleic acid bases per amino acid in each amino-acid-peptide chain... Plus, with very few, partial and minor, exceptions (the mutational paths to which are not clear) they use the same sequence of three DNA bases to code for each respective amino acid.

I have a large warm spot in my heart for the concept that the entire Earth can be understood as a living organism, an idea that is sometimes called the Gaia Hypothesis. But that is a story for another time...

While we cannot stall the apparent flow of the time we have, it might be worth spending some of it to try to understand how it is that we go about doing what we have been doing, how it is that we do our business of be(com)ing human.

Where *Do* We All Come From?

Most of us know that storks and cabbage leaves have very little really to do with how new humans come to be added to the population. In its home countries in northern Europe, where there really are storks nesting on roof ridges (and more recently on power line pylons and cell phone towers), that particular children's fable about new babies being delivered by the storks is a wonderful story, and it has served heartwarmingly worthy purposes.

I like the myth of reincarnation, especially in the Buddhist tradition, where everyone always can get another chance, but I can't help thinking there must be more to it than that. With only a little bit of checking around, it is easy enough to learn that the total number of living humans has been increasing at an increasing rate for a very long time. New arrivals have long been more numerous than contemporaneous deaths over the population as a whole, so the constantly increasing accumulation of new humans cannot be fully explained by any kind of recycling program for previously-owned sacred immortal souls...

> Unless... maybe... I never thought of it this way before — perhaps, if all those prenatal failures (which outnumber live births by at least three to one and quite possibly eight or nine to one or more) really are immortal souls and they get do-overs like everybody who actually ever gets to be alive...? Naaah; there still would not be near enough; the only way

would be to divide them. ....perhaps it is better not to think of that as a
matter of fact ...

Every human population that ever became known to the rest of us has one
or more of such creation myths, explanations of the origin of the world in which
they live and of themselves and of all the other (kinds of) lives with which they
share their world. Some of those stories serve only to entertain or pacify curious
children, and there are others that adults are expected to believe in order to be
trusted with full membership in the group.

The idea that only one such origin story or creation myth can be true, and
that anyone who cannot believe that particular one can just go to Hell is a pitiably
ordinary human perversion of an old parallel idea in the long process of becoming
human as the species we are today: It seems that every human population ever
discovered and reported, besides having a creation myth to explain the presence
of their kind on Earth to begin with, also has one or more understandings of how
their kind is different from — separate from, better than, more important than,
to be protected from, destined by their creator god/s to dominate, etc. — the
members of other such groups.

Some understanding of that sort is a fundamental component of each group's
boundaries, their definitions of who they are and are not. Such boundaries are
essential components of every system at every level. No system maintains its
integrity and functionality without boundaries. Boundaries are essential to the
maintenance and function of any system, including human life. In the system
that is human life, just such boundaries are, for example, the basis for every form
and instance of homicide.

Where did we come from? How did We get Here? The stork is every bit as
good an answer as most until the child is old enough to ask "So ... from where,
then, comes the stork?" Beyond that point, we have the stories adults are supposed
to believe. "Who made you?" ... how does that go? "And why did [that Person,
Maker, person-maker] make you ...?" No doubt some of you remember ... that's
the Catechism. No problem with that until and unless somebody tries to tell you
that those questions are questions of fact from which the questioner may expect
the answerer to provide real and rational answers.

In most such structures, origin fables and creation myths account for the
first few of each species of organism, and the rest of the story acknowledges

that, beyond that beginning, the nature of any given individual living thing is a function of the nature of its parentage.

The understanding that "like begets like" is a very old one. When we begin to teach the most basic elements of Biology in primary or middle school, the little darlings usually do not have "like begets like" in so many words on the tips of their tongues. Not one of them, however, even at that age, will admit to not already knowing that mares sometimes drop foals but never ever have kittens. They know that bird eggs hatch out little birds and not lizards. They know that the mother 'possum has only baby 'possums, bless their nasty-ugly little hearts. They know that acorns will sprout little oaks if they sprout at all and will never ever sprout rosebushes.

Exactly how all of that happens in molecular mechanochemical detail is so incredibly complex that it may well never become altogether clear. We have become reasonably certain that the observed continuity has something to do with genes and the developmental regulation thereof, and we are getting smarter every day about all of that and how it happens.

We have become smart enough for example to know that all living organisms have many genes in common, implying and defining a single kinship — with, therefore, a single ancestral origin somewhere very far back in time. We are not yet, however, smart enough to know in exact molecular detail how differences among genes and/or the regulation thereof produce the different structures of pigs and peonies and people and parrots.

It makes all the sense in the world that there might have been, must have been, many emergent beginnings of living systems that lacked coherence and led to oblivion. On the other hand, however many times Life on Earth might have started, it succeeded only once as far as we have been able to learn. All of the outrageous diversity of forms of Life that we know about are of one "kind," biochemically speaking, at the most basic level.

> "Life as we know it" is a single coherent complex emergent system.
> and "coherent complex emergent system"
> means, more or less exactly ... what?

I have mentioned "system" and "emergence" before. I went right on as if I fully expected you intuitively to understand these fundamental concepts that are at least right up against the limits of anyone's understanding, definitely including mine. The paragraph just above seems to be the first place I have said "emergent system." I even said "complex emergent system." That phrase is internally redundant, because emergence is a function of…, arises from…, is driven by… complexity, and "systems" in this context, at the level of living systems, are by definition marvelously complex and by definition emergent. Even the simplest of living systems, even the single-celled ones, are of and in a level of complexity well beyond our current understanding in molecular detail… but we will not be giving up soon.

Emergence is what happens when…, is the process whereby…, is the phenomenon through which … patterns and systems arise out of interactions among large numbers of multiple kinds of simpler components. "Large numbers of multiple kinds of components" is where the essential complexity in there is to be found. A pile of parts that are all the same, more or less regardless of how many of them there are, is simple, and may be expected to stay that way. For example, when thousands of identical bricks are used to build a house, a modicum of complexity may arise from a deliberate externally applied organization and well-placed whacks with the edge of a trowel to change the sizes of some of the bricks, but that is nothing like the spontaneous self-organization we need to understand here for living systems.

On the other hand, when there are only a few pieces, no matter how different they are, not much in the way of complexity can be expected to arise. Complexity depends on both number and diversity of parts.

The Emergence of "Emergence"

The concept of "emergence" has been in use since at least the time of Aristotle. Whenever a collection of components becomes — because of interactions among its (… large number of multiple kinds of …) components — something greater than or other than a predictable resultant of its components, then emergence is in effect.

The emergent thing, that which emerges, by virtue of inevitable spontaneous self-organization occurring among multiple heterogeneous components by way of their complex interactions (what Stuart Kauffman calls "order for free"), has properties not present in, not predictable from and not reducible to, the properties of its components (and quite possibly not seen before, ever, anywhere).

Integration ... Biogenesis

The "coherence" I attributed in that same sentence to the complex emergent system that is Life, might also be called integration or integrity. A coherent, integrated emergent system is one which can maintain itself in its specific highly ordered structure over an extended period of time. It can grow, by accretion if by no other means, until interactions among its components evolve to include metabolism and facilities to reproduce the components. It can reproduce, as a whole, by fission and duplicated and renewed growth if by no other means, until its structure and functions evolve more sophisticated mechanisms like genes and gametogenesis. It can adapt to changes in its environment, and it can evolve/ acquire capabilities for responding to such changes, by making changes in its structure or behavior ... at which point it becomes appropriately and necessarily incumbent upon us to acknowledge its sentience in at least a rudimentary sense — its ability to detect and recognize and respond appropriately to changes in circumstances. Thus, "biogenesis": the inevitable, spontaneous, self-organizing origin of Life from non living matter. There really need not be any more to it other than further details ... which we may reasonably suppose will be forthcoming.

This has been called "strong emergence" to distinguish it from situations in which the change is not so great, with which we have no real present concern. It has also been called "uncomfortably like magic" and "illegitimately getting something from nothing." Stuart Kauffman calls it "order for free" and got a MacArthur Fellowship for providing examples accessible to those of us who have not navigated up from the ground floor into complexity theory.

It is easy enough to appreciate how it is that many people might be satisfied to call it a miracle, and commence their search for The Maker, The Worker of such miracles. However, as I have already told you and now remind you — to call anything magic or miracle without even trying to understand it is at best no fun, and arguably, from at least any scientific viewpoint, irresponsible.

Just a couple more pieces need to be attached somewhere near here, and then we can move on, namely: the relationship between scale and complexity and scale-invariant fractal self-similarity, and the role of chaos. As scale increases, i.e., as things get larger, complexity is likely to increase. Our brains are bigger than chimpanzee brains because ours have more cells (which may have been caused by a relatively simple mutation that allows a few more divisions of the primordial cells of the embryonic brain). However it got that way, the linearly greater number of cells generates an exponentially greater number of cell-to-cell connections, because the complexity of the cell-to-cell connection network scales as/to the power of N = number of cells. That means, for example: to *triple* the number of cells (assuming about the same density of connections per cell) is to *cube* the number of cell-cell connections, which probably already had millions or even trillions of zeros. If, for example, there are two conditions each — such as on/off — for 10 switches, then there are 2^{10} possible arrangements of the switches = 1024, about a thousand. For three times as many of the same switch units, the possible arrangements thereof are now 2^{30} in number. From a threefold increase in the number of component parts, there arises a million-fold increment in complexity.

As the scale of the system increases, the patterns descriptive of the structure in general need not change, due to the fractal scale-invariant self-similarity I have mentioned, which is to be expected in such structures. If the structure, for example, is like a tree, the new additions by which the structure is enlarged will in general repeat the branching shapes of the earlier and now larger structure out of which they grow. For a reasonably convenient illustration, cut a head of broccoli in half from the stem through the head and carefully disassemble it from that starting point. Cut each successively smaller piece you pull off in half from the stem through the head, and it will look very much like a smaller version of the structure of which it was a part, from which you removed it.

As the complexity of the system increases, the emergent properties of the system as a whole may change; perhaps to intensify, perhaps to include novel, even drastically novel, properties. At each level of complexity entirely new properties may appear, and the whole may become not merely more than, but even quite different from, the sum of its parts (cf. Anderson, 1972).

The Emergence of an Appreciation for Complexity and Emergence in Biology

Biology has not in general, over most of its history, spent much of its time and energy on thoughts of systems emergence and integration such as these. That is beginning to change. We are attempting to climb above our (still rapidly growing) understanding of the genome, and of the combinatorial codes of transcription factor gene products and other such effector molecules required to control transcription and translation, in turn to control cell differentiation. Our primary means of moving in the right direction is our steadily increasing ability to visualize those interactions using such wizardry as "microarrays" and ongoing prodigious expansions of computing power and algorithmic insights to analyze the data we collect from them.

Working our biggest and best and fastest computers, limited as they are by the dazzling but finite creativity of our best programmers, as hard as we can, we are just beginning to gain an appreciation for the complexities of the systems that constitute the biochemical functionalities operative within living cells. You might want to look up "genome," "proteome," "interactome" and "systems biology" for orientation and opportunities for further insight.

The Developmental Perspective

This writing is mainly about the kind of biology called developmental biology, taking to the cellular and molecular levels the study of embryos and how their shapes and their parts and the shapes of their parts change during their development from a single cell to their mature body structures composed of many different kinds of cells performing the body's many different functions. One fundamental thing we know about all that is that all of it is done by the molecular, mechanochemical functions of cells.

You may find, if you look, several different statements about — estimates of — the number of specific cell "types" in a mammalian body such as yours or mine. Most of those counts or estimates are in the range between 200 and 300, but I have seen some over 400 (including about 150 different kinds of nerve cells alone). I have never felt any need to believe any such number, and have long wondered how such numbers might be arrived at, but never bothered to find out until I felt this present need to explain it. The answer is cytological, or histological. In plain words, those estimates are about cell structures, and how the different "types" of cells look different in the microscope.

I am comfortably certain that the number of cell "types" defined by genetically and epigenetically differentiated biochemical function must be far greater than a few hundred, but obviously the answer would vary depending on the criteria for "cell type." We now know, for example, that there are at least several histological "types" of cancer cells that can be divided into more specific types on the basis of differences in gene function, which is much closer to the "real" basis of cell differentiation … (and much closer to a realistic approach to improving specificity and effectiveness of cancer therapies).

> Between your scalp and your toenails,
> everything that happens [in the Organism that is You] is done by cells.

That certainly and necessarily includes everything that happens between your ears and between your legs. You may think that sounds absurdly over-simplified, but in fact the universe of human knowledge and understanding contains only a very few, if any, ideas that are more profound than that one. Life is cellular. Everything alive works as it does because the cells of which it is built work the way they do.

Things like viruses, the living-ness of which has been subject to debate from some corners, are not full-fledged cells capable of performing all the functions of life, but they are in all cases descendants of living cells and they remain able and constrained to commandeer the functions of living cells to further their survival and reproduction — they hold their positions in the system that is Life as satellites of their host organism species.

> I personally do consider viruses to be living things, because they have genomes that "need" to be reproduced, and the "means" to get that done and pass them to new generations. They mutate and evolve by means of and for the sake of their own genomes.

All of what we are and all of what we do, have ever done, will ever do … is done by cells — rarely acting alone in organisms as complex as we are, almost always as networks of cells within tissues within organs within organ systems within organisms. Plus: No cell ever can do anything it does not "know how" to do.

The functions of which each cell is capable depend entirely and absolutely and exactly on what molecular structures and reactions the molecules inside the cell are capable of generating, and those depend in turn on a combinatorial code of control elements that specifies which DNA sequences are to be read or not read, where in the body, at what time in development, and how fast times how long equals how much. That is all there is (and, from only slightly different viewpoints, nowhere near all there is) to cell differentiation and the development of complex organisms like each and every individual instance of *Homo sapiens*.

Plus, again, the whole unfolding is probabilistic, stochastic, governed in substantial part by chance variations in the outcomes of the great many component processes, because cell functions respond to finely tuned biases which cannot be absolutely specified or predicted. Even (the system that is) the single living cell is sufficiently complex that there is always a practically infinite list of things that might go wrong, or might go about being right just a little bit differently and thereby produce a very different and still functional result.

It has been proposed, and I don't know anyone who knows enough to contradict the proposal, that the number of cell-to-cell connections in the average human brain is larger than the number of stars in the Universe. The number of possible alternative states or configurations of the system of networks made by all those connections has been said to be greater than the number of subatomic elementary particles (the hadrons and gluons and muons and quarks and such stuff ...) in the Universe.

There are apparently at least a few hundreds of billions of cells in a normal human brain, and there are reasons to believe that each of them is connected with every other, not all at once, not all directly and not just one-to-one but in many dimensions simultaneously. Most of the activity of the human nervous system is electrochemical in character, and all of those connections are switches, in a way. Many of them are rather more like rheostats or dials or dimmer switches than on-off switches. Many — maybe even most — of those connections are not only capable of being turned on or turned off, but capable of occupying, if not an infinite number of states in between, then at least several-to-many intermediate states.

The number of different things "the brain" — your brain, my brain, any other brain — can "do," or "be," is a function of the number of possible variations in the state of the brain as a whole. That number in turn is the product of the number of states available to cell-to-cell-connection number 1, times the number of states available for cell-to-cell-connection number 2, times ... etc., all the way through to the *grand product* of all the values from the whole very large number of cells and cell-to-cell connections. (Common math notation uses the Greek letter Π, upper case π (pi), as an "operator" indicating the result of applying the algorithm: "multiply together all the numbers in the set following and governed by this operator Π.") If, for example, all the cell-to-cell connections of the brain were to have only two states, the complexity of the full set of connections scales as (is proportional to) 2^N, two to the Nth power, where N is the number of those connections.

In fact we are certain that the variability in most of those innumerable connections is far greater than just two to the power equal to the number of on/off alternatives, simply because many and maybe most of those connections can occupy multiple states other than on and off. However, even if each cell-to-cell connection had only two states, on and off, the number of the possible states of the average human's three pounds of thinking-pudding as a whole would be breathtakingly, uncountably large, a number with at least hundreds of millions of digits.

Some people have argued that the human mind can never, will never, understand itself. I have never been able to grasp fully what they are thinking about as their reasons for that assertion, perhaps in part because most such statements are not what we might call paragons of articulateness, but here is my take on it: To understand a thing is, in a way, to surround and contain the thing to be understood. I believe that the things I understand are somehow "inside" some part of my Mind, my Self — that they must somehow become *parts of me*. Don't you feel — believe — imagine the same? Who you are, Who I am, your "Self," My Self is somehow a function of every thought you/I have ever had.

> We are the sum of the stories we tell ourselves,
> and those stories are necessarily rooted in our experience,
> and by how we choose to interpret the experiences of others.
> What we choose to remember is critical
> since the narratives that play in our heads shape everything.
>
> *Jon Meacham*

If a thing I want to understand is as big as or bigger than, as complex as or more complex than, the only tool I have available for use toward understanding it with, the only vessel in which I might contain it ...

... by the way ... there is an idiom I never heard before coming to this part of this country, about "wrap my (your, her, his) head around it" ... Maybe if we can make-a-mark-and-move-the-ruler, enough times? ... that might actually be a very good metaphor for our capability for symbolic abstraction ... there may in fact be no way to know or to say where the limits are, and there is good reason to suppose they are still expanding.

For years I carried around, in some strange place in my head, an idea the origin of which I could not remember or discover. It goes like this:

> "...the Ultimate Reality in the Universe
> is Mind contemplating Itself" ...

I did then and still do think that is a truly marvelous thought, and I do wish I could take credit for thinking of it first, but I was always fairly certain that I had not composed that line as an original idea, had not been the first one to think it, and I wondered often where I had gotten it, where it had come from ... who said that? Apparently I had not paid proper attention when I first found it. It would not go away.

As it turns out, that statement is a good English translation from Aristotle's native Greek of the nature of his grand ultimate god of all gods (they were numerous in his spacetime ... by the way, did you know? ... the concept of atheism and the word "atheist" were first applied to the early Christians, who claimed only one god and did not recognize the many gods of the neighboring religions).

It could easily have been what Aristotle meant to say in the first place if he had to say it in English. It makes all the sense in the world to me, and I have never discovered any reason to make any effort to argue against the idea. In my opinion, it may be Aristotle's very best line:

> "the Ultimate Reality in the Universe
> is Mind contemplating Itself"

... more biocentrism ...

Once one becomes aware of the idea, it can be found in many places, from Zen to quantum mechanics.

The (human) mind (it IS NOT the only kind) remains a mystery, commensurate for now with the mystery of life itself, but I believe we can argue quite effectively that neither life nor mind is in fact a miracle — unless your definition of miracle is simply "anything we don't yet fully understand in atomic detail." They are not violations of natural laws. They are not even highly improbable. Life and mind are not even unlikely results of chance events (if only because the age and scope of the Universe continue to provide effectively infinite opportunities, in time and space and physical substrates, for any possible occurrence). And quantum mechanics provides plausible reason to suppose that every imaginable version of any reality is in fact possible, and has, beyond statistically reasonable doubt, already happened.

Beyond certain appropriate levels of increasing complexity, even much less than the complexity in which we are presently immersed and on which we routinely feed our minds, the emergence of Life and of Mind, and their continuing, ongoing unfolding, have been and continue to be inevitable. Those are not things I have come here to prove. I consider these ideas convincingly demonstrated elsewhere (Stuart Kauffman, Douglas Hofstadter, Steven Pinker and Matt Ridley have provided the best explanations I know about), and those ideas serve as fundamental background and target for this work.

So developmental biology, of the human body, and of the mind that is of and in the body, is still embryology, broadened and deepened to think more in terms of the specific cells and molecules that are building and shaping the parts of embryos and the functional relationships among them. What we want to do here is develop an understanding of how adult humans are built from embryos, and where embryos come from and how embryos come to be what they need to be to go on eventually perhaps to become real human persons. We will need, along the way, to do some more thinking about what it takes for a thing to be realistically considered a "human embryo" to "begin" with.

We — that is, apparently the majority of people in very nearly every human culture in this world — tend to think of being human as something very special — "in God's image" and all that — holding dominion over all the Earth and all that is within and upon the Earth, and so on, and so forth. Some of the many variations on that theme are not so foolishly destructively arrogant and aggressive. Some do not include a God, some have many gods, and some have the good sense to omit the dominion thing in favor of a more respectful coexistence with the rest of Life.

Even in the most pacific of those cultures we have ever imagined, whether or not they actually exist, the human remains very special by virtue of having the choice, and most especially by being aware of having the choice.

We also generally tend to think that "thinking is what makes that so." That is not exactly the same assertion as "thinking that is what makes it so," and that is not where we need to go right now, but you may find it worth some time to think about what the differences between those two thoughts might really be (it IS important).

The language in that last paragraph is a bit playful, but it does not quite slide to the level of being facetious. Thinking that the human is the only organism with any sort of mental life is a very common bit of species-chauvinistic nonsense, and thinking about thinking is absolutely anything but trivial.

I therefore and hereby propose that the most powerful and wonderful (the most "in-God's-image" or "god-like?") thing about human thinking, which might most closely approximate Aristotle's Ultimate Reality, which might in fact constitute the kind of evolutionary advantage that our human species has derived from — or which has driven — the dramatic evolutionary expansion of our mental abilities, is reflection.

Reflection is perhaps most readily represented as *the ability to think about thinking*. Remember, "the ultimate reality in the universe is mind thinking (about) itself"

It is indeed a special and wonderful thing, not just to know, but to know *that* I know, what I know. The structure of the human mind's consciousness of itself is widely imagined to be among the species' greatest distinctions from other species. There are those who believe that consciousness of Self is absolutely peculiar to the human species. I find it instead entirely credible that that distinction, and most especially that idea of absolute peculiarity, may be imaginary.

One bit of behavior that has served as an experimental criterion for self-consciousness is the ability to recognize Self in a mirror. Some of the other great apes, elephants, dolphins, and birds have all been reported as having given credible evidence of coming to know that the image in the mirror is of the own body (cf. de Waal, 2008; Prior *et al.*, 2008).

With the ability to think about, to reflect upon, my own behaviors — I can, for example, rehearse offline for dangerous deeds that might become necessary for my survival, which is in its turn necessary for maintaining any prospect I may have for reproduction. Thus it would seem quite clear that gaining a capacity for reflection would be to gain an enormous survival advantage.

I can think about how I came to know (this, that, or some other thing), and about how I might get some more of any kind of understanding I happen to enjoy. Some have called that consciousness of self the defining characteristic of human-ness. It can reasonably be argued that *the human Self comes into, and maintains, its existence only in and by virtue of its consciousness of itself.* Along the same vector, it is very difficult, if possible at all, to distinguish Self from "soul," by whatever meaning one may give it …

> The human SoulSelf builds itself from its lifetime of accumulated experience; sensations, perceptions, responses, all the things that have, in and prior to the present moment, happened "inside" it. It is always different, from every other instance of its "kind," and from itself in every instant but this one.
>
> The human individual body:brain creates/builds its SelfSoul by becoming conscious of its own functions, crucially including gaining consciousness of its consciousness, and by learning the peculiarities that distinguish its Self from the other entities it recognizes as similar but not the same.
>
> Every emergence of Mind from matter is unique (the matter of each individual human is unique; the mind of each individual, being a function of the matter, is unavoidably unique), and the structure of that uniqueness defines the Self.
>
> The Self IS its consciousness of itself, and has no existence apart from that.
>
> > "I" is an idea having an idea about an idea.
>
> There is a body, of which the brain is part, with which that idea of Self is individually uniquely congruent.
>
> > A joyful life is an individual creation
> > that cannot be copied from a recipe.
> > *Mihaly Csikszentmihalyi*

Instinct? Sure ... but: To build a system that can perform a particular function is one thing. To build a system that can learn how to do the same job, and then decide appropriately and accurately whether and when and where to do that job ... that is something else entirely.

The projection of the consciousness of self and of mind and its nature onto others may be seen to constitute what is called "theory of mind" (my ability to suppose that you have a mind that works more or less like mine, that you have intentions and emotions and all such things happening in your Self like those that seem to keep happening in my Self).

I find it compellingly straightforward to suppose that any organism that can, for example, recognize the intention on the part of another organism "to make a meal of me" ... will have a decided survival advantage over any individual organism lacking that ability. That individual organism will be more likely to survive long enough to reproduce, and will thereby be naturally selected to pass on its genes and thus increase the representations of its alleles among the genes of the population of its species.

The ability to think, in the comfort of my own bed by my own fire, about how I might go about doing something dangerous allows me the safety and comfort of rehearsing offline, while, for example, the beast I think about eating is not around to interfere with the development of my vision of the relationship between his demise and my dinner.

Reflection is recursive. Since I just wrote about thinking about thinking, then: obviously, I can think about thinking about thinking ... because I had to think about that to write about it. That statement in turn constitutes an output from thinking about thinking about thinking about thinking!

In math, we often use parentheses to help keep such things clear. I repeat, with parentheses: Since I just wrote about (thinking about thinking), then: obviously, I can [think about (thinking about thinking)], which statement in turn constitutes an output from {thinking about [thinking about (thinking about thinking)]}! And so it could go on forever, or at least until I get tired of it. Come to think of it, there is surely some limit to how many layers of that I can keep in mind at one time ... but! if I write it down, I can just add one more level and keep on re-reading it till I run out of paper to write another line or run out of time to reread it ... does that count? Of course it does! Just like making a mark and moving the ruler, as mentioned above. The ability to abstract it to a symbolic representation on paper is a very important part of the package. Reflection is recursive, infinitely so in the abstract, as in symbolism, but limited in practice by its physical, cellular substrate.

This particular fundamental characteristic of the human capacity for thought is a bit like one of those great palace ballrooms with great mirrors at each end. The capacity for reflection is, seriously, my best guess as to how the kind of mind that humans have has managed to give itself the feeling of being so much more powerful and versatile than any other kind of mind in the world. As long as we cannot really know exactly what any other mind, let alone any other "kind" of mind, is thinking, that will probably be as close as we can get.

When I think about thinking, or about anything else I can do or think about, that requires reflection … one manifestation or function of my mind is being considered by (is subject or object of the functions of) another part. When we can't keep straight which part is which, things can get crazy … people call that someone else's "auditory hallucinations" — or their own "voices."

Lithosphere/Geosphere, Biosphere, Noösphere

Eduard Suess (1831–1914) was an Austrian geologist who discovered, for the benefit of posterity's collective consciousness, the extinct supercontinent Gondwana and the Tethys Ocean, of which the Alps were once part of the bottom. He was an expert, you see, on the "geosphere" or "lithosphere" — the inorganic, mineral world. He published, around 1901, *Das Antlitz der Erde* (German, "The Face of the Earth"), which is said to have been a popular geology textbook for many years. That work may also reasonably be considered the first treatise on ecology, because it was there that Suess introduced the concept of the biosphere (the organic, living world) and discussed at some length its relationships with the geosphere.

Vladimir Vernadsky (1863–1945) deepened the idea of the biosphere to the meaning it has carried into the 21st century, and first popularized the concept of the noösphere (the cognitive world, the sphere of human thought) and discussed its interactions with the biosphere.

In Vernadsky's theory of how Earth has evolved, the noösphere is the third major stage in Earth's development, after the geosphere and the biosphere. As Life emerged from and transformed the geosphere, forming and becoming the biosphere, in just the same way the emergence of cognition arose from and transformed the biosphere, forming the noösphere. Vernadsky saw the emergence of Life and then of Mind as implicit essential features of Earth's evolution.

Vernadsky's noösphere concerned itself primarily with the lithosphere, where he foresaw humankind's mastery of nuclear processes bringing the power of the creation of resources through transmutation of elements. The closest present reality to that concept would seem to be the nuclear generation of electric power, with controlled thermonuclear fusion still "in the pipeline."

The Jesuit paleontologist Pierre Teilhard (tay'-ard') expanded his concept of the noösphere from Vernadsky's, to a system of concepts for which he was punished by his superiors in the church because of the threat his evolutionary concepts were thought to present to Augustine's Doctrine of Original Sin. His construction survived that suppression and has not gone away, any more than did Galileo's re-conception of the solar system ... "nevertheless, it moves."

Teilhard (according to at least one source, "de Chardin" was not his family's proper surname, but an affectation to ancestry in an extinct French aristocratic line) has been called the "Patron Saint of the Internet," because of his conception of the noösphere as a sort of "collective consciousness" of human-beings — emergent from the interaction of human minds, more or less exactly like that "collective mind" I mentioned just above.

Teilhard's noösphere, emergent from the self-conscious complexity of the entirety of self-conscious human thought, continues to grow in an exponential relationship with the number of interacting self-conscious nodes in the system-that-is-the-human-species. As discussed above, the number of component logic elements in that collective Mind approximates the compound product, over all of its nodes, of the number of inter-nodal connections per node ("individual mind"). This scales as (connections per node) to a power equaling the number of nodes in the whole system (that latter now fast approaching seven billion individual human minds, of which perhaps not all may be considered "active" and/or "on-line" at any given instant). Can you grasp the idea of a number that large, or tell me how to?

Teilhard foresaw, in his "Law of Complexity and Consciousness" (his description of the nature of evolution in the universe) most importantly the tendency in matter toward increasing complexity and the corresponding growth of consciousness that he saw as inevitable. Teilhard further envisioned the noösphere growing towards an ever greater integration and unification, culminating in the transcendent "Omega Point," where consciousness will rupture through time and space and assert itself on a higher plane of existence, from which it cannot be brought back down ...

Because Teilhard's Law of Complexity/Consciousness applies and operates always and everywhere, and because of the practical infinity of time and space, and the corresponding practical infinity of opportunities for matter to take steps toward increasing complexity, it is beyond a statistically reasonable doubt that our Earth-bound version is not all there is of Life.

I surmise [perhaps I should say "Let us suppose for the sake of argument ..."] that the Collective Consciousness here described has not yet achieved Self-Consciousness and therefore presently remains without a Self/Soul. But ... wait a minute! That would be "Personhood As We Know It," wouldn't it? Consider ants; the individual ant ... and the ant colony ... there may be in here a plausible comparison: One at a time, individual ants have very limited behavioral repertories, but a healthy ant colony is a complex emergent living system which can and must — for the sake of survival and reproduction of the species — perform many different quite sensitive and intelligent interactions with/in Life System. This cannot be achieved by the individual ant any more than by any single cell of a human body — the queen is a very "high-maintenance" individual — she cannot lay the eggs without the ongoing support of the whole, healthy colony system to keep her alive and well. My surmise about the Selfhood of the Collective Consciousness might well have all the scope and power of the individual ant's "thoughts" on the structure and function of its colony and its species and its place in the System that is Life.

Even bacteria — single cells, mind you, to all appearances more or less entirely self-contained in their ongoing life-functions — communicate (by "quorum-sensing," and perhaps other means) and cooperate to do things as biofilm colonies that they cannot do as free-swimming cells one cell at a time. Within their colonies, there is specialization, and division of labor — different individual single cell bacteria have different functions in the biofilm colony — one of which seems to be self-sacrifice for the good of the colony, to be eaten by the protozoan predator and thereby gain access to destroy the predator from the inside.

The changes between those different, free cell **vs** biofilm, behavioral repertories occur in the same sorts of ways as developmental differentiations of cell function in multicellular organisms. Some genes get turned on, some get turned off, and others show differences in the levels of their activities. If we define an organism by its structure and its behavioral repertory [as in fact we generally do], and define a complex system as emergent when its complexity assumes or acquires properties not present in and not predictable from the properties of its component parts [as in fact we do], then between the individual ants and bacteria and their respective colonies we must declare the emergence of complex supersystem organisms, superorganisms. The colony is an organic supersystem of the individuals … just like the human population? Just like the Internet? Just like the noösphere?

In the context of this understanding, the emergence of Life and (then?, later, or necessarily simultaneously?) of Mind still may be seen as mysterious or miraculous, but need no longer be seen as supernatural. The fact that we do not fully understand a thing or a process does not mean we need the imagined machinations of a deity to stand in for an explanation. Vernadsky, Teilhard, and more recently Stuart Kauffman (in *Re-Inventing the Sacred*) have argued, in other than so many words, for the 'creativity of evolution,' a concept offered to the collective consciousness at least a century ago in the work of Henri Bergson [1859–1941, who won the 1927 Nobel Prize in Literature for *l'Evolution Créatrice*]. The work has been published in English with the title translated as "Creative Evolution," but my understanding of what his title in French actually says is "Evolution the Creatrix (Creator, feminine)" — which notion is reinforced by reading it.

The number of cells participating in the brain's functions is huge, and the number of cells may or may not be the most important, but is certainly by far the simplest, part or measure of the brain's complexity. What really seems to matter most is the number of cell-cell connections and the complexity of those connections. The number of possible states of the brain, of that whole system of connections, is on the order of the product, over all of that huge number of cells, of the number of variable states available to each one of those cells. … I have said this before. Some serious estimates of the number of cell-cell connections in the human brain suggest that the number may exceed the number of subatomic particles in the Universe. Is that wild hyperbole? To the extent that I have been able to understand and investigate it, I don't think so, but I can't swear to it, because I haven't had time or other wherewithal to count them and I don't believe anyone else has either.

Surely no two brains have the same number anyway, no one brain has the same number from one day to the next, and the true total number would change many times before anyone could finish counting them anyway. You could take my word for it, OR you could look up what others have had to say about it for your own edification. If you go about the search openly and honestly, you will find yourself facing a choice among several numbers. Just remember I clearly labeled it "estimate" and made my uncertainty clear. I believe that the calculations I have seen, leading to the impression I have explained above, are plausible.

To bring it to a point, the point is: The human brain is a tissue, an aggregation of cells, so hugely complex that we cannot even make a confident statement about exactly how complex it is, and we have no reason to suppose that any specific thing anywhere is more complex. Is your mind the same as your brain? No. No more than "frying" is the same as your skillet. Is your mind IN your brain? In a way, yes, but not in the sense that we might expect to find some specific piece of brain tissue that is or contains the mind.

Brain is a thing — about three pounds of pink, white and gray pudding-looking stuff with some dark maroon lines and some plump rolling-hills-and-valleys sort of wrinkles — a taupe sort of color when pickled in formalin, the most likely way for most of us ever to see one. Brain is a thing, a discrete organ, a simple non-abstract noun. Mind is a concept, not a thing but an idea, an abstract idea that cannot be seen or touched or physically moved or manipulated. It can be "changed," but that is at a different level of abstraction in an altogether different direction. Mind is something that is done by (and perhaps reflexively to) the brain. Brain is structure; mind is function and behavior, of and from and dependent upon that structure.

Your brain is the cellular substrate of your mind; your mind is a system of functions of your brain. Yes, even while no one can tell us exactly how, that much I certainly do consider proven. For example: I could change the temperature of the cells of your brain by just a few degrees, either way, and thereby cause your mind to work a little less well, then much less well, then not at all for a little while, then not at all ever again ... depending on exactly-how-many-is-a-few degrees I change the temperature.

> ... sign this informed-consent form right here _____,
> and we'll get started in just a few minutes ...

We can do the same sorts of thing, and achieve the same sorts of results, with any number of drugs and other "chemicals," natural or synthetic — including things as ordinarily benign as water and salt, or with small fractional changes in the concentrations of mundane simple substances that are always there, such as ammonia, urea or carbon dioxide. We can find evidence for the understanding that brain is substrate of mind and mind is function of brain in any number of developmental anomalies where we see specific disorders of mental function every time we see certain regularly corresponding anatomical anomalies or traumatic disruptions of brain structure.

One may argue, but there is no need to argue, and there is no future in attempting to make any argument ... Resistance is futile. Mind is a function of (brain as part of) body ..., as opposed to having any other relationship with brain, or being particularly and essentially a function of any other subsystem of the living body.

The necessity of including emotion as part of mind invokes the necessity of broadening our understanding of the physical substrate of the mind to include — really, I believe, the whole body, but at least the rest of the nervous system and the musculoskeletal system. "Feelings" are physical, as witness ages of considering the heart as the seat of our emotions.

Besides size and complexity and connectedness, and the complexity of the connectedness, and emergent properties of that complex connectedness, which all work together to make it much more than the sum of its parts and give it properties well beyond anything present in or predictable from any or all of its parts ... (you should recognize the definition of an emergent system) ... there is another physical feature of the human brain that is fundamental and crucial to its human-ness and its human functionality ... and (only) at first glance, apparently much simpler.

There is a physical feature of brain tissue, other than cell number and connectivity, anomalies of which are significantly associated with at least very nearly every anomaly of human brain function. It is simple in concept, obviously a feature of and a variable in embryonic development, and crucial to every known more-or-less-specifically-human feature of human brain function.

The one particular crucial element in the complexity and functionality of the human brain that I have in mind is its functional asymmetry. The left and right halves of the human brain are built differently and they function differently. It all works together, and any damage to any part of it diminishes all of its functions at least transiently, but its many and various subsystems do show some degree of specialization to specific ones of all the brain's functions.

The best more-or-less straightforward indicator of this human variation in brain function asymmetry is handedness.

In point of practical fact, handedness — neuromotor hand preference — is arguably no big deal in and of itself, at least most of the time, at least for most people. Some tasks are harder to perform with a wrong-handed tool (a tool designed and built for use by the hand opposite the one you would prefer to use). Happily enough, however, people in the nonrighthanded minority are on the average more flexible in their behavioral functionalities, and do not have nearly as much trouble using a majority-handed tool as the righthanders do when things are the other way around.

It may also be considered a good thing that there are not all that many tools for the use of which handedness makes an important difference. Scissors and golf clubs and guitars are the worst I can think of offhand. Humans vary a great deal as to the importance they perceive those particular implements to have in their lives — if, for example, I never pick up a golf club again, I will almost certainly be okay, but I cannot comfortably say the same about scissors or guitars.

Handedness is important to this story because it is the single most accessible indicator of the fact that a not-strictly-righthanded minority of people have brains that are built differently and which function differently from those of the fully-right-handed majority. The establishment of brain function asymmetries begins in embryogenesis, in cellular events and processes that are different for twins from what they are for singletons. When humans, like any other organism, do something in an unusual way, we are given opportunities to learn more about how the usual way of doing it works.

The distribution of handedness in the human populations is not random. The estimated fraction of nonrighthanders in the population varies a bit depending on the exact sampling methods used, and differences among sampled subpopulations may be real, but ... 7:1, 8:1, and 9:1 (which fairly well cover the range of published handedness sampling results) are each and all very definitely not the same thing as 50:50.

Some of the scientists who have worked on these questions have been supposing recently that random directionality is the true nature of the minority brain-function-asymmetry status. In other words, randomness is the default pattern expressed when the genetic mechanism/s responsible for the characteristic asymmetry of the "normal" pattern are not working for whatever reason. For handedness it has been argued (not compellingly demonstrated) that there is no genetic determination of nonrighthandedness, just right-handed versus random. If this is correct, then we can only detect about half of the members of the minority because the random other half of the members of the minority look just like the majority.

Even if the minority really has twice as many members as we can detect, the largest fraction we could justify supposing from existing data is about one-fourth, and 3:1 is still definitely not statistically the same as a random 50:50 in any sample larger than 12. Clearly, human development within normal limits includes a strong bias in favor of one direction over the other for multiple physical asymmetries of the individual human body, very specially and specifically including that of the motor functions of the brain. I will hope to show you that the development of those asymmetries is set in motion by the first cell division of the single cell (zygote) that sometimes forms from the fusion of egg and sperm.

We no longer suppose — as we once did — that brain function asymmetry is a strictly, specifically human thing, but we do still believe that functional asymmetry in the human brain is stronger and more complex than the asymmetries in all other kinds of brain, and that this physical and functional asymmetry is an important part of the changes in brain structure that have driven, or followed, our development into human-ness as a species.

Essentially every significant deviation of human mental and behavioral development is associated with anomalies of brain function asymmetry, as represented adequately for most purposes by unusual motor hand preference.

It cannot be reasonably imagined that unusual handedness causes academic giftedness or retardation or alcoholism or seizure disorders or psychosis or any of a number of other anomalies of behavioral development — but nonright-handedness has significant statistical associations with each and all of those departures from the usual. In all of the major mental illnesses that have been studied appropriately, the underlying brain dysfunction is asymmetric, as are the underlying normal functions.

The original question that drew me into this research, and has sustained my curiosity for decades, is: Where in development does this system of left vs right brain function differences come from? How does development establish all the structural and functional differences between the left and right halves of the human body, especially in the brain where it is so central to what we think of as making us human?

How Does the Embryo DO Left vs Right?

I thought that might be a piece I could bite off and chew, large enough to be of great interest, but small relative to any way of considering the mind as a whole and perhaps simple enough that I might be able to learn something about it and explain what I found.

What first made me think I might be able to make progress toward an answer for that question was reading published observations that twins are more often nonrighthanded as compared with members of the general (primarily single-born) population. Twins do embryonic development differently and develop brain function asymmetry differently. What are the connections? By studying the ways in which twins and chimeras do the building of the embryo differently from the ways most of us do it, we might expect to learn a great deal about how it usually happens.

> "To me the most astounding fact in the universe, even more astounding than the flight of the Monarch butterfly, is the power of mind that drives my fingers as I write these words. Somehow, by natural processes still totally mysterious, a million butterfly brains working together in a human skull have the power to dream, to calculate, to see and to hear, to speak and to listen, to translate thoughts and feelings into marks on paper which other brains can interpret. Mind, through the long course of biological evolution, has established itself as a moving force in our little

corner of the universe. Here on this small planet, mind has infiltrated matter and has taken control. ...

"It appears to me that the tendency of mind to infiltrate and control matter is a law of nature. Individual minds die and individual planets may be destroyed. Or, as Thomas Wright said, "The catastrophe of a world, such as ours, or even the total dissolution of a system of worlds, may possibly be no more to the great Author of Nature, than the most common accident of life with us." The infiltration of mind into the universe will not be permanently halted by any catastrophe or any barrier that I can imagine. If our species does not choose to lead the way, others will do so, or may have already done so ... The universe is like a fertile soil spread out all around us, ready for the seeds of mind to sprout and grow. Ultimately, late or soon, mind will come into its heritage.

... "I do not make any clear distinction between mind and God. God is what mind becomes when it passes beyond the scale of our comprehension. God may be considered to be either a world-soul or a collection of world-souls. We are the chief inlets of God on this planet at the present stage of his development."

Freeman Dyson, in Infinite in All Directions

Chapter Three

Genetic Twin Studies of Schizophrenia

Mind/Brain Asymmetry, Multifactorial Inheritance and Epigenetic Variation

Nothing hurts a new truth more than an old error.

Johann Wolfgang von Goethe

In the beginning of the work I am describing here, even though my second child officially saw light of day only two minutes after my firstborn, I knew nothing more about twins than the fact that people call children "twins" when there are two of them born from the same pregnancy.

I developed a curiosity about the genetics of schizophrenia when a close relative of mine was given that diagnosis in error. Important genetic twin studies had been carried out for decades prior to that in the effort to find the cause/s of schizophrenia. It was that curiosity about schizophrenia, and that approach by way of twin study genetics, that put me on a path toward a life's work of curiosity about twins and about twinning as an unusual version of the process of building and growing into individual human lives.

At the beginning and for a while thereafter, I was ignorant of the common knowledge — what I later learned that some people said "everybody knows," "obvious facts" about the biology of twins and twinning. I did not know it, and that was definitely a good thing. I did not know that what I would decide to do could not be done, so I set about doing it. Normal variations in asymmetry of brain functions, and asymmetric abnormal variations in brain functions, soon made themselves known to me as issues in schizophrenia and in twinning — at first by way of seeing clear differences from "the general population" among schizophrenic twins. I knew even less at that beginning about handedness than I did about twins, and less still about the statistical methods I would have to learn to help me make sense of what I would find as I got farther into it.

In the beginner's mind there are many possibilities;
in the expert's there are few.

Shunryo Suzuki-Roshi

We know accurately only when we know little;
with knowledge doubt increases.

Johann Wolfgang von Goethe

I had already figured out that there is no story in the world more complex or more compelling than the story of the business of becoming human. There is no story more worthy of our attention than the one about how human life fits in with Life in the Universe. There is no better way to approach that than to learn about how each of us came to be all of what we are by way of unpacking the contents of a single cell, putting things where they belong and setting up housekeeping.

I had begun to know that everything about all of what we are and all of what we do is done by cells, never working alone but always in complex systems of cells, of subsystems within systems and supersystems of cells working together in tissues and organs. Remember the ant … and the ant colony … and the bacteria …

I had begun to know that no cell ever does anything it does not "know how" to do. Each cell must have the mechanochemical wherewithal — the means — to do what it needs to do. It must have the genes to make the proteins to make the necessary mechanical systems of cellular, subcellular, tissue, organ and organism structures. It must have the chemical systems of enzymes to catalyze the great many different chemical reactions that will be necessary from time to time in development and from place to place in the body. I had then, and still have now, much more to learn about that. I do not believe a week has passed in the recent past few years without some new piece of developmental genetics appearing in the scientific literature or the newspapers and thereby becoming available for learning.

In part because at first I knew no better than to 'believe' what the literature said about the biology of twins and twinning, I thought I could see a logical way to learn things from twins about how the business of becoming human unfolds.

Everybody already knew that monozygotic twins (There are no "identical" people … have I told you that already?) must have some very odd sort of embryogenesis, to make two embryos out of a cluster of cells that would otherwise, if everything were to go as usual, become only one embryo. The "splitting" idea always sounded goofy to me, but I really did have to believe that there had to be something odd happening ... We thought we knew that dizygotic twins, in a process quite distinct from that of monozygotic twinning, with nothing odd about it, must come from two egg cells.

'Everybody' assumed that had to mean separate and independent double ovulations, in the same way it seemed to happen in all the litter-bearing mammals. Those species generally seem to have more or less exactly one ripe follicle and then one follicle scar (corpus luteum) per fetus. No credible evidence exists as to numbers of corpora lutea in human multiple pregnancies.

We thought we knew this had to mean that dizygotic twins always go through two perfectly ordinary (separate and independent) embryogeneses (that is the proper plural of embryogenesis, which is the proper name for the process of building an embryo from the substance of a zygote). They should develop just like singleton embryos, but in the same womb at the same time, in parallel. That was what the literatures of twin biology had to offer at that time.

I had, by then, a good bit of prior experience working with mutant bacteria and viruses, learning, from the ways the mutants do things wrong, how the normal bugs do those same things right. From that background, I felt certain that I could learn some things about human developmental biology from developmental differences that I was certain I would soon find between monozygotic and dizygotic twins.

Unlike "genetic twin studies," my thinking here is not about within-pair differences and whether monozygotic co-twins are more alike than dizygotic co-twins. I wanted to know about any differences there might be between monozygotic and dizygotic twins as groups of "different kinds" of people with different embryogenic histories. I wanted to find and understand any differences between all monozygotic twins as one kind of people and all dizygotic twins as another kind of people, for the prospect of understanding the developmental consequences of those very different paths through embryogenesis that "everybody" visualized for them.

My perspective at that time, based on all of that "common knowledge," was that the differences between monozygotic and dizygotic twins as groups should be the same as the differences between monozygotic twins and singletons as groups, except perhaps for any effects of spending gestation sharing a womb. In other words, the dizygotic twins were assumed to be the developmental equivalent of singletons, but they might be affected by some experience/s of twin conception, gestation and/or delivery, and could therefore serve as a control group for any possible effects of those twinship experiences peculiar to going through twin conception and pregnancy.

In fact, as it turns out, the effects of sharing a uterus during twin gestation are negligible, with the possible exception of challenging uterine resources late in pregnancy when the total mass of baby bodies in the uterus to feed and clean up after may go well beyond the expected products of a singleton pregnancy.

As I would soon begin to discover, as the results of my studies would show, that system of underlying premises I had set up for my work is very importantly wrong. The why's and how's of its being wrong turn out to be the most important parts.

Twinning and Brain Function Asymmetry and Schizophrenia

In the beginning, there were twins and there were left-handers, and they were the same people more often than the appropriate joint probability statistic would suggest that they should be. That is to say, people who are born as members of a twin pair are lefthanded, or — more accurately — nonrighthanded more often than are people who are born single.

Twinning and departures from the human population's most common motor hand preference are not statistically independent — they are strongly associated. But I did not know that part right away. I learned that part from poking around, under and through and behind, the literature about twin studies on schizophrenia.

I have even quite reasonably suggested that every lefthander in the world may be a twin. At least one eminent senior worker in the field liked that idea in my work well enough to publish it elsewhere as if it were his own ... but that will fit much better into the puzzle after we learn about all the twins who are born single. It is also, in fact, entirely possible that all of the victims of midline/fusion malformations and all of the schizophrenics in the world are products of twinned embryos. … we'll get to those farther along, as well …

Genetic Twin Studies in Schizophrenia

Schizophrenia is a horrible disease, and to this day, even (!) with my trying to help for 40 years now, no one out there knows (well enough to prove it to the world) how schizophrenics come to be so sick, or even how to identify and explain the things that are different in their development, that cause their brains to function in the ways that make them so sick. Demonic possession became unsatisfying as an explanation for schizophrenia a long time ago, but it is still very nearly as useful as anything else that has been proposed.

Theories about "schizophrenogenic mothers," who created "double-bind" life situations and kept at it until the child had nowhere to go but crazy, also had their day and fell by the wayside along with most of the other blame-the-mother theories about the causes of distressingly unusual forms of human behavioral development.

A dozen or more different "environmental" contributions and sources have been considered and corresponding hypotheses tested and rejected. Not one of them has ended up fitting well with all of the available data, and not one has helped much with the effort of understanding where schizophrenia comes from or how it does what it does to its victims. Furthermore — all of the family studies and adoption studies and all the finest, fanciest, newest correlational path analysis twin family studies agree: *whatever is important about "the environment" in the development of schizophrenia is not about the shared family environment, not about the people with whom one shares a (life at) home.*

The genetic parts of the family connections explain a greater fraction of the whole system of relationships than everything else put together. These genetic analyses are basically about the statistics of correlation; genetic relatedness to a person previously diagnosed with schizophrenia correlates very highly with the risk of another such diagnosis in the family.

The fact that twins believed to be monozygotic are not always both schizophrenic has been considered for decades to constitute evidence — that there must be important "environmental" contributions to the causes of schizophrenia ["environmental" here meaning "not genetic in a plain Mendelian sense" and therefore imagined to mean something about life experiences]. This is one of the more promising of all uses for our developing understanding of epigenetics — about ways in which much of variation in development, and in the control thereof, has more to do with controls on the expression of gene sequences than it does with the sequence of base pairs in the DNA. More about that later.

Some people still do not want to accept the concept of mind as function of brain. Because of the astounding complexity of its structure and its functions, we cannot tell exactly *how* the cells of the brain work together to put on the show that is the mind. There is plenty about the mind that still appears to be somewhere between mysterious and miraculous. There is accordingly plenty of enthusiastic sympathy for those who suppose that the human mind will never be able to understand and explain itself.

> Any Universe simple enough to be understood
> is too simple to produce a mind able to understand it.
>
> *John Barrow*

… does that also mean … any mind simple enough to be understood is too simple to understand itself …? … just wondering …

As long as science cannot provide a system of detailed, concrete mechanisms, some people will prefer to hang on to believing in some sort of ineffably transcendent thing that they can think of as justifying our history of behaving as lords over all we survey. That, in turn, is apparently just an idea that most people will never want to outgrow or give up.

It is hard enough to conduct the business of being human with any grace or skill without ever giving any thought to the great many qualities from which human life cannot be separated that ensure its final inevitable absurdity. Not a single one of us is going to get out of this world alive. You are absolutely responsible for every single instant of your time on this Earth. You cannot absolutely control even a single instant of your time on this Earth. No one ever had a nervous breakdown as a result of not trying hard enough to control his or her trip through this world.

> Everything has been figured out except how to live.
>
> *Jean-Paul Sartre*

Schizophrenia and the affective psychoses (the latter including major depression and manic-depressive psychosis, now most commonly called unipolar and bipolar affective disorders, respectively) together have been called the "functional psychoses," and schizophrenia has been called "the graveyard of neuropathology," because no one has been able to find what is wrong with the brains of schizophrenics to cause their minds to go so badly out of line with those of most of the rest of us.

"If we can't find anything wrong with the structure," so the reasoning seemed to go (assuming that our failure to find whatever might be wrong with the structure means that there is nothing wrong with the structure), "then the problem must be in the function ..." I am willing to bet that at least one major flaw in that logic is immediately more or less obvious to you, but apparently they were not always obvious to the scientists on the scene at the time. Honesty and humility are virtues in science, too; maybe even better or at least more important than in the rest of life. Remember the more or less universal assumption about the accuracy and significance of our observations?

Schizophrenia is a devastating chronic mental illness. It has been responsibly estimated to cost the USA over 60 billion dollars per year in health care costs and lost productivity — and that estimate is several years old. Fewer than 10 % of its victims are able to hold full time jobs, and what jobs they can hold are almost always substantially below any expectations the victim might have had before becoming psychotic. Although some other mental illnesses can generate more drama and uproar from time to time, over the long haul schizophrenia is the worst. (From young adulthood through the rest of life is unarguably a long haul, to be shortened on the average only by a substantial excess of violent deaths among schizophrenics, just under half of which are suicides.) There have been reports of some "physical" illnesses which are significantly rare among schizophrenics, but apparently none that are significantly more common.

By the way, when any work like this one says "significant" like that, unexplained and unqualified, it means "statistically significant." "Statistically significant" means an observed difference is large enough that, according to an appropriate statistical measure, it would be hard to believe that a difference that large could have happened purely by chance in the course of taking the samples for the comparison.

At least manic-depressive (recently, most commonly called "bipolar") psychosis can be exciting for an occasional while, when the last depressive crash fades far enough into the past to be forgotten for a while and the next one cannot be seen coming from way up there in a manic phase. Schizophrenia is never any kind of fun for anybody in it or in reach.

As I said a page or two back, I first became interested in schizophrenia because someone I love had recently gotten that diagnosis — because the doctors knew nothing about all the amphetamine and chemically related substances that he had been using. According to numerous apparently sound reports, anyone who takes enough "speed" long enough (the necessary dose and duration are variable among individuals) will become psychotic. Without knowledge of that particular kind and history of drug use, only with extended observation can mental health care personnel distinguish the presenting psychosis from schizophrenia.

In those days, about 25% of the hospital beds in this country were occupied by people with a diagnosis of schizophrenia. That number has since then dropped dramatically, but not because the frequency of schizophrenia has changed. We just don't lock them up much anymore, certainly not for decades at a stretch or the rest of their lives whichever comes first, as was once the case. That change seems to have begun in California during Mr. Reagan's tenure as governor.

I was in graduate school in California then, and I remember myself and many liberal-minded others having some sympathy for arguments to the effect that society had no right to restrict the freedoms of people who did not present immediate danger for themselves or others. I could not, however, see what was happening to them because of those changes as liberation. The civil rights argument made a more or less noble cover story, but, in some other layers of the reality, there appeared to be at least as much concern as to the costs to the taxpayers of the state for maintaining a decent system of state facilities for the mentally ill. Many of them had been involuntarily committed entirely because they did not have sense enough to know how badly they needed to go there and stay voluntarily.

All of this has had the result that, in 21st century USA, our schizophrenics today are mostly to be found among our homeless. Many of them can from time to time be pretty scary up close, but for the most part, they cannot collect themselves enough for long enough to conjure up clear and present danger to themselves or others. And "clear and present danger to themselves or others" has always been and still is the only circumstance in which our system of laws gives society a legal right to deprive anyone of their freedoms in order to segregate them for the comfort, convenience or safety from embarrassment of their own or the rest of us. Besides, schizophrenics are many times more likely to do violence to themselves than to anyone else, and it costs a lot of money to keep them locked up and to house and feed them. They cannot pay it, so taxpayers had to.

Nearly half of all schizophrenics attempt suicide at least once in the course of living with their illness. In most situations today, in the absence of very substantial family wealth, there is in most public facilities a specified maximum time, generally no more than about 72 hours, during which to diagnose, prescribe and release. It can readily be seen as a sort of existence adverse to most human preferences and to all the things most of us like best about being human.

Schizophrenia damages the most particularly human things about the victim, the highest-level functions of the human mind. It is a thought disorder. Its victims cannot in general hold places in society, cannot hold jobs, cannot maintain safe and productive relationships even with themselves.

Drug therapies are believed by some to have improved over these recent decades, and they are said to make a modest approximation to a normal life available to some of the patients. Even with very good luck with drugs, life prospects for the victims are almost always irremediably reduced from the lives that they seemed to have been headed for before their breakdowns. Further, there has always been a substantial fraction (odd, but it seems usually to be about one-third) of patients who simply do not respond in any useful way to any given drug. Happily, that is not the same third for every drug, but a sizable fraction do not respond well to any of the drugs.

Each and all of the antipsychotic drugs (including the new 'atypical' antipsychotic drugs, which in spite of strong promotion surrounding their introduction are making no significant difference in the range of outcomes) also have such unsavory side effects that many of the patients decide it feels better on the average day to stay crazy. At least on the surface of it, willful noncompliance with antipsychotic prescription drugs is still, and seems likely to remain, a large and enduring problem.

Schizophrenia "runs in families." As long as I have been acquainted with the problem, it has been said that about 1% of the human population worldwide will become diagnosably schizophrenic sooner or later (most commonly as young adults). Some recent work indicates that the perennial "about 1%" statement of lifetime risk is on the high side. A recent "critical review" of international statistics says that the number should probably be more like seven or eight per thousand; that is, seven- or eight-tenths of 1% — and still more or less uniform worldwide.

That uniformity is a profoundly fascinating part of the picture, but, in the several decades that we have been telling that to each other, we have yet to understand anything that we might stand to learn from that observation.

Given the apparent huge variation in efficiency and effectiveness among the national public health statistics organizations of the world, I have always wondered how any confidence could be placed in the precision of any such estimate, anyway. The primary importance of the number, it seems ... the best reason for wanting to know it more or less exactly, is as the reference point against which to compare frequencies in various groups of relatives of already diagnosed schizophrenics. Comparisons such as that really require that diagnosis should be uniform across groups to be compared. So it would seem that consistency without great precision may suffice if the accounting is the same in the groups to be compared.

Through it all, through one decade of research work after another, by far the single best predictor of any person's individual risk of becoming schizophrenic is the closeness of his or her genetic relatedness to someone who has previously gotten the diagnosis — relatives with whom one shares genes, not so much housing, meals, homework or after-dinner games.

Genetic relationship accounts for a larger share of the schizophrenic-vs-nonschizophrenic variation among family members than any other factor or combination of factors. But these pedigree studies persistently show us that none of the known single-gene patterns will fit the data. No simple Mendelian model will explain the way schizophrenia segregates in families.

Adoption studies are another very important parallel approach. Those studies show no change in risk in either direction as a result of the changes in relationships following on adoption:

— Children of schizophrenics adopted into normal families show no reduction in their genetically elevated risk as a result of living in a normal home.

— Children of normal parents adopted into families with one or more schizophrenics or future schizophrenics show no increase in risk over those who stayed at home.

Neither schizophrenia nor its normal alternative can be transmitted socially within any reasonable range of behavior. Perhaps it can be taught, or induced psychologically, but probably not unintentionally.

Any "environmental" contribution, therefore, is supposed/imagined to arise from individual, "non-shared," life history. The parents and the home environment they provide may be good or bad in many dimensions, but together they contribute nothing we have yet been able to demonstrate that might help in explaining the development of psychosis, to give us any hope of preventing or reversing it. Numerous widely varying kinds of "environmental" variables are more common among schizophrenics than in the rest of the population (including but not limited to: being born in winter, head injuries, "birth stress" from difficulties in delivery, viral infections suffered during pregnancy ...), but not a single one of those "environmental factors" affects every schizophrenic or even a majority of them, and not one of them affects only schizophrenics.

Sorry, but in fact parents-as-environment really never score many points in such analyses of any form of mental and behavioral development. Peer groups are in general much more influential, so that the most likely aspect of child-rearing for parents to influence significantly seems probably to be in steering the child to spend time with the "right" people as peers. In my personal experience, it seems that nothing any parent ever has to say to a child makes any real difference until and unless it is confirmed for the child by another arguably responsible adult who has no apparent personal interest in the results. An apparently offhand poke from Uncle Buddy or Aunt Geneva can often mean more to a child than years of harangues from Mom or Dad.

This is where Genetic Twin Studies come in. There was at least one earlier writing on the subject, but Sir Francis Galton (a cousin of Charles Darwin), from his writings in 1875, is generally given credit for the idea of using twins to ask questions about the tendency of any given trait to be inherited. Galton made many solid contributions to genetics and psychology. It was Galton who coined the phrase "nature versus nurture," and he can indeed fairly be called the founder of the science of behavioral genetics.

Galton did not understand with cellular precision the different origins of the very-similar twins **vs** the clearly-different twins that he studied. We have to give him some slack there … if anybody had that one nailed down, even now, I might be writing a rather different kind of book, if any. Galton was, after all, roughly contemporary with Mendel and ignorant of Mendel's work along with the rest of the world. He did his best work a few decades before "genetics" even got its name. He did, however, clearly demonstrate the utility of the basic concept that still underpins the genetic twin study method … that is: when similarity between siblings in one given trait under investigation correlates strongly with within-pair similarities in all other traits examined, the similarity in the relationship is to that extent probably mediated by heritable variations.

If twins who closely resemble each other in "all" of their visible and measurable traits also tend to resemble each other with respect to a trait being studied, and twins who never did resemble each other so much also turn out less likely to match with respect to this trait being studied, then the trait being studied probably has some important genetic contributions to its causes. This idea of comparing within-pair similarity between monozygotic co-twins, vs dizygotic co-twins — as groups of pairs — has long been considered one of our most powerful tools for analyses of the sources of variation in human mental and behavioral development.

> I have … sought some new method by which it would be possible to weigh in just scales the effects of Nature and Nurture, and to ascertain their respective shares in framing the disposition and intellectual ability of men. The life history of twins supplies what I wanted.
>
> *Francis Galton, 1875*

Epidemiology (the study of factors affecting health vs illness of populations) is most widely understood with respect to its concerns over the spread of infections, infestations and environmental toxins in the population. More basically, in fact, it is about how this person or this group of people has come to have this trait or this health problem and those other people over there do not have it. Infections and environmental toxins are always high on the list of usual suspects, but they certainly do not make up the whole list of proper epidemiological concerns.

Even with respect to pathogens and toxins from the environment, epidemiologists have long known about "host factors" — individual human variation in sensitivity to one or more of the threats from the environment. Chickenpox, for example, is without question an environmental cause of disease. It comes from the environment in the form of a virus named *Varicella* that is always around somewhere nearby. But there are "host factor" individual variabilities in the ease of acquiring the chicken pox infection, and in how generously virus particles are released into the environment from the infected person, and in whose body will or will not allow the virus to lie dormant in nerve junctions near the spine until it later erupts as shingles, in patches following the distribution of that nerve's branches through the skin. I am given to understand that no one but a shingles victim can fully appreciate just how nontrivial and inconvenient that last kind of individual variation can be.

Genetics within medicine — "medical genetics" — is a part of epidemiology, and a kind of epidemiology. The analytic face of medical genetics is about learning how this person or group of people got this particular trait or got sick in this way when most other people do not. We would still very much like to know how anybody becomes schizophrenic. When pedigree-based studies of the sorting of a trait among members of families are not compatible with a simple one-gene answer, when there is no credibly simple pattern of variation in repeat risk as a function of degree of genetic relationship to affected family members, then, for a long time now, a favorite next step has been to set up a genetic twin study.

The Multifactorial Liability/Threshold Model of Inheritance

If we cannot explain the distribution pattern of a familial trait with a simple one-gene model, we go straight to "multifactorial inheritance" — stated more exactly, that is the multifactorial liability/threshold model of quantitative inheritance.

In theory, we know how to analyze the distributions expected for two or three single genes, but we seldom have the necessary handles on their separate manifestations to follow their interactions through families. The "multifactorial inheritance" hypothesis sets up a model of many genes, of small, equal and additive effect, plus environmental effects. Yes, that is technical, and yes, it is jargon, and this is one of those places I warned you about, where just a little too much simplification can make the idea unrecognizable and useless. I drag a few dozen students through this patch of woods every year, but I expect them to learn how the calculations are done, so that effort requires the use of rather many more words than I really want to try to squeeze in right here. But … let's give it a bit of a try; to make it as simple as possible and still keep it true:

Multifactorial means "involving many factors." Not one gene as in the simple Mendelian distributions, but many genes (there is also "polygenic," but that is not the same as "multifactorial" because that concept does not include environmental factors). The multifactorial model includes many genes *plus* "environmental" factors, which are imagined to be modifying the influences of one or more of the "many" genes involved.

Genetic analysis by way of the multifactorial threshold model is an exercise in the fractionation of phenotypic variance. … see, there; we are already off the rails! — what that little four-word pile of polysyllabic jargon really means is an effort at estimating the contributions of various inputs to the final range of individual variation in a particular trait. In any given sample of people or families, the members vary in their phenotypes (the way they look, with respect to the trait the cause/s of which we are trying to analyze). Some, for example, are schizophrenic and some are not. We want to understand the origins of that variation/variance. We want to quantify it and analyze what we have, to see what other information we have that reliably goes with it.

We use the multifactorial model for genetic analysis of two rather different groups of traits. Most of my students find it fairly straightforward to learn it with respect to all the "bell-curve," continuous, "metric" kind of traits: height, weight, blood pressure … variables that can take any of a continuous infinity of values, extremes of which are pathological or indicate dangerous tendency toward likely future pathology. A value usually near the peak of the curve — near the most common value — is the average, the value that most of the values tend to cluster around. The variance is a measure of the dispersion of the values in the sample around the mean (the width of the bell curve at half its height is a good proxy). Some bell curves are pointier and some are rounder on top, depending on how more or less closely the values in the sample tend to cluster around the mean. A distribution with a narrow, pointy peak has a smaller variance than a broad, softly rounded one.

Every person in the sample can be scored by the deviation of his value from the mean value for the whole group. The multifactorial/threshold model of inheritance simply supposes that every individual value is the mean value plus a component of deviation (plus or minus) caused by his or her genes. Because it WILL show up AND assume an important role, we might as well go on and write it down: we also need a component representing the imperfections in that relationship. That is the component of the variation that is not genetic in a Mendelian way, at least partially independent of the Mendelian genetic component, and which therefore gets called "environmental." It is in general more accurate to call it the error term, because it collects everything that does not perfectly fit the idea that all the variation is caused by (a certain limited class of) genes.

At the end of the exercise, we will have quantitative estimates of what fraction of all the observed variation belongs to each of those sources. The model can be, and has been, made much more complex and sophisticated over the decades of its use, but not really much more powerful or accurate.

"Environmental" factors in the basic variance-fractionation models of genetic analysis actually include all forms and components of statistical lack-of-fit. All values of the trait in question among the members of the sample population that are not exactly where their genetic score says they should be are showing some lack-of-fit. The "environmental" components are parts of the total phenotypic variance (variation in the appearance of the sampled individuals with respect to the trait in question) which cannot be accounted for by the main prediction of the model, in this case the only prediction, namely simple Mendelian genetic relationships.

Some of that non-fitting variation may be truly environmental, as in virus infections and such, but ... that "environmental" (lack-of-fit) term in the mathematical model also invariably includes the effects of:

a) any *genes* segregating in the sampled families that (we don't know about and which therefore) are not specifically included in the model — they are assumed not to exist and therefore to have zero effect

b) any *interactions between genes* whether or not included in the model — interactions cannot be part of a linear (all first-order) model, so are assumed not to exist and therefore to have zero effect

c) any *interactions between genes* whether or not included in the model, *on the one hand and non-genetic factors on the other* — again, interactions are non-linear, they are assumed not to exist and therefore to have zero effect, plus

d) all random *error in sampling, diagnosis or measurement*.

The model assumes and tests the assumption that all the variation is genetic and behaving in perfect accord with Mendel's Laws, and that the genetic component tracks through family relationships as if it were a block of genetic information accounting for a single trait. The single trait in question can be height, weight, blood pressure or — not schizophrenia itself, but *liability to schizophrenia*. Variation from "environmental" causes — all variation other than that due to the genetic component in the model — does not fit that single-variable model, and shows up as lack-of-fit.

How many is "*many* genes"? It might be quite many, or it might be only two or three or four. We just barely know how to make predictions for two simple, additive, non-interacting genes, let alone several or a great many, to say nothing of interacting genes regardless of the number thereof. We cannot usually analyze even a two-gene situation effectively because that requires that we must somehow realize and measure a separate component of growth or behavior or appearance representing variation due separately to each of the two genes. If we are looking for the influence of two or more genes on a single phenotypic trait, there will not in general be two or more sub-phenotypes or components of the phenotype that we can follow separately through pedigrees. There are numerous other difficulties as well, including some serious ones, but they will never matter at all until we get to the other side of that one … It is not clear how many is the minimum number to constitute "many."

"Small" effect is arbitrarily set to mean that no one of the involved genes is responsible for as much as 25% of the observed variation. It is also not at all clear how equal "equal" may be expected to be. It is fairly clear that it means no epistasis is allowed in this model. Any and all effects of epistasis are assumed to be zero. That means that every copy of every gene involved contributes its effect/s regardless of the status of any other genetic component of the system. We know some gene systems where, for example, a system of several genes determines how numerous and how large the black spots on the animal's white coat will be if there are any spots. Regardless of the versions of each gene present at those spot-number-and-size gene loci, there will not be any black spots unless another gene is present in a particular version, to allow that there may be any spotting at all. That plus/minus spotting gene is epistatic to the set of genes that determine variation in spot size. Its effects on the coat-color phenotype (the appearance of the animal with respect to coat color) are not equal to the effects of the other genes involved.

Another example: human ABO blood types are determined by gene-product enzymes that either do or do not change the type-O cell-surface-marker into the type-A and/or type-B cell-surface marker. Cells of blood type O have neither of those enzymes, and therefore will change none of the type-O antigen into either the A antigen or the B antigen on their surfaces. A mutation (called "Bombay") that causes the type-O surface marker (called "H substance") not to be made, so that there is none of it available to be made into A or B, will cause an individual expressing that mutation to type as O (as if lacking both A and B) whether or not the gene products that make A and B are present. A mutation in one gene prevents the expression of either of two others; this is another example of epistasis.

When the expression status of one gene controls the expression of one or more other genes — that by definition is epistasis, and it means that one gene pair has more influence than others in the system. That does not fit the definition of "equal effect" and it does not fit in the multifactorial model. So we must assume epistatic contributions to variation have a value of zero and no effect.

"Additive" presents a similar sort of problem. The restriction to "additive" effects means that no dominance is allowed as part of this model. To say that a particular version of a gene (a particular allele) is "dominant" means that one copy of the dominant version makes the organism look just as it would with both copies being of the same dominant version. The dominant version of that DNA sequence is expressed regardless of what allele occupies that locus (DNA sequence site/location) on the other chromosome. That means the presence of one copy of the dominant version can mask the presence of the recessive version on the other chromosome of the pair. So, in the presence of dominance, some allelic versions of a gene will have more influence than others. That does not satisfy the "additive" restriction of the model.

In all the world of living things and all our studies of their genetics, we know of only a handful of situations where the behavior of a genetic system clearly and simply yields a very good fit to a multifactorial or even a polygenic model. Even so, we more or less insist on using it as an approximate fit for all the situations in human development where we are convinced that the process is genetically controlled but we cannot make sense of any single gene model. Many of our applications of multifactorial genetic analysis have to do with variation in elements of growth — height, weight, blood pressure, etc. A large fraction of it is about features of behavioral development, such as handedness and psychosis.

"Many" genes? Actually, with a perfectly ordinary amount of variability in their expression, three or four gene pairs can easily generate a distribution of their combined manifestations that cannot be distinguished from the same sort of distribution for three or four dozen gene-pairs, in samples of usable size for credible statistical analyses.

There are, in all of known genetics, a very few traits that fit that multifactorial model more or less exactly in all of its particulars. Schizophrenia is not one of them, but it fits in several ways and we have had no better way to understand the aspects of its distribution which do not constitute a good fit.

Genetic twin studies have been the favorite, and have been considered a very powerful, way to study the genetic bases of complex, especially behavioral, traits such as the functional psychoses.

The cardinal element of the genetic twin study is the sample. Before we can do anything else, we must find a number of twin pairs sufficient to give statistical credibility to our results, in which at least one member has a solid, reliable diagnosis of schizophrenia. For purposes of genetic analysis, we must come as close as we can to certainty about the uniformity and reliability of the diagnosis. If, for example, the diagnostic range for our sampling happens to include two or more different ailments which are similar in every way we know how to measure, but which arise from different mechanisms, our effort to identify "*the* mechanism" is doomed.

We can feel fairly safe in supposing that all cases of a particular illness in a single family line of descent are arising from the same one mechanism, but we can seldom accumulate statistically useful numbers of members of/from any single family. When we have to combine families to reach statistically useful pedigree sample sizes, we lose the comfort of supposing that all cases in the group have arisen from the same genetic causal contributions. This has been proposed as the most likely reason that many very well executed genetic linkage studies, up to and including whole genome scans, have been contradictory or fruitless with respect to schizophrenia.

When I started this work, the range and reliability of diagnoses of schizophrenia were subject to roaring debate, with the diversity of honest and well-informed opinions including the frequent and vigorous assertion that the diagnosis of schizophrenia was so diverse as to be meaningless. Even two psychiatrists trained by the same mentors might disagree about a given case.

It has been said (and still is, but apparently somewhat less often) that there is no such thing as schizophrenia beyond a label we have used to justify restraining people who behave too oddly to be tolerated comfortably by most of us others in society. The debates were never really settled, but my psychiatrist colleagues tell me that most of today's medical insurance programs will pay for treatment of schizophrenia only under the hypothesis that it is a biochemical disorder of brain tissue, to be treated accordingly with antipsychotic drugs. Psychotherapy is disappearing.

The study on which I really got started took a deliberately broad approach to diagnosis (every twin who could be found in the records of a very large mental hospital in the suburbs of London who ever got a diagnosis of schizophrenia anywhere), with a major effort at diagnostic precision in the form of requiring consensus diagnosis by at least four members of an international panel of six highly-regarded clinical-research psychiatrists working from detailed case histories. In my opinion, it stands as the definitive Galton-style twin study of schizophrenia (as diagnostically defined in that sample).

An interesting additional feature of that study was that one of the principal investigators had previously been involved in a twin study of handedness, the sample for which had included some of the same twin pairs now counted among this sample of twins with schizophrenia. In that smaller sample, it had been found that twin pairs the members of which differed in handedness usually also differed with respect to the diagnosis of schizophrenia.

In their sample, monozygotic pairs with one twin lefthanded and one righthanded usually did not both have schizophrenia, while the pairs with two righthanders usually were concordant for schizophrenia — both sick. Schizophrenia and handedness had no statistically significant association; there was no evidence that nonrighthandedness and schizophrenia had any sort of causal relationship. The right-handed twin in such a pair, and the lefthanded twin, were, within the range of sampling error, equally likely to be the single diagnosed schizophrenic.

In their new larger sample selected for the presence of at least one schizophrenic in each twin pair, they extended that analysis of handedness. They collected and included handedness data from all but one pair of their subjects. The answer did not change. Monozygotic pairs with different handedness were usually discordant for schizophrenia, but there was no consistency in the handedness of the twin who was the schizophrenic in the pair.

I now recall this as my first encounter with the excessive frequency of nonrighthandedness among twins. I would soon learn that several other people had known about it and published such results several times in the course of several decades before I found it again among this sample of schizophrenic twins. Since my work, several other studies have confirmed it — twins are lefthanded more often than single-born people are. To know how that comes to be so became a very important part of the quest represented by this writing. There will be more about this farther along.

In that sample of twins collected because at least one member of each pair was schizophrenic, the lefthanded twins clearly were not lefthanded because they were schizophrenic, nor were they schizophrenic because they were lefthanded. In that sample, the excess of lefthandedness was especially high among the monozygotic twins. Because, that close to the beginning, I still had some residual tendency to believe what I was told, I was still under the illusion that this made sense. Clearly, monozygotic twins must have an odd, asymmetry-disturbing sort of embryogenesis … and everybody knows that dizygotic twins suffer no such thing. From separately ovulated, separately fertilized, egg cells, they should have separate, and independent, and therefore "normal" embryogeneses. The literature was more or less common-knowledge-everybody-knows unanimous in supposing that the excess of any anomaly in twins must be expected to belong only to the monozygotics.

I changed the perspective from that of the previous investigation, by removing from any further consideration all of the pairs that included *any* nonrighthanders (just exactly as I might do if I found some of the members of the sample with histories of head injuries, on the chance that that history or anything associated with it might contaminate diagnosis).

I set those aside, as pairs because I thought the source of the anomaly might be — no, must be — somewhere in the "splitting" of their shared embryonic cell masses. That difference in the circumstances of their embryogenesis, more or less obviously, must be capable of affecting both members of any pair affected in that way. I supposed that the important feature was not that co-twins differed in handedness, but that the determination of brain function a/symmetry was unusual in the embryogenesis of either member and presumably therefore of both because it was necessarily the same occasion of embryogenesis.

With only this apparently small difference in my way of looking at the subjects, with the setting-aside of this minor fraction of the subjects, the results of the classic genetic twin-study considerations for the same sample changed dramatically. Monozygotic pairs in which either member was not strictly right-handed were much less likely to be both-schizophrenic. Because of that difference, taking out of the sample those pairs that included nonrighthanders also took out most of the pairs that were discordant, and made the result much stronger in favor of a genetic basis for the schizophrenic psychoses in this sample.

Most of the monozygotic *pairs with either member nonrighthanded* had milder, atypical cases in their schizophrenic members, including many of the cases with less-than-unanimous consensus (minimum four of the six diagnoses) among the diagnostic panel, and those with fence-straddling diagnoses like "schizoaffective." Most of those pairs with either member nonrighthanded included only one twin with a diagnosis of schizophrenia.

One of the strangest notes in the whole song was this one: Among five of the monozygotic pairs with discordant handedness, the lefthanded twin in each pair had a diagnosis of "schizoaffective" psychosis, while their righthanded co-twins were all consensus-diagnosed with "inadequate personality." Don't forget that every case history being used for diagnosis was considered in isolation from every other case history in the sample — all of the diagnosticians were blind to the twin pairings.

So ... at least among twins, the reduction of normal brain function asymmetry that is represented by nonrighthandedness appears to act both as a predisposing factor for schizophrenia (being even more frequent in twin pairs with schizophrenia than in normal twin pairs) and also as an ameliorating factor (being concentrated in pairs with only one schizophrenic, and usually one of the milder cases at that).

When I purged the sample of twin pairs that showed the causes or consequences of unusual brain function asymmetry in their shared embryogenesis, the monozygotic/dizygotic casewise concordance ratio changed dramatically. There was now near-perfect agreement between monozygotic co-twins, far stronger than between dizygotic co-twins. Ninety-six percent of cases of schizophrenia among monozygotic twins in this all-righthanded sub-sample are in concordant, both-schizophrenic pairs, vs 6% in dizygotic pairs sampled and sorted the same way.

Many found my interpretation of this result odd, dismissed it and went back to arranging and performing their twin studies, but these observations did then, still do and always will cause me grievous doubt about the appropriateness of the twin study method for questions about schizophrenia — or any other behavioral phenotype. Given some special relationship between twinning and brain function asymmetry, and also a special relationship between schizophrenia and brain function asymmetry, then it is seriously unlikely that the distribution of schizophrenia among twins can be straightforward. This takes nothing away from twin-study evidence to the effect that there are genetic contributions to the causes of schizophrenia, except for any pretense of precision in quantifying those contributions, or of sorting "genetic" vs "environmental" contributions in any meaningful way. [… with those features gone, there is actually not much left from the usual paper about such results …]

There is an essential, fundamental, but usually unmentioned and presumably therefore unrecognized, assumption of the Genetic Twin Study: To suppose any logical soundness in the analyses, the trait in question must be the same trait in monozygotic twins, in dizygotic twins, and in the rest of us in the singleton majority about whom we hope to learn something from the genetic twin studies. As discussed above, the mathematical basis of the genetic twin study is the "fractionation of the phenotypic variance." When the trait in question has different phenotypic variance structures in the comparison groups, the analysis is seriously flawed.

A couple of groups have published claims that they have failed to replicate those results of mine on handedness and schizophrenia in twins. This is generally supposed to be the kiss of death for scientific work — if someone else does the same experiment and gets a different answer, that is a bad sign. The fact is that they really did not even try an accurate replication effort. In each case their diagnostic range was substantially different. I had written in the meantime to point out that the then-proposed newer diagnostic criteria would have excluded, from the sample I had used, those twins with the atypical or mixed diagnoses. That group includes most of those I set aside because of their unusual brain function asymmetry. Along with them go the most interesting findings.

During my stay with Robert Elston's group in Chapel Hill from 1975–1978, I gathered substantial experience with multivariate statistical methods. By the use of those methods, simultaneous variation in multiple measurement dimensions can be used with particular effectiveness to classify members of a sample into groups, or to decide on the size and importance of differences between groups sorted by any of the various criteria.

Also available to me at that time were MMPI (Minnesota Multiphasic Personality Inventory) data from most members of the same sample of schizophrenic twins I had been studying. The MMPI is a rich, complex questionnaire that can be used to quantify multiple dimensions of emotional and cognitive function, with built-in consistency checks and extensive analyses of large diagnosis-normative groups to support interpretation.

Those data were lent to us in the hope that we might be able to find a multivariate MMPI indicator for the genetic predisposition to schizophrenia which must be lurking in the (so far) nonschizophrenic co-twins of the monozygotic schizophrenic twins. The idea was that the nonpsychotic co-twins of monozygotic twin schizophrenics must have the same genes that caused the schizophrenia in their co-twins, and perhaps the MMPI could show us a psychological manifestation of the genes that caused schizophrenia in one twin of a pair, and which are present but not (yet?) (fully?) expressed in the other twin.

I could not, did not, find that. That result is not there. Nonschizophrenic co-twins of schizophrenic twins could not be sorted into zygosity groups by any linear combination of MMPI scores, giving us no way to identify the carriers of an unexpressed schizophrenic genotype. The nonschizophrenic monozygotic twins are all supposed to have the gene/s which caused schizophrenia in the co-twin with "identical" genes, but the same should be true of only perhaps as many as half of the nonschizophrenic dizygotic twins.

From other work, by the way, part of the background that required the making of this comparison, the risk for schizophrenia is exactly the same in the children of nonschizophrenic co-twins of monozygotic twin schizophrenics as it is in the children of the schizophrenic co-twins. If that unexpressed schizophrenic genotype that we assumed would be present in the nonschizophrenic monozygotic twins was manifest in the MMPI, our analyses should have been able to separate them from the nonschizophrenic dizygotic co-twins that we could expect were nonschizophrenic because of not sharing the necessary gene/s.

While exploring relationships among those data, I did find a few other things that struck me as being at least as important, if not more so:

I found a statistically-significant difference in the MMPI structures of schizophrenia-in-a-monozygotic-twin *vs* schizophrenia-in-a-dizygotic-twin. In other words, according to my results, when twins have schizophrenia, monozygotic and dizygotic twins have different "kinds." Monozygotic twin schizophrenics and dizygotic twin schizophrenics have different kinds of psychosis as expressed in MMPI structure. This is not a good sign for people who want to perform Galton-style twin studies.

I also found a significant difference in the MMPI structures of schizophrenia-in-a-twin-from-a-pair-with-lefthander/s *vs* schizophrenia-in-a-twin-from-a-pair-with-only righthanders. When twins have schizophrenia, lefthanded twins and righthanded twins have different kinds. Lefthanded twin schizophrenics and righthanded twin schizophrenics have different kinds of psychosis as expressed in MMPI structure.

According to these results, the logic of the genetic twin study as applied to the analysis of schizophrenia is intolerably stretched and strained and twisted. This is worse, logically and practically much worse, than comparing apples & oranges. Monozygotic twins and dizygotic twins are not the same, and in neither group are the twins the same as singletons with respect to the psychological structure of schizophrenia. This is about comparing relationships within-pairs of apples against relationships within-pairs of oranges, in hopes of learning something about the rest of us who are … maybe peaches, maybe bananas, but apparently neither apples nor oranges.

The same is certain to be true in analyses of almost any variable of human mental and/or behavioral development, because apparently all such variables are intimately related to human brain function asymmetry, and twins do the development of brain function asymmetry differently from singletons (more about which later).

So, it seems that those results took the genetic twin study of schizophrenia as far as it could go, and then dropped it over the edge, demonstrating its complete lack of appropriateness for the task. Those MMPI results of ours were never published. The people who had collected the data became seriously concerned about the quality of the data for purposes of multivariate analyses and denied our use of the data for publication of those unfortunate results.

We cannot do much more with schizophrenia right here, but schizophrenia turns out to involve left-brain functions more than right-brain functions, and my results in twins contributed to the impetus for the work going on around me that eventually went on to make that clear. Exactly how or why the relationship behaved the way it did in the twins is still not exactly clear, but it certainly seems that most of the oddity is about embryogenic asymmetries in building the brain and the odd ways that twin embryos do those jobs.

At any rate, it was abundantly clear that twins in the sample I studied were much more often nonrighthanded than figures from general population samples. As I studied it further, though, what people had already written made it sound as if perhaps many or most of them might be left-handed because they were twins.

The fact that twins are more often lefthanded than singletons is what we call a statistical association. Handedness and twinning are statistically not independent; they are associated. If blue eyes and blond hair each occurs in half of the population, and if those two traits are statistically independent, then there will be about 25% of the population with both those traits, 25% with neither, and 25% with either one and not the other. [If half have blue eyes, and half of those (half of everybody including equally the blue-eyed ones) are blond … then one-fourth are blond and blue-eyed.] If, however, we find, for example, one-third with both, one-third with neither, and one-sixth each with one and not the other, then those two traits are strongly associated in the population at hand. If there are one-sixth each with both and neither, and one-third each with one and not the other, then the traits are strongly negatively associated.

That is rather like the way things are with twins and handedness. Twins are half-again to twice as likely to be nonrighthanded as singletons are. This is a correlation, or association. It says nothing whatever about cause and effect.

Nothing. < that's a period. Nothing. Seriously. And, the difference between correlation and cause is crucially important. No doubt you have seen correlations and associations in print or other news media without this warning about the difference between correlation and cause.

Remember this one?: People who regularly drink one glass of wine with dinner have fewer heart attacks than heavy drinkers or tee-total-abstainers. That statement does not offer any evidence to the effect that you should expect to reduce your risk of heart attack if you add a regular glass of wine to your routine if not already there. Nor, in fact, does it provide evidence that you will reduce your risk of a heart attack if you have been routinely drinking rather more than that and you now bring your alcohol consumption down to one daily glass of wine. Neither this nor any other correlation as such tells you anything at all about cause and effect. A pair of traits may be correlated — even perfectly — not because of any causal relationship between them, but because of a relationship each one shares with something you have never heard of, let alone looked at and properly measured.

If we were to divide a large crowd of tee-total-abstainers who have had one heart attack each, into two groups, one of which groups changes nothing but to add a glass of wine to dinner daily and the other group changes nothing, and if we should go on to find that the frequency of second heart attacks is quite different between the groups ... then we might start to move a bit closer to information about causality.

Here is the point, set apart: We have no reason to imagine that twinning causes nonrighthandedness or vice versa, but we can be certain that the two phenomena have something important in common, and finding out what that is seems likely to be important and worthwhile.

At the beginning of my involvement with this story, it was customary to blame the excess lefthandedness (just like every other developmental anomaly among twins) on the monozygotic twins. Monozygotic twins, you see, must somehow build two bodies from the single mass of cells that has descended from a single zygote by several successive cell divisions. One might easily imagine that something in the process of forming two embryos from one embryonic cell mass — forming two body symmetries: two different head-tail frameworks, two different back-belly alignments, two different left-right lines, and sorting all available cells into one or the other framework — might cause something about left-vs-right-handedness to show up looking different from the usual.

It does not matter as much to your development to have your stomach and your liver switch sides as it does to make sure head and tail directions are appropriately defined. Acephalus-acardia might be interpreted as a significant failure of the head-tail axis. An embryo with no head and no heart lives to become a fetus, and a fetus with no head and no heart survives to term — only when a normal twin inside the same amnion (inner gestational membrane) can supply blood circulation to keep the lump alive. The rest of them, if any others occur, must fall aside before anyone knows of their existence.

Nor have I ever seen or heard of any product of conception lasting as long as maternal recognition of pregnancy (six to eight weeks of gestation) with any condition that could be seen as an anomaly of dorsoventral (back-belly) patterning homologous to such (uniformly lethal) anomalies that we have seen in, for example, fruit flies.

Left-right is apparently the last and the least of the embryogenic structural axes, the most variable, the least stringent, the least fundamental and the most subtle of the axial definitions. It must apparently be modular (specified more or less separately at several different levels) along the head-tail axis, and apparently the decisions at various levels are at least to some extent independent.

From all the work surrounding the twin studies and the adoption studies, in particular, it came to be generally accepted that the origins of individual cases of schizophrenia include important genetic components. There has been a tendency to think "if genetic, then of course it must be biochemical," so grand and expensive efforts have been exerted to discover any substance in the tissues or body fluids of schizophrenic that might give us a clue to the biochemical basis of schizophrenia. Several suspect substances have been studied in some depth at considerable expense but with no lasting or widespread success, and new ones are still arising.

I am convinced, from these findings in twins and other related contributions, that schizophrenia is a developmental disorder. It is a disease of embryogenesis, primarily concerning the mechanisms under which the cells of the developing brain grow, divide asymmetrically, and move into their positions in complex arrangements suitable to their complex connections and interactions in the adult brain.

I am further convinced that the "(genetic, therefore) biochemical" variations that cause the deviations from normal development that result in schizophrenia perform their functions from start to finish during embryogenesis of the brain. There need be no reason to expect anything out of the ordinary in the tissues or body fluids of an adult schizophrenic, because the differences may have all been made in embryogenesis and the responsible gene products are no longer being made. I am confident that any differences found in tissues or fluids of adult schizophrenics are consequences and not causes of the disorder.

I have always found it right up against the threshold of the miraculous that any drug can make the average schizophrenic feel any better, let alone perform better. The idea that — because schizophrenia clearly is substantially genetic in origin, therefore it simply must be biochemical in mechanism, absolutely does not mean that there is any biochemical anomaly to be found in the adult schizophrenic, let alone one that can be fixed with a prescription, and certainly not necessarily one that is the same for all schizophrenics.

If the cells of certain areas of the brain are not put into the proper arrangements during embryogenesis, every developmental process or tissue function farther along is at risk of failing when the ongoing focalization of brain functions brings its demands to bear more and more sharply on a poorly organized patch of brain tissue.

Because the brain substructures and functions involved in schizophrenia are asymmetric, related to other brain function asymmetries, and subject to anomalous distributions among twins and between co-twins, I am further convinced that schizophrenia is a disorder of epigenetic[a] functions that are involved in defining such asymmetries in cell functions and in cell movements during embryogenesis.

---

[a] epigenetic functions are not yet understood very widely or very well, and we will have some discussion of their many variations and implications at some length farther along here, but, for now: In this context, epigenetic functions have less to do with any given DNA base pair sequence and its genetic functions than with control of the expression of the genetic DNA sequences. An epigenetic variation ("epimutation") is usually heritable among descendants of a cell in which it occurs, even though there has been no genetic mutation, no change in DNA base-pair sequence. In some organisms, some epigenetic variations are also heritable from one generation to another.

It must also be noted, however, that all epigenetic functions are performed by gene products. I have not seen that considered in the large and growing epigenetics literature, but it must eventually fit in there somewhere.

X-inactivation is probably our longest-known and best-known example of a major epigenetic function. Mammalian females have two X-chromosomes and males have one X- and one Y-chromosome. The X is a fairly large chromosome and it carries many important genes. For some of those genes, dosage (the number of copies of the DNA, and thus usually the amount of gene product that can be made) is part of the genetic differences responsible for sex differentiation.

For most of the genes on the X, however, dosage is not part of the sex differences, and something must be done to balance the dosage. In all mammalian females, most of the genes on one of the X-chromosomes are shut off in each of the cells. So, in some cells, a mammalian female expresses all of her mother's X-chromosome and only a small part of the one that came from her father; and vice-versa in other cells.

On the average, the distribution over her body as a whole will be half-and-half, and variation around that number appears to be small in general, but occasionally a skewed distribution will arise, with some interesting consequences. Normally, which X gets shut down is randomly-chosen in one of the cell divisions very early in embryogenesis. It is random, but it is permanent, and the change is passed down to all descendants of that cell for the rest of her life. All of the DNA sequences are still there, the same in all of her cells, but some of her cells have most of father's X shut down, and some have most of her mother's X turned off.

Many of the cell divisions of embryogenesis are asymmetric, with one of the daughter cells maintaining a relatively undifferentiated stem-cell-like functional configuration while the other goes on to its more highly differentiated function. Those asymmetric divisions of functions between the daughter cells absolutely depend on some asymmetry of *chromatin* (DNA + scaffolding and packaging and control proteins in the chromosomes) structure which changes the functional coding of gene expression.

Asymmetry in the formation of a multicellular embryo is one thing, and asymmetry in the division of a single cell may well be something else — but probably necessary for the multicellular version. An obvious candidate is to be found at the level of the asymmetry of DNA strands and of the histone proteins that govern packing up of the DNA for movement in cell division and that control the availability of specific DNA sequences for duplication or for expression via transcription and translation.

The universe is asymmetric and I am persuaded that Life,
as it is known to us, is a direct result of the asymmetry of the universe
or of its indirect consequences.

Louis Pasteur

DNA is inherently, fundamentally asymmetric. The two strands are different in many ways, and every kind of living cell appropriately examined to date has shown an ability to read and to use those differences, as well as differences between the "old" and the newly replicated copies of the leading and trailing strands. There is room therein for another whole level of combinatorial code by which gene expression and thus cell differentiation might be regulated.

From the vantage point of this understanding, some discordance between monozygotic co-twins for schizophrenia (and every other disorder to which we have attributed "multifactorial inheritance") is to be expected. In spite of repeated clear demonstration by the results of adoption studies that variation in physical and emotional home environment has no effect whatever on the outcome of genetic vulnerability to schizophrenia, the protagonist genetic analysts still know no name other than "environment" for the source/s of that discordance.

The differences between co-twins for diagnosis of schizophrenia, for example, are now generally assigned to "non-shared environment," the personal and private variations in individual life history that are presumably never allowed to affect or be affected by the rest of the family. The failure to identify the "environmental" contributions that so many analysts want to believe are indicated by discordance between monozygotic co-twins for these disorders is therefore also to be expected.

The first and most pervasive elements of the "environment" to affect the function of any gene or gene product are the configuration of the DNA strand from which it might be transcribed and the ensemble of the products of other genes in the same cell and interacting cells.

There are many malformations and disorders of mental and behavioral development which are considered to be inherited by a "multifactorial" mechanism, and which show greater discordance in twins than might be expected from the results of adoption studies and of pairs-of relatives analyses not involving twins. I find it of compelling interest to note that all of them involve asymmetries of embryogenesis. We will spend more time on that when we get deeper into thinking about malformations and their relationships with twinning.

Chapter Four

Reviewing/Reciting The Orthodoxy

Think wrongly if you please, but in all cases think for yourself.

Doris Lessing

To make a mistake is only an error in judgment,
but to adhere to it when it is discovered
shows infirmity of character.

Dale Turner

To remind me, and to let you see, what we are up against here, suppose we go through the story of the biology of twins and twinning as it has been told and retold over the most recent past several decades, and perhaps play a little game of "What's Wrong with This Picture?"

Watch closely, and try to see what parts of the story should perhaps cause doubts or second thoughts. If you find something you do not understand, that fact absolutely does not imply that YOU are at fault. If something sounds stupid to you, do not just go away shaking your head. A lot of the stuff people have told you all through your life and your education is in fact absolute nonsense. If it does not make sense to you … that very well might not represent any deficit in you or your efforts or your faculties, but you will get nowhere if you never challenge anything.

Once you begin to discover how very often The Last Word is the wrong word, the Latest Greatest Idea is not even a good idea, the Received Truth is at best a fable however long it has been circulated and translated from however "high" its origin. Depending on how honest you had hoped your sources would be, you might easily shed much of your fear of making mistakes and have no end of fun uncovering those made by those who have gone before.

Here We Go (The Order is Largely Arbitrary)

Everybody knows:

There are two kinds of twin pairs.

Twins of those two kinds arise by two completely unrelated mechanisms in the formation of the embryo.

Some pairs of twins are identical.

Identical twins come from splitting a single embryo, so they have the exact same set of genes and an increased frequency of developmental anomalies.

Identical twins have exact copies of the same set of all their genes.

Identical twins can therefore be expected to have all the same genetic traits, good, bad or indifferent.

Twins have excess frequencies of many things that can and do go wrong in development. This is more or less exclusively due to embryo splitting among the identical twins.

The later in embryogenesis that splitting of the embryo happens in identical twinning, the more likely it is to screw something up.

If a particular trait (a malformation, a disease, or any specific element of an individual's structure or behavior) is "really" genetic, all identical twin pairs should match for that trait — whatever one has, the other also must have if it is "really" genetic, because they have identical genes!

Twins are lefthanded more often than singletons. This is because of "mirror-imaging" among the late-splitting identical pairs.

Difference between the members of a twin pair in handedness means something special about embryogenesis.

Twins who differ from each other in handedness are "mirror-image" (and therefore) "late-splitting" "identical" twins.

There is an excess of malformations in twins, which likewise arises from the strange embryogenesis of identical twins, from disturbances in embryonic development caused by their splitting and all the associated rearrangements.

(only) Identical twins are perfect tissue transplant donors for each other.

All twin pairs born inside a shared gestational membrane (chorion) are identical … all monochorionic twin pairs are monozygotic. All dizygotic twins are dichorionic.

> Doctors who deliver twin babies often tell the mothers that the twins are "fraternal" if they "have separate placentas" or if they are dichorionic (came with separate chorions, the gestational membranes, the "bag of waters"). In other situations, some baby-catchers say the twins must be "identical because there was only one placenta" if their dichorionic placentas have fused.

Some twin pairs are clearly "non-identical." Maybe one is a boy and one is a girl, and we are therefore pretty sure they are not "split" from a single fertilization event by way of a single embryo. Maybe the members of a twin pair are both boys, or both girls, but they just don't look much alike beyond being of the same sex. It seems that most people call twins like that "fraternal." The French call them *faux jumeaux* (false twins). From some perspectives, all newborn babies look exactly alike. Therefore, sometimes we have to wait a little while to be sure of this particular outcome.

Everybody knows that fraternal twins arise from double ovulation.

> There is a remote possibility that some pairs of otherwise separate and independent egg cells have come from one and the same single ovarian follicle. (The ovarian follicle grows to look like a big blister on the surface of the ovary as the egg cell inside ripens through its last stages of development in preparation for ovulation. The follicle gets about as big as a marble or a grape, about 15 millimeters, about six-tenths of an inch in diameter, right before it bursts to release the egg cell.) When we see ovarian follicles containing two developing egg cells, we call them "binovular follicles," but we do not see those often enough to have any reason to imagine that they could account for anywhere near all the dizygotic twin pairs. Besides, no one has ever reported natural human

twins (or anybody else for that matter) who was born alive from either of the egg cells in any human binovular follicle.

Coming, as we are supposed to believe they do, from two different egg cells from two separate ovulations from two separate follicles, dizygotic twins must be expected to develop independently and just like singletons, with no excess of anomalies of embryogenesis.

Fraternal twins are therefore neither more alike nor less alike than the members of non-twin sib-pairs. Just like pairs of non-twin siblings, some fraternal co-twins are very similar to each other in appearance, and some are quite different.

It is easy to diagnose the zygosity of boy-girl twins — if the boy and the girl are both normal, they are always dizygotic twins.

> If you have boy-girl twins, and someone you meet says "Ooh, twins! Are they identical?" say "Yes, of course. Thank you for asking. But I really have to go now," and leave — promptly or even more quickly if you can. There can be no good outcome and it is not worth the effort.

Very rarely, one of a pair of monozygotic twin boys loses his Y-chromosome in a cell-division accident early in embryonic development, and he goes on to develop as a Turner syndrome female, with only one X-chromosome — no Y, no second X.

> The great majority (95% or more) of Turner syndrome embryos are never born; they fail quite early in pregnancy. Those Turner syndrome girls who survive to birth seldom reach five feet tall or grow normal breasts or pubic hair unless they get complex hormonal therapy. They are generally happier people than most people would think they have any right to be. They are almost always within a useful normal range of intelligence as long as they minimize their involvement with mechanical things, and they are able to live useful lives, but very few will ever ovulate. Any twin pair containing a normal boy and a girl with Turner syndrome is not the same thing as the boy-girl-both-normal pairs mentioned above.

With only those criteria explained so far, we cannot tell all the identical and fraternal pairs apart in any representative sample, but we can compute how many are identical and how many are fraternal in any given group of twin pairs by the

Weinberg Difference Method. This is easy: We assume that half of all sperm cells carry a Y-chromosome, which will cause any normal, viable conception initiated by that sperm cell to develop as male. We assume that the other half of all sperm cells carry an X-chromosome, any normal, viable conception initiated by which will develop as female.

Wait... That Is Not All...

We assume that all dizygotic twinning events are independent, so that sex-pairing should follow the expectations of the binomial distribution. The simplest appropriate explanation of what "the binomial distribution" means right here and now is that the joint probability of a certain pair of events or outcomes (probability of both of them happening) is the product of the separate probabilities of the two. Both twins of a pair should be male one-fourth (½ × ½) of the time, both should be female one-fourth (½ × ½) of the time, one-fourth will be male-female (½ × ½) and one fourth (½ × ½) female-male, so half will be same-sex pairs and half will be boy-girl pairs. This is exactly like flipping two coins.

If we simply subtract the number of boy-girl pairs in the whole sample from the total number of same-sex pairs in the sample, then ... the remainder, the "difference," according to the Weinberg "Difference" Method, is the number of monozygotic pairs. The total number of the same-sex pairs includes all of the identical pairs and half of the fraternal pairs. The other half of the fraternal pairs is composed of the boy-girl pairs. If we subtract the number of boy-girl pairs from the total number of same-sex pairs, that is the same as subtracting the number of same-sex fraternal pairs, and therefore the difference is the number of identical pairs. See? I told you it would be easy!

It is also by the use of the Weinberg Difference Method that we have convinced ourselves that the frequency of monozygotic twinning is constant the whole world over, exactly as we should expect of such a developmental accident just like any one of the malformations that are excessively frequent among the monozygotic twins. On the other hand, because the fraction of girl-boy pairs varies over racial groups, we assume that the fraction of dizygotic twinning varies in frequency over racial groups as would be expected for a trait that is subject to genetic variation.

Because fraternal twins come from double ovulations, and fathers hardly ever ovulate, the genetic variations in frequency of fraternal twinning can only represent nothing other than variation in a genetic tendency for double ovulation and must travel through the generations within families only by way of the females.

Results from application of the Weinberg Method have also given us our common knowledge understanding that the excess of developmental anomalies among twins (symmetry malformations, nonright-handedness, fetal and neonatal mortality, etc.) is due to the monozygotic twin pairs, as we said above that we would expect from their odd embryogenesis.

> That, too, is simple, by way of the simplistic Weinberg machination. Those anomalies (and more-or-less all other difficulties, interestingly enough) are less common in live-born boy-girl pairs than they are in same-sex pairs of either zygosity. Therefore (because we insist on supposing that the boy-girl pairs are perfectly representative of all the dizygotics, so any difference between them and the whole same-sex crowd must be due to the "identical" members of the same sex crowd); the Weinberg calculation always comes out showing that these problems are practically exclusive to the monozygotics — always — there is no way it could be otherwise. Any problem that is not substantially more frequent in boy-girl twins than in same-sex pairs *will*, as the night follows the day, be assigned to the monozygotics by Weinberg calculations. It could never be otherwise.

This excess of anomalies in twins, all assignable to the same-sex pairs, has been for decades presumed to be caused by the weird embryogenesis of the monozygotic twins, i.e. splitting of the embryo thus disrupting the ongoing consequences of developmental decisions at the cellular level.

Some folks have suggested a "Third Kind of Twin," and that prospect remains a cause of some debate. Maybe, if dizygotic twins sometimes come from one egg cell and two sperm, they should be intermediate between identical and fraternal, that is, "sesquizygotic" (sesqui- = 1.5, as in sesquicentennial celebrations of 150 years since the town was founded …), or "half-identical," because all the genes they get from that one single egg cell from their mother should be identical.

All of the above is what you will find in every other book *about* twins and twinning that you can presently put your hands on. Have you discovered What's Wrong with This Picture? Some questions, even on my Human Genetics course exams, have no wrong answers. This one does. Wrong answers here include any answer that does not include "at least very nearly everything." This chapter could be subtitled: "Falsehoods that the World of People Who are Interested in Twins have been Telling Each Other about Twins and Twinning AS IF They Knew It Were All True …" but that is too long for a title and does not get everything across anyway.

It is true that the members of dizygotic twin pairs resemble each other just about like non-twin siblings do. Some co-twins are very similar, some are quite different. Otherwise, there is not a single complete sentence above that has been, or ever could have been, shown with solid direct evidence to be true.

Oh, well … yes … actually, it is also true that fathers very seldom ovulate, but as far as I know nobody ever wrote that down before. Until I wrote it down there just a minute ago, my impression of the whole history of the idea is that it has only been implied, but it has always been implied. A more to-the-point statement of that part of the story would be the *assumption* that there is nothing that the male partner or his sperm cell/s can do to cause double ovulation, which is required by the common knowledge for the occurrence of dizygotic twinning. If, however, it was never really about double ovulation in the first place, that idea is of course irrelevant …

Systematic analysis of this doctrinal system reveals that the core belief is the one about double ovulation as the origin of dizygotic twinning. Most of the rest of the errors arise from that one, primarily by way of results from the Weinberg method — because the Weinberg calculations require several assumptions that are consequences of the assumption of independent double ovulation.

Every conclusion gained from application of the Weinberg method is logically false, because the logical and mathematical foundations of the method absolutely depend on several assumptions which are known to be false with respect to actual human twins. The worst of those failures is that the logic is inherently and irredeemably circular, because the expected conclusion is always imbedded in the premises. Because of the absolutely required (but never stated) assumption that boy-girl twins are entirely representative of all dizygotic twins, we must also assume that any differences observed between boy-girl twins on one hand and all same-sex twins as another group must be due to the monozygotic twins among the same-sex pairs.

Many people over the century since the Weinberg Method has been in use have been aware of The Method's stated assumptions that X-bearing and Y-bearing sperm are equal in number, that any one event of fertilization would have no relationship with any other, and that sex-pairing in twins should therefore conform to expectations of the binomial distribution.

Most of those same people have not been aware that the logic of the Weinberg Method also absolutely requires them to assume that there is no relevant difference, in any part of embryonic or fetal development, between boy-girl pairs and same-sex dizygotic pairs. (If we are going to say that all differences between boy-girl twins and same-sex twins are due to the monozygotics, that cannot be true unless the same-sex dizygotics are just completely and exactly like the girl-boy twins.)

What if? ... what if that just is not so? It would wreck the entire logical structure and render the whole process false and disruptive, much worse than useless. What if what "everybody knows" is wrong, and has been wrong all along, throughout the entire history of these questions about twins and twinning? What if ...

What if ... ? What if the boy-girl pairs are absolutely not developmentally representative of all dizygotics, and it was the same-sex dizygotics who were having most of the problems? Nothing but the circular logic of the Weinberg method says otherwise, and that is what I found when I tested it with regard to fetal and neonatal mortality — the same-sex dizygotic pairs were at least as bad off as the monozygotics, and much worse off than either member of the boy-girl pairs. Much more to come on that question ...

Every piece of the story built upon conclusions derived by application of the Weinberg method is logically false, and biologically false where properly questioned. Proper questioning, in turn, has depended on deciding not to believe anything without evidence just because somebody tells you "everybody knows."

By the way, in case I have not made it clear already, there is no such thing as any two identical people. Monozygotic co-twins differ from each other in a great many ways. Also, only about one-fourth of all dizygotic twin pairs are brothers and might therefore properly be called fraternal. The other three-quarters include at least one sister. About one-fourth are pairs of sisters and about half are brother-sister pairs.

Human *chimeras* are individuals whose bodies contain two or more genetically different cell populations. Chimerism is of course another major thread of this story. I hope we will come to agree that chimerism and twinning are as intimately related as parts or aspects of any one process that has been erroneously thought to be two could ever be.

Chimerism has not been given more than the occasional fleeting thought over the ages during which the Orthodox Common Knowledge Canon of Twin Biology as outlined above has been cemented into place by repetition. Thinking of chimerism in terms of "people made from parts of two people" makes it just entirely too weird for most people even to try to understand — their minds wander all the way off the page and leave the building before they even hear the rest of the explanation. Before we get all the way finished here, we should be able to get beyond that.

Human chimerism is unquestionably real, absolutely not mythical and not magical, and not even rare — nevertheless, it presents quite a weird concept to most people when they first become acquainted with it, so I will do my best to help you deal with it by explaining how it must happen the way it does, and how easy and ordinary that might be.

According to the literature, there are several kinds of chimeras, differing by the (imagined) mechanism of their formation:

Artificial Chimerism: chimerism resulting from medical artifice includes all successful organ- or tissue-transplants involving any donor other than a monozygotic co-twin. This may include blood transfusions, which may occasionally include stem cells capable of establishing permanent colonies.

Spontaneous chimerism, arising without medical intervention, occurs by two mechanisms: 1) Fetal cells routinely enter the mother's bloodstream during pregnancy, and those may include stem cells which can form permanent colonies. 2) "Twin chimeras" are (imagined to be, and consequently in much of the older literature defined as) chimeric in blood (only), due to connections between the placental circulations of dizygotic twins in the uterus. This has been reported and discussed almost exclusively in cases where only blood has been examined, and there is in fact no reason to suppose that any significant fraction of spontaneous embryonic chimeras are chimeric in blood only.

"Dispermic," "generalized," "whole-body" and "tetragametic" (from two eggs and two sperm cells) are labels that have been given to chimeras who are believed to have arisen from fusion of dizygotic twin embryos, each from separate sperm and egg.

What's wrong with this part of the picture? The statement about spontaneous chimerism occurring in women via fetal invasion is true. Those other distinctions are imaginary, not at all based on observation, repeated or repeatable or otherwise.

Connections between placental circulations of dichorionic twins have in fact been found — <u>at less than one-thousandth of the lowest frequency of chimerism found in direct surveys</u>.

There is no reason to suppose that dizygotic twin embryos from separate and independent ovulations can ever come into embryonic cell-to-cell contact. By arranging for the fusion of separate embryos or parts of embryos, several laboratories have produced many thousands of artificial, experimental chimeras for decades of inspiring work in developmental biology and genetics.

To do that, they have always had to remove the zona pellucida (the tough, elastic, non-cellular "eggshell" that surrounds the egg cell and all of embryogenesis until after "hatching" and implantation), or else work only with embryos after they have "hatched" out of the zona.

Clearly, however, spontaneous embryonic chimeras, born with two genetically different cell populations, can only have arisen from dizygotic twin embryos. This is how it comes to be that the two cell lines are genetically different — in exactly the same way that siblings differ. Having somehow spent time in a single shared embryonic cell mass is the only way they could have gotten into a situation from which to end up sharing some of each other's cells. Exactly how that must be happening, and how extremely unlikely independent double ovulation is as an explanation, will be an important part of what we will discuss in all of what is yet to come.

All of the work to be discussed in what follows, and all of the results thereof, grew from my original realization that — quite contrary to the revelations and traditions of the twin biology of the past — the excess of nonright-handedness among twins belongs quite equally to both monozygotic and dizygotic twins.

That is where my thinking along the lines of the orthodox stories of the biology of twinning first got derailed.

That might easily be overlooked or dismissed as trivial, but it has fundamental things to tell us about how we all got here — about the business of becoming human, as individuals and as species. In ways we have only just begun to understand, differences in the structures and functions of the left and right halves of the brain are fundamental to whatever is human about the human brain, and it is a matter of hugely fortuitous serendipity that unusual motor hand preference (handedness) shows up as a clue to how things are being done differently.

Nonrighthandedness may be subtle, but it is our best indication of membership in a minority of people whose brains are put together differently. Their differences in the distribution of nonrighthandedness tell us that twins do the embryonic building of the brain differently. The fact that monozygotic and dizygotic twins both do it differently from singletons, in the same ways and to the same extents, is nothing short of cataclysmic.

Whatever is odd about the embryogenesis of monozygotic twins is equally odd in dizygotic twinning, and in the same ways. From any point of view, properly considered, within the developmental biology of the human, this is earth-shaking. Stay with me, now ... If you have made it this far, you can handle the rest. There is more where that came from, and it is not about double ovulation.

Truth is mighty and will prevail.
There is nothing wrong with this, except that it ain't so.

Mark Twain

Chapter Five

On The Zygosity of Twins

Nine times out of ten, in the arts as in life,
there is actually no truth to be discovered;
there is only error to be exposed.

H. L. Mencken

It ain't what you don't know that gets you into trouble.
It's what you know for sure that just ain't so.

Will Rogers

It is at least a tradition, and, as far as I know it may even be a Law somewhere but I have not been able to find the reference to where it is written … Any and every book that is to any important extent *about* twins must include a chapter, with a strong preference that it should be quite close to the first chapter, on the zygosity of twins. The purpose of such a chapter is to explain what we believe [and have been repeating for decades until most of us have convinced ourselves and each other that it is "what we know" instead of just some long-ago somebody's opinion that has hung around in the literature too long without ever being mindfully considered or challenged] about the different "kinds" of twin pairs, their different cellular origins and how we might come to know how to parcel twin pairs out into their proper sorts for purposes of other or further analyses.

A partial list of such chapter titles as are to be found in other books is as follows: Kinds of Twins; Types of Twins; Zygosity Determination and the Types of Twinning; Types of Twinning and Determination of Zygosity; Identical and Fraternal Twins: Living Laboratories …

Related topics that are sometimes mixed in with such chapters, and sometimes stand alone, are: Placentation; How are Twins Formed?; The Causes of Twinning; The Epidemiology of Twinning; Factors Which Influence Twinning Rates; Factors Affecting Twinning. Those of you who have done some previous reading about twins probably already know what all those chapters about zygosity say: 1) there are some monozygotic twin pairs, that come from one zygote that develops into a cluster of cells that somehow "splits" to become two embryos, and 2) there are some dizygotic twin pairs, that come from two separate zygotes, which arise from two (presumed to be separate and independent) ovulations.

Mono- is to One as *Di-* is to Two

The "common knowledge" has it that there are those two kinds of twin pairs, and that they arise from entirely different embryogenic mechanisms — on the one hand (presumably mechanical) "splitting" for monozygotics and on the other hand "double ovulation" for the dizygotics. Chapters about zygosity in books about twins are there to explain, in furtherance of that belief system, what we claim to know about all that and about how it comes to be the way that it is, and to a far lesser extent (in fact, very nearly not at all, anywhere) to explain how we get to where we claim to know it.

No evidence? No need! Everyone knows this is how it is.

In the traditional characteristic false simplicity of those writings, the protocol for determination of the zygosity of any individual twin pair goes like this: There are two of them, yes? Okay, that's what "twins" means … they're twins. The French will call them "false twins" if they are not monozygotic …

> Les faux jumeaux résultent
> de la fertilisation de deux ovules distincts en même temps.
>
> Margaret Freda
>
> ("… the false twins result from the fertilization
> of two *distinct egg cells* at the same time …")

Does the pair consist of one normal boy and one normal girl?

If so, they are surely dizygotic.

If they are of the same sex, then they might be monozygotic, or they might still prove to be dizygotic. According to the traditional protocol, one must now check the placental membranes. If they are inside a single chorion, then the twinning event by which they became twins did not happen until after the trophoblast cells that would become their shared chorion differentiated, therefore, they must be monoembryonic (both bodies coming from a single embryo — clearly, because they share at least one tissue (the chorion) that differentiated during embryogenesis!) and we have long presumed that can only mean monozygotic. We now know that presumption to be unfounded and often false. According to two independent studies a couple of decades apart, in Scotland and in Taiwan, it is wrong one-quarter to one-third of the time. More about that later when we discuss chimeras and chimerism at greater length.

If the twins are of the same sex and they arrived in separate sets of gestational membranes (dichorionic), then they might still be either monozygotic or dizygotic, and we will need similarity testing to decide, ideally strong genetic tests.

By the way, about half of all dichorionic twin pairs, whether they are dizygotic or monozygotic, appear at delivery with their two placentas fused to form a single placental mass. Therefore, the number of placental masses provides no information as to zygosity. If your obstetrician has proclaimed "your twins are fraternal, because they have separate placentas," or "they are identical, because there is only one placenta," then buy an extra copy of this book and carry it or have it delivered to your OB's office … If I have done my part of this endeavor well, an obstetrician willing to take the time and put in the effort should be able to understand it.

Many of those chapters or papers on zygosity go on to discuss various sets of genetic markers and laboratory and mathematical protocols by which to assess the probability of monozygosity.

Dizygosity is considered to be proven by any difference between the co-twins for any appropriately tested genetic marker. Monozygosity can only be stated as a probability. Monozygosity can never be considered absolutely proven because — regardless of how many markers we may test — there might always be a difference we have not found, with respect to a marker we have not tested. For each marker used in testing, there is a finite probability that we will find a difference between the twins if there is one. (This is a function of the mixture of different versions of that marker's DNA sequence in the population. We choose, to use for testing, marker sequences that exist in the population in numerous versions, to maximize the probability that relationships will be informative.) Each additional genetic marker tested without finding a difference between the co-twins decreases the probability that there is any difference to be found, and increases our estimate of the probability that the co-twins belong to a monozygotic pair.

Although we cannot consider monozygosity absolutely proven from any genetic result, we can fairly efficiently reach the position of being able to declare a vanishingly small probability of dizygosity. This is essentially the same question as the forensic matching of a sample from a suspect with a sample from evidence. Proof of difference is easy. If the samples are the same, it is straightforward to reach a vanishingly small probability that the genotypes came from randomly-paired different genomes, putting the status of the question "beyond a reasonable doubt." [Two samples from the same person may be different; more later.]

Sound knowledge of the zygosity of each and every pair of twins is, more or less obviously, crucial to the purposes of genetic twin studies, where the object of the endeavor is to measure and compare within-pair relationships separately among monozygotic and dizygotic *groups of twin pairs*.

The people who do those genetic twin studies (after the manner of Galton as described earlier) want to know if, and if so by how much, the members of monozygotic pairs resemble each other more than the members of each particular pair of dizygotic co-twins do for the trait in question.

They cannot do that very well at all if they are not sure which twin pair is of which "kind," to know which questions to ask of each pair and in which column to count them.

Knowledge of the zygosity of a pair of twins has also been thought to be very important if ever one of the twins needs a tissue transplant. It is generally presumed that "identical" co-twins will always be perfect tissue donors for one another.

This has been shown not to be true. Recipients of such transplants often need immune suppression for success. It seems reasonable to suspect that those pairs are not in fact monozygotic, that the donor has some antigens the recipient's immune system sees as foreign. A reasonable explanation must depend on knowing exactly how zygosity was decided, to understand how the asymmetric chimerism (in which the donor twin has a second cell line not shared with the recipient) was missed. More later.

It is generally presumed that dizygotic co-twins, on the other hand, will have the same prospects of matching of cell-surface antigens that non-twin siblings have, no better than 25%. This latter proposition, because of the overwhelming strength of the "common knowledge," has not been well tested, but there is good reason to suppose that it is also not true.

The really interesting thing about this is that both of those presumptions are probably not true, for the same reason, having to do with chimerism.

Some twin pairs that have been diagnosed as "identical" because of arriving at delivery in a shared single chorionic membrane are not in fact monozygotic. Many members of dizygotic twin pairs have incorporated some of the co-twin's cells since embryogenesis. One or both members had both kinds of cells present when their immune systems established immunological self-definition and tolerance and they may therefore be mutually immunologically self-tolerant.

Realize that chimerism may be asymmetric, with one twin having some cells from both embryos and the other having only one cell line and therefore not chimeric. Should that occur, the non-chimeric member of the pair could not be expected to accept a transplant from the chimeric twin.

For purposes of understanding the biology of human twinning, as part of the larger purpose of understanding human developmental biology, zygosity has been considered an issue of great importance because of the common knowledge understanding that monozygotic and dizygotic twins arise from different mechanisms, with different genetic consequences.

If indeed they do arise from different mechanisms, then indeed we do need to know which pairs are of which kind so that we can study their separate patterns of occurrence and development, and the similarities and differences between the zygosity groups, so that we can figure out things about how the different kinds of twinning happen differently, if indeed they do happen differently.

It has generally been supposed that we do not need to know the zygosities of each individual twin pair for these population-level understandings of the biology of twinning. Rather, it has been supposed, we might expect to learn about the different biologies of monozygotic and dizygotic twinning by knowing the respective proportions of the two kinds of twins in populations under varying circumstances that might affect their relative abundance. For that, we have the

Weinberg Difference Method [an algorithm according to which the difference between the numbers of like-sex pairs and unlike-sex pairs in any given sample of twin pairs is believed to equal the number of monozygotic pairs among the same-sex pairs].

The common knowledge has always been put forward as if "monozygotic" (arising from a single zygote) means the same thing as "arising from a single egg cell" and "dizygotic" (arising from two zygotes) means the same thing as "arising from two egg cells." This leaves the obvious (and profound and fundamental and much more complex than the orthodoxy would have us suppose) implication that those two egg cells from which dizygotic twins might develop must arise from two ovulations from separate and independent follicles.

This whole history, at least 13 decades in the making, has had most of us feeling compelled to suppose, and requiring each other (and students, and twins and their parents) to suppose, that there must be two different systems of events and processes, constituting two different meanings, or kinds, of "the biology of twinning."

What is history but a fable agreed upon?

Napoleon Bonaparte

To follow by faith alone is to follow blindly.

Benjamin Franklin

… a myth's power does not depend on its plausibility …

George F. Will

The time over which this idea has continued to be passed down as if it were fact indicates that it has held a certain plausibility — many people have found it easy to accept those stories as credible.

I have held the conviction for a long time now that the truly most interesting contribution to the understanding of human development that can come from the study of the biology of twins has little if anything to do with the genetic relationships within pairs.

I have wanted to know about the causes and effects of any differences in embryogenesis between monozygotic and dizygotic twins and the progeny of single conceptions. What are the developmental consequences of changing one embryo-in-progress into two embryos? What can we learn from that about the building of normal, single embryos?

I have tried very hard, for over 30 years of studying twins and the literature relevant to the biology of twins and twinning, and I have not been able to find any sound evidence that any naturally conceived pair of human dizygotic twins, ever, arose from independent double ovulation.

That really turns out to be the heart of the issue. Nor has anyone responded to my published challenges to that hypothesis by offering any such evidence.

Double Ovulation Is Not a Fact

It is not a repeatable observation, as required to constitute scientific evidence; it is not an observation at all. Double ovulation has never been observed and reported in direct association with an event of spontaneous human dizygotic twinning.

It is not a self-evident first principle.

It is not supported by evidence. It is a guess … at best, the loftiest possible characterization of the idea of double ovulation as origin of natural dizygotic twins is to call it "hypothesis." If we give the idea the status of hypothesis, so that it might be thought of and dealt with as if it is a matter of science, then it matters that that hypothesis makes certain predictions.

Those predictions must be tested by further observation, the results of which observations must support its predictions if we are to move toward elevating the hypothesis to the status of theory and continue to believe it and to teach it. The double ovulation origin of dizygotic twins has always been considered an obvious fact, rather than a hypothesis the predictions of which need to be tested. The upshot of all that has been that the necessary predictions that in fact this hypothesis makes were never tested before I decided that the issue needed to be raised.

Twin-Prone Women

Mothers of naturally conceived dizygotic twins are considered to have demonstrated themselves to be "twin-prone" women.

Those mothers and their sisters (the maternal aunts of naturally conceived dizygotic twins), have been reported to do double ovulations — as recognized by ultrasonography — more often than a control group of other women who had never shown any inclination to twinning.

In some arguments I have heard, this result has been offered as proof of the double ovulation origin of dizygotic twins. So far, so good, and that is all very nice, but ... the number of women in the sample studied was too small to generate statistical significance for the results of the comparison, even at the level of "the members of this group did this thing more often than the members of that group did it," to say nothing of "how come?"

In that small sample, the mothers of dizygotic twins and their sisters did seem to do "double ovulation" more often than the untwinning women, but the observed difference was well within the possibility of random sampling error for samples of the size gathered for that work. The abstract of the paper as published says the result is statistically significant, but the data in the body of the paper clearly contradict that. Even if we assume the apparent result as fact without need of statistical support, not one of those (or any other) observed double ovulations was reported (in that study or any other, ever) to have resulted in the conception and delivery of twins.

That was obviously difficult and probably expensive evidence to obtain, involving frequently repeated ultrasonography, but the world awaiting the science of the argument owes no discount for that. It is indirect evidence at best, and it is statistically negligible in the bargain, but it has been considered by motivated believers to be quite telling.

So, zygosity is not simple. It is not nearly as simple as the old stories would have it. How twins got whatever zygosity they have, and how we can tell which one they got ... neither story is as simple as the orthodoxy, the common knowledge, the tradition, would have it.

Zygosity and Chorionicity

There are plenty of participants in this field of work who must be very upset to hear it, and may well yet refuse to believe it, but we really do know for certain that not all monochorionic twins are monozygotic, and not all dizygotic twins are dichorionic.

Even as they must grudgingly admit that monochorionic dizygotic twins do exist, most students of twin biology will still maintain that such pairs are extremely rare anomalies. There is no logical basis for supposing that monochorionic dizygotic twins must be very rare, simply because the acknowledged and reported discovery of such twin pairs has been very rare. We have no sound evidence for supposing anything about the frequency of monochorionic dizygotic twins.

> … thinking anecdotally comes naturally …
> thinking scientifically does not …
> … anecdotal associations are so powerful
> they cause people to ignore contrary evidence …
>
> *Michael Shermer*

I cannot reasonably assume, and I cannot honestly tell you, that monochorionic dizygotic twins are rare, just because they have only rarely been discovered. They have been only very rarely discovered in largest part because very nearly all of the people who might have discovered them were quite certain all along that they would never discover any such thing because no such thing ever happens, or happens only with extreme rarity, so they never did the work that would have been required for the discovery of something they just knew was not there anyway.

We almost never find what we are certain we will not find.

I have, for example, seen girl-boy pairs found in datasets, marked "monochorionic," the records of which were, thereupon and forthwith, deleted. That was considered certainly and obviously to be an error on the part of the delivery team that recorded the data, or someone who had transcribed it. They would seem unlikely to escape the notice of anyone paying proper attention, but the failure of the paying of proper attention is not at all rare, and — even if they are noticed — they are likely to go unreported, being too weird to believe, or for lack of appreciation of their profound significance.

Increased Twinning and Chimerism in IVF Conceptions

Monochorionic girl-boy pairs have been found and reported more commonly among products of IVF (*in vitro* fertilization) pregnancies than in natural conceptions, and those sets have therefore been reported as if their occurrence obviously represented some sort of pathology deriving from their technologically assisted conception.

If we could believe that the accounting to date might be correct, then we might go on to suppose that girl-boy monochorionic twin pairs would, or should, represent half of all monochorionic dizygotic pairs, but I do not believe I can go there with you. Boy-girl twin pairs are very special in a great many ways, especially with respect to differences in their growth rates in embryogenesis. Experimental mixed-sex mouse chimeras almost always appear at birth as functional males, and we have no basis for assuming whether human mixed-sex chimeras would or would not behave differently.

In embryogenesis, male cells do very nearly everything faster than female cells, in mouse and man and every other mammal properly tested to date.

The major continental subpopulations (that have been known as "racial groups") differ in frequency of twinning, and in the fraction of twin pairs that include one member of each sex. Those same groups consistently differ also in the fraction of males among their newborns. The fraction of females at birth and the fraction of twin pairs that are opposite-sexed are correlated; for each of those variables, those group values are in the same rank order. I have argued elsewhere that the mechanism in common may be the slowing of female embryogenesis by the paternally-imprinted X-chromosome.

Boy-girl twin pairs are most frequent in populations where the paternal X-imprint is "weakest," as evidenced by the lowest fractions male at birth.

There is good evidence that the manipulations required for all of the artificial reproduction technologies have some probability of damaging the imprinting of gamete DNA and other aspects of their epigenetic functions.

According to the way we have always understood the embryogenesis of monochorionic twins, they must be monoembryonic, arising from a single inner cell mass inside a single trophoblast. We have no reason to imagine that monochorionic twins could arise from independent oöcytes or from separate embryos by fusion of separate trophoblasts to form a single shared precursor to the chorion and a single shared precursor to the placenta. The normal timing of these differentiations would require any such supposed fusion to take place before differentiation of the trophoblast, when the zona pellucida would still prevent any contact.

In the production of many thousands of experimental embryonic chimeras, no fusion between two differentiated trophoblasts was ever reported — always the cells of the morula (before trophoblast differentiation) or cells of the inner cell mass were used (after they have differentiated from the trophoblast). Only when they were placed together inside a single trophoblast did they fuse, and never did that fusion result in twinning — the result was always either chimerism or failure to continue with development.

One thing we know for certain about the cellular event that initiates monozygotic twinning, the change-of-plan that sets one embryo on a path to build two sets of back-belly, head-tail, left-right body symmetries from cells that usually generate just one body symmetry, is this: It must happen after syngamy[a] if the DNA complements of these twins are to carry very nearly exact copies of the DNA genome that is first established in a single zygote nucleus.

Monozygotic twins are embryogenic descendants of a single zygote cell with a single nucleus — therefore necessarily having undergone their twinning event after syngamy, after that single zygote nucleus has been formed.

The central figures in syngamy are the pronuclei — the paternal pronucleus released from the sperm head after it has penetrated the secondary oöcyte and unpacked, and the maternal pronucleus just formed in the secondary oöcyte by the completion of the last step in the preparation of the maternal half-genome — when the second meiotic division leaves the second polar body just outside, and the maternal pronucleus just inside, the oöcyte membrane.

Each pronucleus contains the substance of all 23 chromosomes, in the form of one copy of each chromosome, one member of each of the chromosome pairs that were in the normal adult nuclei of the parents. After the sperm penetrates the egg cell, and its nucleus is unpacked and reorganized by enzymes in the

---

[a] Syngamy is the "final" stage in the fertilization process, when the maternal and paternal pronuclei fuse to form the zygote nucleus, and the fertilized egg cell becomes the zygote — the single cell from which an embryo might be constructed if all goes well enough.

egg cell cytoplasm,[b] each pronucleus starts from its position near the membrane of the egg cell and moves to the center. There they meld (details unknown) to establish the zygote nucleus with a new, single nuclear membrane surrounding a full diploid[c] set of chromosomes.

The chromosomes of the newly constituted full set replicate their DNA and assemble all the RNA and protein accessories of replication, to begin the series of "ordinary" mitotic cell divisions necessary to build the embryo and grow it to adulthood … again, if all goes well.

The mechanical ability to conduct cell division in the zygote and in all of the cells descended from it depends upon the pair of centrioles brought in by the sperm to form a microtubule organizing center (MTOC) from which to conduct chromosome segregation in mitotic cell divisions — the egg cell having lost its centrioles during the meiotic cell divisions of oögenesis.

This dependence on the sperm centrioles for the establishment of a functional zygote is a cardinal feature of human embryogenesis, not done the same way in all mammals, and it must be kept in mind for some parts of what follows. [This fact, together with genome imprinting during the building of sperm and egg cells, would seem to assure the impossibility of parthenogenesis (embryonic and fetal development from the egg cell without fertilization, in other words "virgin birth") in the human.]

One thing we know for certain — by definition, actually — about the cellular event which initiates human dizygotic twinning is that it requires the presence of two different zygote nuclei. These are necessarily the products of — there absolutely must have been — two maternal pronuclei and two paternal pronuclei assembled by way of two processes of syngamy into two zygote nuclei. [This says nothing at all about the membranes and thus the number of "cells" separately defined by their membrane boundaries.]

---

[b] The cytoplasm is the 'fluid' content of the cell. In fact it contains more water than anything else, but it is so full of cytoskeletal structures and membrane-enclosed organelles that it is really more of a gel than any kind of liquid.

[c] Most complex organisms have their chromosomes contained within a nucleus, in two copies each. This is the normal diploid state, the normal "ploidy." A triploid has three copies of each chromosome; a tetraploid has four copies.

People who study such things have told each other and their students for a century or two that the dizygotic twinning event is an occurrence of independent double ovulation. There is in fact no evidence for that hypothesis, no evidence whatever that any pair of natural human dizygotic twins ever came from two separate and independent ovulations. There is, however, evidence against it.

Splitting? What Splitting? How?

Because of the old belief in "splitting" as some sort of mechanical separation of parts of a single embryo to form monozygotic twins, the world has long appeared to be comfortable with the idea that monozygotic twins might show signs in their development of having passed through an odd sort of embryogenesis. So, the excess nonrighthandedness in twins (and of fusion malformations, and of virtually every sort of birth or developmental anomaly) was attributed to the monozygotics alone for decades. Publications about handedness in twins continued for decades to include the results from the 1920s and 1930s, which made certain that the result continued to come out that way, apparently because that conclusion "made sense" and the truth did not.

The idea that dizygotic twins might have more or less exactly the same frequency of nonrighthandedness as the monozygotics have did not then and still does not make sense under the assumption of double independent ovulation as their origin. That IS, however, exactly what the data demonstrate if one excludes data collected in the 1920s and 1930s before blood typing was available for an objective diagnosis of zygosity.

The rapture over Spemann's discovery of "splitting" newt embryos to cause monozygotic twinning had everyone thinking that any twin pair that included one nonrighthander and one righthander simply MUST be monozygotic, specifically "late-splitting" monozygotic … even if they did not look at all alike, even (according to some stories) if they were of different sex. That "late splitting" could mess up anything, you see?

It simply is not so. While it may seem obvious that monozygotic embryogenesis could be expected to generate or foster anomalies of a/symmetry in development, and monozygotics might therefore be expected to have unusual features in the many dimensions of a/symmetry in development, then, of course, dizygotics, if they are assumed to be derived from two independent ovulations, by way of two independent embryogeneses, should develop exactly like singletons, with no excess of such anomalies.

It simply is not so. Dizygotic twins and their family members have the same frequencies of nonrighthandedness that the monozygotic twins and their families have, not only among the twins themselves, but also among their siblings and their parents.

Dizygotic twins have the same excess frequencies of symmetry-related malformations in their sibs and offspring as the monozygotic twins have ... except for the excess frequency of twins among the parents of children with certain neural tube defects, which is in fact clearly more strongly associated with dizygotic twins than with monozygotics. More about both of those farther along.

Traditionally, monozygotics get the credit for the whole excess among twins of the symmetry-related malformations: In particular the neural tube defects, the congenital heart defects, and the orofacial clefts. That just is not the way things really are.

Everything in the literature blaming monozygotics for the excess malformations in twins draws that conclusion from "Weinberg Method" estimates, which are profoundly flawed logically, quite circular and inappropriate, and generally wrong for any sample of twins other than normal twins of white European ancestry. More about that later, too.

There are probably more errors and false conclusions about zygosity than any other part of this. The belief that there are two "kinds" of twin pairs and that they arise by unrelated mechanisms, is the core of the whole mythology. That simply is not the way things really are.

Chapter Six

Most twins
(products of twin embryogenesis)
are not twins
(born as members of double pregnancies)

What we see depends mainly on what we look for.

Sir John Lubbock

Twins born alive but alone, the sole survivors of twin pregnancies, are much more numerous than twins delivered together as live twins. According to my best estimate, which was arrived at by critical and conservative calculations, which appears to be the best available estimate in the literature, and which has not been responsibly challenged in relevant literature, there are about 12 times as many sole survivors from twin conceptuses as there are live-born pairs of twins.

For our most important purposes here, it really does not matter if that number is not 12, but in fact eight, or 15, or five, or 30. That number, that proportion, is free to change as a function of variation in the circumstances of pregnancies in particular and in general. The point, and what you will not find changing, is that significantly more twin conceptions and pregnancies result in the delivery of sole survivors than of liveborn twins. That, you may comfortably believe. Most human conceptions fail, among multiple conceptions at least as often as among singletons, and the members of twin conceptions can fail independently at least some of the time. Some sole surviving twins may be recognized as such because of being delivered with some residue of the former co-twin, but most of them bring with them no easily recognized way for us to know that they are who they are.

At least three-quarters of twin conceptions fail completely, resulting in no live baby delivered, much like the fraction of single pregnancies that are lost, under optimal conditions. The data I used for the core components of those analyses were all from highly privileged pregnancies, under "optimal conditions," which conditions include: Couples of proven fertility, with healthy mothers of reasonable reproductive age, under research-level medical attention. The majority of pregnancies in the human population do not occur under conditions that good, so survival to term in the human population as a whole is almost certainly not as good as this. It would not, I think, be an unreasonable estimate that survival from conception to term among all conceptions in the whole human population is not more than 10%.

In the mid-to-late-1970s, serial ultrasound imaging of ongoing pregnancies emerged from being a university medical center research curiosity toward becoming the routine standard-of-care that it is today in most obstetrical practices in the developed nations of the world.

As the data that were collected from repeated ultrasound imaging of pregnancies began to accumulate, it became apparent that many twins that were found in ultrasound images that were collected in the first trimester of pregnancy were no longer twins at delivery, or even in another ultrasound image collected a few weeks after the previous one.

Some twins (products of twin embryogenesis)
… actually, most of them …
"vanish."

From an obstetrical viewpoint, and throughout the twin family and twin research communities, this realization was more than a little bit of a big deal! There are many more twins early in pregnancy than there are to be found at delivery! Many twins vanish.

The point in time when a pregnancy can first be detected on ultrasound is sometimes called "early in pregnancy" — but only by obstetricians. Please understand, in the flow of the full story of human reproduction, that stage of development is "early in pregnancy" ONLY from an obstetrical viewpoint — it is early *only* in *clinical* pregnancy. Most of the developmental biology and cellular differentiation is over and done with by that time, and over two-thirds of all pregnancies initiated at the same time have failed.

From another perspective, an interesting number of people doing psychological therapy or life-counseling of various sorts were working with clients who just knew that the causes of their distress centered on the consequences of having lost a twin. Some had survivor's guilt over having somehow caused the demise of the co-twin, or at least for being favored to live after the other had died. Others had a variety of emotional deficits or distortions in their lives. With not very many exceptions, they had no physical evidence for the twin they believed to be missing, and some of the most compelling stories are the ones in which there was said to have come a revelation, decades down the road, that such a belief had a basis in fact. These individuals were excited to learn of this reinforcement of their stories — to know that the loss of a co-twin is a biological possibility that is not at all uncommon. Some found in this a very welcome vindication after years of being considered deluded or worse.

In a broader context of human developmental biology, this was not, and should not have been seen as, news. Many people still have trouble believing this particular piece of the truth, primarily because of some very common difficulty in understanding the kinds of mathematical inference that make it clear, but:

> It is a fundamental fact of human life that
> most human deaths occur before birth;
> most "human" "lives" end before they begin.

If these are to be counted among human lives, then the majority of instances of human life end before they begin. If they are "created" "for a purpose," then, most of them must have been "created" for the purpose of dying before they are born.

> In the best of circumstances,
> in the most privileged of pregnancies,
> fewer than 25% of human conceptions reach full term.

Over two-thirds of "human embryos" fail and disappear even before embryogenesis is finished. Their lives are finished even before anyone ever had any idea that they might have been considered to have begun.

What may be the hardest part of dealing with this realization is another of those differences I warned you about, the many differences between all of what is happening and the parts we can easily see. Here, the problem concerns understanding of the words "conception" and "pregnancy." For the most part, when people talk about pregnancies and the survivals and losses thereof, they are thinking about *recognized pregnancies* and limiting their considerations to only a fraction of the reality of human reproduction — at most about a third of all conceptions.

At the level of ordinary conversational discussion, it makes all the sense in the world to think of pregnancy in terms of recognized pregnancy — in ordinary conversation, a woman becomes pregnant when she comes to know that she is pregnant. This corresponds closely to the concept of clinical pregnancy, because it is at approximately that same time that the presence of the pregnancy becomes detectable by clinical observation.

Hardly ever does any woman arrange to visit her reproductive health care person without thinking that there is some sort of reproductive thing going on that might require some health care. For no more complicated reason than that, clinical recognition usually comes after (at least tentative) maternal recognition. From the obstetrical viewpoint, clinical recognition of pregnancy is the beginning of the pregnancy — pregnancy begins when the obstetrician has been chosen to look after it. It is, however, a long way beyond the beginning of the developmental biology of the business of building a new human, which has been and will remain my focus throughout this writing.

Take-home biochemical pregnancy tests can make a difference of sometimes as much as a few weeks in a woman's understanding of the onset of a pregnancy. These devices are used primarily by women in technically-sophisticated societies who are either eagerly hoping to begin a pregnancy or who are ominously concerned about the possibility of having failed to avoid beginning a pregnancy.

For a majority of women even in technologically sophisticated nations, and for nearly all of them elsewhere, pregnancy is still recognized the old-fashioned way … If the woman has been pregnant before, she may recognize a particular way of feeling different from normal or usual, and suspect the possibility, but two consecutive missed menstrual periods is very widely considered to make the fact of pregnancy fairly clear.

Ovulation is generally said to occur "near the middle" of an average cycle. That idea is not quite entirely meaningless, but it is very nearly so. The length of "an average cycle," and thereby its "middle," varies among women, and declines with a woman's age, becoming three days to four days shorter on average between ages 20 and 40. This variation in cycle length among women, and among cycles for any given woman, is concentrated in the part of the cycle before ovulation, the "follicular phase" — when the egg cell is maturing and being prepared for ovulation, inside the blister-like bubble that we know as the ovarian follicle on the surface of the ovary.

The "luteal phase," the part of the menstrual cycle after ovulation, is rather less variable in length than the follicular phase. In most cases, the next cycle begins very close to two weeks after the ovulation in the current cycle, with much less variability than there is in the length of the follicular phase before ovulation. Aside from continuous variation, give-or-take a day or two between cycles for one woman, or between women, this age-related reduction in average cycle length is a discontinuous variation when seen on an individual basis.

The nature of this change in cycle length distribution with age reveals that there are two "kinds" of human menstrual cycle, differing qualitatively by their average length. It is not the case that all or most cycles gradually get shorter.

In the most common kind of cycle, the average cycle day of ovulation is about the 14th. The average length of the menstrual cycle over the whole population of women and cycles declines because of an increasing frequency of the shorter "kind" of cycle, in which ovulation is much more likely to occur on the 10th cycle day than on the 14th.

The probability distribution of cycle lengths does not move from its peak at 14 days, to 13, to 12, etc., to a new peak at 10 days. It does not slide; it jumps. At all ages, follicular phase lengths cluster around 14 days and 10 days. As women get older, the peak of ovulations at 14 days becomes smaller, and the peak at 10 days rises. The average of all cycle lengths in the whole population decreases as we mix in more of the shorter type of cycle.

For example, this group has an *average* of 13.8:
[14 14 14 14 14 14 14 14 14 14 14 14 10 14 14 14 14 14 14 14 14],

and this group has an *average* of 10.2:
[10 10 10 10 10 10 10 10 10 10 10 10 10 14 10 10 10 10 10 10 10 10 10],

and neither contains any values other than 10 or 14.

The egg cell deteriorates very rapidly after ovulation, such that more than half of all egg cells are no longer capable of normal development if penetrated by a sperm cell beyond about 12 hours after ovulation. By 24 hours after ovulation, the probability of fertilization and normal development is essentially zero.

> The *most fertile time of the cycle*, when insemination is most likely to result in a successful fertilization and a viable zygote, is *the day before* ovulation.

This gives sperm cells time to reach the ampulla (wider section nearer the inner end) of the oviduct, where they can briefly loiter about, finishing their biochemical preparations ["capacitation"] for the prospect of penetrating an egg cell; waiting, still fresh, for the appearance of a freshly ovulated egg cell.

That timing actually works out very much like the process used by mammals with reflex ovulation (rabbit, cat, minks, ferrets, voles, shrews, etc.), where progress through the final stages of ovulation is triggered by copulation, giving sperm and egg cell about equal time to get to the ampulla of the oviduct, the best place for fertilization to occur.

There is good reason to suppose that the human may be a *facultative* reflex ovulator. If there is no insemination to trigger reflex ovulation at the egg cell's best time, then ovulation proceeds spontaneously — like most writings on the subject have always claimed it does. [Actually, it is the spontaneous ovulation that is facultative — by definition non-obligate, available as "Plan B," a backup, secondary process.]

You see, pretty much everything we think we know about the human menstrual cycle, everything you can find in all the textbooks and journal articles, *very nearly everything ever written about the human menstrual cycle is about cycles in which conception did not occur.*

Think about it ... When conception occurs, the behavior of the female reproductive system, from the time of the last normal menses before a conception to the next normal menses after the consequences of that current conception are resolved, is by definition a "cycle" but it is not much like the usual "cycle." After insemination, if fertilization has succeeded and a pregnancy has begun and is continuing, then that cycle does not end in menstruation and the start of a new round of the ongoing cycle, but in delivery, somewhere along the continuum between spontaneous abortion, miscarriage, premature live delivery, stillbirth or full term live birth.

If embryogenesis continues to near completion, a second consecutive period goes missing, and a visit to the obstetrician is likely to confirm pregnancy. It is now about six or eight weeks since fertilization, usually eight to 10 weeks since the last normal period. What has that little parasitic interloper been doing all that time?

Very nearly all the hard work of developmental biology, the building of the embryo, embryogenesis, is complete by about six to eight weeks, about the length of a cycle and a half after fertilization. If all has gone well, the embryo is by that time about the size of a "baby" lima bean, or an average adult fingernail, and … it has its own fingernails! Primordial forms of all of the tissues and organs required for a fully functional body have been formed by the successive divisions and differentiations of the cells descended from the zygote. Sex is not yet obvious to external observation because, although all the parts are there, they have yet to grow into their final relative sizes and move into their final relative positions.

The most salient point is this: By the time in development when the presence of a pregnancy achieves maternal and clinical recognition, most of the things that can go wrong with the development of that pregnancy have already gone wrong. Fewer than one-third of all conceptions that started at the same time have survived to the stage of being recognized. Between maternal/clinical recognition and time for term delivery, another 10% to 20% of those that have survived long enough to be recognized will yet fail, as spontaneous abortions ("miscarriages") of recognized pregnancies.

Against this background, it can hardly be considered surprising that many twin pregnancies fail. According to the best available estimate, fewer than one in 50 twin conceptions result in the delivery of a live pair of twins. Like single pregnancies, over three-fourths fail completely, resulting in no live deliveries.

The most interesting part of this for many people is the part in between the live birth of a pair of twins at one extreme vs no survivors at the other extreme; the in-between story of the sole survivors, the single-born twins. Ever since we have known that some twins vanish, it has been obvious that some other twins — the sole survivors of the vanishing, that have developed from embryos that began as twins — are born alone. To assemble all the data, and parse out the patterns they contain, so that we might realize more or less accurately how many twins are born alone, is another thing entirely.

It turns out, according to the best estimate I have been able to derive and which has not so far been contradicted or improved upon, that for every live-born twin pair there are about 12 sole survivors, single babies who are products of twin embryogenesis. For the most part those survivors show up totally alone, with no evidence that the unborn co-twin was ever there. Just as most deaths occur before birth, and for the same reasons, most twins are born single.

Most twins (defined as products of twin embryogenesis)
are not twins (defined as members of a multiple delivery).

According to the best estimates available, conservatively about one pregnancy in eight begins with a twin conception, and conservatively about one in eight of all liveborn individuals began as a member of a twin conception (including live-born twins and sole survivors).

Twin embryogenesis seems likely to have left the sole survivors with more or less everything that is importantly different about the prenatal development of twins. This has not, however, been demonstrated. No one has yet shown a reliably feasible way of identifying the sole survivors of twin conceptions who are at large in the world, just going about the living of their lives, having no idea, and giving no one else any idea, that they are not "ordinary" people.

Some have called this "The Vanishing Twin *Phenomenon*,"
but it is entirely too common to be considered phenomenal.
Some have called it "The Vanishing Twin *Syndrome*,"
but it occurs in entirely too many different ways, for entirely too many
different reasons, to consider its aggregated characteristics to be indicative
of a specific morbid entity (as required for the definition of syndrome).

It is a fairly simple matter that most human pregnancies fail, most human deaths occur before birth. Most "human" "lives" end before they are in general considered to begin. Most of the failures occur before the existence of the pregnancy has been recognized. Twins do all of those things in the same ways and for all of the same reasons as singletons, and with perhaps a slightly less favorable distribution of outcomes.

The most important parts of the story for present purposes are:

1) Most of the products of twin embryogenesis are born single, usually with no evidence readily visible at delivery to indicate that the other twin was ever there.
2) Twins as a fraction of individual live births approximately equals twin pairs as a fraction of conceptions. The current best estimate of both those fractions, in highly privileged conceptions under optimal conditions for pregnancy, is about one in eight.

Because the samples I used in those calculations were mostly USAmerican and British, these values may be characteristic only of that population of primarily white European ancestry, and may differ from results of similar efforts with samples from other populations. During the time when most of those data were collected, it was not considered proper to notice and record the ancestry of individual members of research samples, lest differences be discovered which might be considered to reflect poorly on one or another subpopulation. The results are likely to be different in detail for twins of African ancestry or East Asian ancestry.

In most such comparisons when properly made, the white European values are intermediate, but closer to Asian than African values.

There is a paper out there in the scientific literature whose senior author later claimed in "the press" to have "debunked the vanishing twin theory." The authors tried to argue that there are nowhere near as many vanished twins and sole survivors as my published results say that there must be. The authors of that paper were coming from an obstetrical point of view, according to which "early pregnancy" begins at clinical recognition, early after the obstetrician has been brought on board. That is not early in any other frame of reference. It is not at all early in the developmental biology, because by that time most of the fundamental cellular work of embryogenesis is completed. Very nearly everything that can go wrong has already done so by the time of clinical recognition, and over two-thirds of all conceptions have already failed. The evidence they presented fit very well among the mass of evidence very closely related to theirs from which the observation of twins vanishing was originally made and from which my estimates had been calculated. Their results showed a loss rate entirely compatible with my estimates for that portion of the duration of pregnancy during which their observations were made.

There have been numerous stories about people who have claimed without physical evidence to be sole survivors suffering the loss of a twin, who have later learned of the death of a co-twin for which they had previously had no evidence. Of all such stories I have heard, I never heard any firsthand, and the idea of memories from before birth is somewhere between mystery and miracle for me, but the data make it clear that the prenatal loss of one twin of a pair IS something that does happen, and frequently.

Chapter Seven

Natural History and Epidemiology of Twins and Twinning

> Believe nothing, no matter where you read it, or who said it
> unless it agrees with your own reason and your own common sense.
>
> *Buddha*
>
> A lie can travel half way around the world
> while the truth is putting on its shoes.
>
> *Mark Twain*

Over a few recent decades, pairs of live-born twins have occupied a rising fraction of human live births. They are still not a large fraction, not even a large fraction of those live-born babies who began gestation as a twin conceptus. Most pregnancies fail regardless of multiplicity, as a result of which sole survivors of twin pregnancies substantially outnumber those born alive together as "twins."

According to sizable bodies of work in several directions over several decades, it has long been considered a certainty that some women are more likely than others to give birth to twins. Women who have previously delivered twins are in fact significantly more likely to deliver twins from a current pregnancy than are other currently pregnant women who have never delivered twins before.

At the other end of the distribution, it may be that some women just cannot conceive twins or deliver twins once conceived, but we cannot yet be certain whether or not that is true, or what the fractions might be.

> Bear in mind that *conceiving and delivering are very different things,*
> regardless of multiplicity, zygosity, sex or sex-pairing.

In Search of a Genetic Basis for Dizygotic Twinning

When we wish to learn whether and how genetic variations may contribute to the distribution of any particular trait, our first step is generally to analyze the pedigrees of families in which there are multiple cases of the trait in question segregating across generations. With large enough samples, we can learn a great deal from such a pedigree study. There is no reason our approach should be different for a study of genetic contributions to the probability of having twins.

When the families of twins are studied for purposes of learning the distribution of twin deliveries over the population, families that find their way into research samples because of the delivery of a pair of twins can be divided according to whether their twins were conceived spontaneously or "induced" by any of the methods which include artificially-induced ovulation.

For the largest such study that I know about, two large groups of families were sampled for pedigree studies of the inheritance of dizygotic twinning.

One sample included the families of women who were known to have delivered a spontaneous, naturally conceived dizygotic twin pair.

The other ("control") sample included the pedigrees of women who had delivered a pair of dizygotic twins after induction of ovulation by a fertility drug.

The control sample might reasonably be presumed to represent that subset of all twinning events which requires no heritable tendency, that fraction of all twinning events which occurred only for "environmental" reasons, in response to the distortions of the normal processes caused by drugs that induce ovulation.

Our favorite and arguably most powerful approach to analyzing data like the results from this pedigree study of twinning is what we call likelihood ratio analysis. We use it to compare the likelihoods of two or more models hypothesized to explain the results. This is highly technical and generally very much loaded with jargon. I cannot get around all of that for you, cannot completely save us from it, but I believe I can explain enough about it for you to have a workable idea of how it fits here:

In likelihood ratio analysis, "likelihood" is the joint probability of all the data that have been obtained [the probability of all these observations happening the way they did, together ... the probability that this one happened this way, times the probability that one happened that way, times ... the next one ... etc., all the way through],

as conditioned on differing hypotheses (according to different guesses about how it works).

The observations are not hypotheses; they are facts; they are what they are. The question to justify all our efforts to collect those data now becomes "How did the results get to be what they are? Let us propose some models of how this might have happened, and see which if any model/s we can reject, and which one of those we cannot reject is the most likely." Under each different hypothesis as to the underlying mechanism, each observation would have a particular probability of being what it has been found to be. This is an objective way of comparing the explanatory values of different hypotheses.

Case by case, person by person in the whole set of pedigree results ... if everything happened according to the "plan A" hypothesis, what was the probability that this person would come out the way he or she has in fact turned out? If this, hypothesis A, is the way this thing moves through families, then what is the probability, under this proposed mechanism A, of a son looking like this one does, or of a daughter looking like that one?

The "likelihood" of this set of results, under this hypothetical mechanism, is the product of all those individual probabilities, case by case, person by person, all the way through the whole data set. ... the joint probability of the entire pedigree as data, if the trait in question is moving through the pedigree according to the model at hand, is the grand product of all the individual probabilities under that model ...

> The uppercase version Π of the Greek letter π (pi) is used
> to indicate the "operator" that symbolizes the instruction:
> "multiply everything after this together"

... Then we do the same calculation again with plan B in mind. If, instead of plan A, it all happened according to the hypothesis behind "plan B," now we score the individual probabilities over the same list, that each person observed and recorded in the database would be as he or she is if plan B is the way this is working.

We go on to do the same thing under each different model till we have computed the likelihood of the whole result dataset separately under each and all of the models that we have thought of that might fit [for datasets large enough to be statistically useful for such work, we would of course be using computers to do all that calculating].

If, for example, we propose the inheritance of a dominant allele that can cause a woman to have twins, then the daughter of a woman who has had twins has a 50% probability of having twins herself ... or less, depending on her age and how many children she has had so far, and how we calculate the age-dependent or parity-dependent penetrance.

Because, in most such calculations, the individual probability values are all less than one, the joint probability, that is, the product of all of the many individual probabilities in a sample of statistically sufficient size, each of the likelihood results will be a very small number, with lots of zeroes between the decimal and the first significant digit.

The farther from "reality" any model is, the less likely each of the outcomes will be, the smaller will be the individual probabilities, and the very much smaller will be the product of all the individual probabilities. If, for example, we have 100 people in the pedigree database, and if every single one of them has a 50% probability of being as they are according to a given model, then the joint probability — the likelihood — of all those results, is one-half to the 100th power, which is roughly one in 10 to the 30th power — many zeroes before you get to a significant digit! All of the likelihoods will be just such very small numbers. Between any two of the models, the model that fits less well — so that specific individual outcomes are less probable under that model — will generate the smaller likelihood.

The ratios between more likely and less likely models can be very large numbers (or the very small reciprocals of very large numbers). It is not unheard of to find one model trillions of times more likely than another, and ratios in the hundreds of thousands or billions or even far larger numbers are not at all unusual with adequate samples. The example I used above, with 100 people, is actually a smallish sample. As applied to analyses of genetic linkage, we routinely dismiss as meaningless a result that shows less than a thousand-fold difference. We would further much prefer to have the same or larger results from several different laboratories with independent samples, if the question is important enough that we really want to have an answer we can believe.

The results of the study at issue here, with those two samples of families with twins, made it abundantly clear that a model involving no genetic contribution to twinning is far less likely than any of the models tested that included any genetic component. All of the models with any genetic component yielded very large likelihood ratios against any model with no genetic component.

There is clearly something genetic about the distribution of twinning events in families over consecutive generations.

Among several different types of genetic segregation mechanisms that could be modeled and tested with those data, the best fit overall was for a single, dominantly expressed mutation. In pure, simple Mendelian dominance, a dominant allele is fully expressed and has its full effect from a single copy of the mutant gene. In this case, the model fit better with the inclusion of age-dependent penetrance (meaning that the mutant gene is not always expressed every time it is present and that older mothers are more likely to show the expression of the mutant gene). According to the results, the mutant allele in question is present in about 30% of the genes in the population. That amounts to a plausible scenario. That means that about 30% of all women are genetically capable of conceiving and delivering twins, but most of their pregnancies will not turn out that way, and the fraction that does will increase with the mother's age.

The obvious next step was a whole-genome linkage scan. If indeed there is a single gene responsible for most cases of twinning in twin-prone families as this result indicates, then we will almost certainly be able to tell where in the genome that gene is located, because closely linked markers will travel through the families together most of the time. When the sequence of DNA, part of the chromosome, that is carrying a mutation that might be causing twinning moves from parent to child, marker sequences that are close by on that same span of DNA will move with it and we will be able to tell which ones — and we already know where they are because we have a decent map of the genome.

We have available many thousands of such markers that we can follow as they travel through families, such that we can have one or two markers that will stay with any particular mutation almost all the time. The larger the number of markers we use, the smaller is the average distance between any two of them, so we can set up a very fine filter. Since we know the precise locations for all of those comparison markers, when we find one of them that hangs tightly together with a particular mutation moving through a large sample of families, we then have a very good idea where the mutation being traced might be.

Now that we have the map of the whole human genome, or at least nearly so, when we find the location of a gene with a whole genome linkage scan, we can simply read the map in that neighborhood for a list of candidate genes, genes whose known functions might be likely to influence the trait we are trying to track down. [There are also still many DNA sequences we know or believe to be functional genes the functions of which we don't know yet. Sometimes we learn the function of such a gene this way.]

As of this writing, those studies have not found evidence indicative of any one single gene that might account for most of the observed variation among families in probability of twin births. It may be that there are several different gene sequences, variations in which may contribute to the risk of delivering twins.

Double-Edged Sword?!
Most Swords have Two Edges;
Sometimes Neither is Really Sharp

Remember that the study in question began with two large sample groups of twin family pedigrees: One group represents the families of mothers of spontaneously conceived twins, while the other is indexed by twins conceived after induction of ovulation. Counting naturally-conceived secondary cases of twin pairs throughout the two groups of families showed that the distribution of twin pairs over the pedigrees was the same in the two groups. The investigators found the same frequencies of secondary cases of naturally conceived twins in every class of relatives on comparison between the two samples.

Although never published, this part of the result is astounding.

Explosive, even.

According to the results of this study, families with twins who were conceived by induced ovulation have other pairs of naturally conceived twins in various relationships elsewhere in their families at exactly the same frequencies as are found in families included in the study sample because of the delivery of a naturally conceived pair of twins.

In both samples of families, included because of ("indexed" by) a spontaneous twin delivery or because of an induced twin delivery as the index case, the question is about genetic relationships with other sets of spontaneous twins in the family. In the two samples, the distribution of those other twin pairs over the sampled pedigrees is the same. *The mothers of index twins conceived after induction of ovulation have exactly the same genetic background with respect to spontaneous twinning as have the mothers of index twins conceived spontaneously.*

That particular observation confused more or less everyone who knew anything about it, and caused the pedigrees that were indexed by artificially induced twins to be left out of publications about the results. As published, this appears to be the definitive study of the inherited predisposition to dizygotic twinning as it was assumed to be occurring, by way of spontaneous double ovulation.

The results (as published, representing only those families indexed by a naturally conceived twin pair) were reported to indicate the existence of a genetic mutation in favor of a tendency to conceive and deliver twins. The version of that gene in question is reported to be present in (i.e., "has a gene frequency of") about 30 to 35% of all the copies of the gene in the population studied.

> The families of the induced twins
> presented exactly the same pedigree evidence
> for a genetic tendency to twinning
> as the families of spontaneous twins.

This fact seems to require the conclusion that (only?) women with that presumed gene in their families — that pattern in their pedigrees — are able *to bear and deliver* twins, with or without artificial ovarian stimulation — and/or therefore without or with spontaneous double ovulation.

Twins born after fertility drug treatment are born only into the same (genetic "kind" of?) families that conceive and deliver spontaneous twins.

It seems to me, therefore, that variation in the probability of delivering live born twins has, for purposes of all reasonable considerations, nothing to do with any imagined inherited tendency toward spontaneous double ovulation, and this tells us nothing at all about the relative tendencies to *conceive* twins.

There is no reason to suppose that the trait being inherited in these families has anything to do with double ovulation. It might much more reasonably be imagined to represent variation in the ability to carry and deliver twins once they have been conceived.

The repeated observation of familial aggregation of twin births remains to be explained. Because powerful whole genome scans have failed to identify a segregating gene DNA sequence to explain the variation, epigenetic variations seem likely to figure prominently in any explanation that may eventually be reasonably assembled. More about that later.

Conception is a verrrry lonnnng way from Delivery

One very important element in these considerations, which is almost universally left out of the published thinking, is the huge difference between the *conception* of twins, that is, beginning the embryogenesis of twins, and the *delivery* of twins. Most of the research efforts concerned with the biology of twinning, such as the pedigree study just mentioned, at face value set about asking questions about the *conception* of twins, and most of the writing sounds as if that is what they are talking about, but … the results being written about universally come from data and studies thereof predicated on the *delivery* of twins, ostensibly as if any change in the delivery of twins can only be caused by a very properly and exactly commensurate change in the conception of twins — as if no twin conceptuses are lost.

In fact, as we discussed in a previous chapter, **delivered twins are a small fraction of conceived twins**, and no one really understands all of the contributions to the large difference between twin conceptions and twin deliveries. It is naïve at best to suppose that everything that happens between conception and delivery is sufficiently constant that variation in the conception of twins generates all of the observed variation in the delivery of twins.

The logical and biological gulf between events of those two kinds is enormous. Delivery of twins absolutely implies and requires not only that there must have been a twin conception and twin embryogenesis, but also that embryogenesis and fetal development and all the requirements for productive gestation have unfolded within normal limits and have succeeded.

The great majority of twin embryogeneses will not result in delivery of twins, without which occurrence the fact that they were conceived as twins in general need never be known.

Subpopulation Differences in Frequency of Twin Deliveries

There are large real differences among major continental subpopulations (which we have been accustomed to calling "racial groups") in the total frequency of twinning (at delivery) and in the mixture of "kinds" of twins. Mothers of African ancestry deliver roughly twice as many twins as mothers of white European ancestry, who in turn deliver roughly twice as many twins as mothers of East Asian ancestry. Those differences are real (strongly statistically significant) and reliably consistent, and have been confirmed repeatedly over several decades.

Also, throughout those observations, the fraction of boy-girl pairs among all delivered twins in each of those subpopulations is directly correlated with total twinning frequency (the higher the population frequency of twin deliveries, the higher the fraction boy-girl. Therefore, the fraction of same-sex pairs among all delivered twins is inversely correlated with total twinning frequency).

Total twinning frequency, and the fraction of all delivered twins that are in girl-boy pairs, are both substantially higher among twins of African ancestry and substantially lower among twins of East Asian ancestry than those values are among twins of European ancestry.

Historically, Weinberg Method results have been used to indicate that these observed differences in total twinning frequency are due entirely to variation in frequency of dizygotic twinning. Because of the observed race differences, it has been considered that the tendency to deliver dizygotic twins is inherited. That approach also gave us the supposition that the frequency of monozygotic twinning is approximately constant worldwide. The general consensus says "of course, this is to be expected, because monozygotic twinning is a developmental accident that can happen to any embryo."

However, on closer inspection, there have always been some very worrisome oddities in those results. The boy-girl pairs in the African samples mentioned above are certainly dizygotic. They were included in these blood genotyping studies only as controls for the zygosity-determining efficiency of the panel of genetic markers that were used. About one-fourth of them were found to match for all the genes tested, just as if they were monozygotic.

If we insist on assuming that boy-girl pairs are, in every way, genetically and developmentally, representative of all dizygotic pairs, and if one-quarter of those girl-boy pairs match for all the genetic markers used in the test, such that the test results might be taken to indicate that those twins are monozygotic, then we must suppose that one-quarter of the same-sex dizygotics must also have scored in the blood typing as if they were monozygotic.

This had to be taken to mean that the twin pairs scored as monozygotic by testing with that set of markers included dizygotic pairs equal in number to one-third of the same-sex pairs that properly scored as dizygotic (one-quarter equals one-third of the other three-quarters).

After this "adjustment" (guided by Weinberg expectations) of their results to accommodate that discrepancy, the results of Weinberg estimation of zygosity fractions in the Nigerian samples were reported to match those of the white European sample.

No such experimental controls were examined and no such "correction" was performed in the analyses of the white European sample.

This adjustment is by no means a small one, by no means is it a minor variation. It is huge, especially when considered against our much more recently acquired certainties that African populations are the oldest populations of the human species on the planet and therefore the most genetically diverse. They have had the most time together in which to accumulate mutations and spread them around. The African populations should therefore be the least likely populations in which to find siblings matching by chance for any one genetic marker, let alone matching by sibling segregation chance for every one of the genetic markers chosen for those zygosity studies. Those markers were chosen for use in those studies because they can, when used together, routinely indicate probabilities of monozygosity in excess of 95% among the European twins.

Commitment to the preservation of the orthodox beliefs founded on the outcomes of Weinberg Method calculations prevailed over these facts. For decades since then we have been told that the Weinberg Method, globally applied, yields acceptably accurate estimates of zygosity fractions, because the Weinberg results agree satisfactorily with results from blood typing.

Therefore, the orthodox story goes on, we should further believe that race-group variation in twinning frequency is confined to variation in frequency of dizygotic twins, and that the frequency of monozygotic twins is constant over those same subpopulations.

We should suppose that the frequency of dizygotic twinning varies genetically as a function of a tendency to double ovulation, and that monozygotic twinning is a developmental accident occurring at an approximately constant frequency worldwide.

These concepts are pillars of the orthodox biology of human twinning. The logic of and by which they were constructed is nonsense. We have no reason to believe any of them, with the possible exception of the unspoken desperate notion that we *must* have *something* to believe. [I have tried it myself on numerous occasions, and I can assure you that admitting ignorance is very unlikely to cause you any real, enduring harm.]

Because of those reports of the constant frequency of monozygotic twin deliveries, we have been expected to suppose that the biology of monozygotic twinning is constant over various populations around the world. However, the facts surrounding that assertion give us another huge little problem! Monozygotic twins of African ancestry are about half monochorionic, monozygotic twins of European ancestry are about two-thirds monochorionic, and East Asian monozygotic twins are about 80% monochorionic. These are independent and repeated results, and the differences are abundantly statistically significant. Funny, but it just does not sound to me as if these populations are all doing monozygotic twinning the same way. You think …?

We have long believed that chorionicity (dichorionic co-twins have separate full sets of gestational membranes; monochorionic pairs share the thicker, outer membrane) represents a fundamentally important variable in the biology of the monozygotic twinning event, particularly as regards the timing of the event relative to events of ongoing embryogenesis. To be specific, we have long supposed that monochorionic twinning events can only have happened after cells that are precursors to the chorion have committed to their differentiation as trophoblast, so that the twin embryos will begin and develop inside a shared future chorion. We have further believed that monoamnionic twinning events must happen still later, after the cells that will become the amnion have committed to their differentiation as such.

According to these results, for mothers of African ancestry, monozygotic twinning events happen earlier in embryogenesis, consistent with their embryogeneses progressing more rapidly, so that they will have a lower fraction monochorionic and a higher fraction dichorionic. Similarly, for mothers of East Asian ancestry, embryogenesis progresses more slowly than for mothers of African or European ancestry, including the embryogenic commitment to twinning, so that twinning events appear to occur relatively later in the progress of embryogenesis, with a resulting higher fraction occurring only after trophoblast differentiation and thus sharing a single chorion.

Whatever the biological bases may be for these very real, reliably repeatable, differences, they have been well hidden by the dead weight of the common knowledge, hidden from understanding and from investigation that might have improved understanding. Several measures of reproductive epidemiology in addition to twinning frequency are distributed in this same [or exact reverse] rank order over these same subpopulations ... where Af = of African ancestry, Eu = of European ancestry, As = of East Asian ancestry:

average age at onset of menstrual cycling:	Af<Eu<As
average age at first pregnancy:	Af<Eu<As
average age at each consecutive delivery:	Af<Eu<As
average parity at any given age:	Af>Eu>As
average age at the last pregnancy in each mother's history:	Af<Eu<As
average gestation length of each pregnancy:	Af<Eu<As
average length of second-stage labor:	Af<Eu<As

average birth weight: Af<Eu<As
frequency of "premature" deliveries: Af>Eu>As
frequency of "low birth weights":   Af>Eu>As
frequency of infant deaths: Af>Eu>As
secondary sex ratio (fraction male at birth): Af<Eu<As

All of these observations of variations in reproductive physiology are in the same rank order over these three "major geographic subpopulation" groups, with mothers of European ancestry intermediate between mothers of East Asian ancestry and mothers of African ancestry.

The overall pattern is that most of the appropriately measured elements of reproductive physiology occur earlier in life, or proceed more quickly, for mothers of African ancestry, and occur later in life or proceed more slowly for mothers of East Asian ancestry, than for mothers of white European ancestry.

Sex Ratio at Birth: It May Mean Something After All

Sex ratio at birth is presently the only one of the above-listed variables for which we have a good candidate mechanism that might explain the variation. East Asian mothers deliver the highest fraction of sons, African and African-American mothers the highest fraction of daughters, with mothers of European ancestry, as usual in all of these considerations of reproductive epidemiology, in between.

In every case where appropriate data are available, the people called "Hispanic" (Americans of Central American, South American, or Caribbean indigenous ancestry) are in the same class as East Asians (from among whom the Americas were peopled, for the most part in the most recent 50,000 years – mostly between 20,000 and 12,000 years ago).

Paternal imprinting of the X-chromosome has been shown to cause slower development of female embryos in every mammal appropriately tested. Only female embryos have a paternally imprinted X-chromosome. This slower development apparently causes a fraction of female embryos to miss time-critical developmental control signals. Greater speed of embryogenesis can account for the greater survival of males through embryogenesis, which can in turn account for the greater fraction of males among deliveries, in apparent spite of male excesses among losses throughout the fetal period.

After the completion of embryogenesis, from the maternal and clinical recognition of the existence of pregnancy to delivery, males are found in excess among all losses of pregnancies. During embryogenesis, however, substantially more female embryos fail, and the male embryos build up an excess of number that they will maintain all the way through to delivery in spite of greater losses through every stage of recognized pregnancy.

(Epigenetic) Genomic Imprinting in Gametogenesis

The apparent origin of the striking pattern among these variations has fascinating and compelling implications for all of human developmental biology.

With a certain few abnormal exceptions, every normal male has one paternal Y-chromosome, and a single X-chromosome of maternal origin. Females have two X-chromosomes, one from each parent. A large number of important developmental control genes are epigenetically marked (imprinted) differently in spermatogenesis vs oögenesis.

As it is generally understood, or as it seems to be in the best-understood examples, the effects of genome imprinting on gene expression are generally negative, such that the imprinted copy of the gene sequence in question is not expressed. This leaves the other allele, contributed by the parent who does not imprint that locus, to be the one that is expressed.

We do know that the effects of genomic imprinting at any given gene locus may not always be negative. We know of at least a few genes with inhibitory functions that are inhibited by imprinting, with disinhibition and up-regulation of their target functions as a result.

Many genes are imprinted in spermatogenesis and therefore expressed only from the chromosome that came from the mother. A substantial number of other genes are imprinted in oögenesis and expressed only from the copy provided by the father.

This imprinting effect was first discovered in mouse embryogenesis: If we remove both the paternal pronucleus (the sperm's half-genome contribution for the zygote nucleus) and the maternal pronucleus from a fertilized egg before the two pronuclei have undergone syngamy to form the zygote nucleus, and replace them with a proper set of one maternal and one paternal pronucleus, many of them will die from the effects of the manipulation alone, but some will develop normally, all the way to delivery.

If we replace the original pronuclei with two paternal pronuclei, every one of those experimental embryos will die, with overdeveloped extra-embryonic support structures and with the failure concentrated in the embryo proper. The paternal imprint reduces expression of some genes the functions of which are required for healthy development of the embryonic body.

If we replace the original pronuclei with two maternal pronuclei, again every one of the experimental embryos will die, but each of these will have a fairly good-looking embryo proper and poorly-formed extra-embryonic support structures. The maternal imprint reduces the expression of some genes the functions of which serve to upgrade the embryo's grasp of maternal uterine resources.

It is to the evolutionary advantage of males in general to maximize the transmission of their individual genomes to the next generation by initiating as many pregnancies as possible. These objectives are furthered by the strongest possible attachment, the most aggressive grasp on the mother's uterine resources.

The long-term reproductive success of the female mammal is not, however, well served by carrying as many pregnancies as the males to whom she might be available might be able to initiate.

Evolution of Imprinting

The normal situation which we have upset in those pronucleus exchange experiments shows a compromise struck by evolution, in which the female's reproductive potential is optimized when her egg-building process controls certain developmental genes by imprinting in certain directions, and the male's representation in future generations is optimized by down-regulating some other functions in embryogenesis and perhaps up-regulating still others.

Female embryogenesis is retarded by paternal imprints on X-chromosome genes installed during the building of the sperm. By that means, sex ratio among live births is set by effects of paternal genome imprinting on sex differences in growth rates and survival in embryogenesis. Because of their developing more slowly, female embryos are more likely to miss developmental control signals and fail more frequently in embryogenesis, setting up an excess of males at the completion of embryogenesis and transition to the fetal period such that there is an excess of males at delivery in spite of excess losses of males throughout the fetal period.

> In skating over thin ice, our safety is in our speed.
>
> *Ralph Waldo Emerson*

Sex ratio at birth varies over these race-group subpopulations, in strong correlation with multiple other indicators of relative speed in female development and reproductive functions. I have suggested that all of those correlated variations in reproductive epidemiology may vary as they do because of variation in the extent or strength of the paternal-X imprint, and its effects on the relative speed of male vs female conceptuses in embryogenesis.

Besides these clear "race" differences, there are other important differences in the likelihood of twinning, as a function of the mother's age and of her parity (the number of children she has already delivered). The effects of age and parity are not trivial to separate, even in theory, and perhaps not even possible in practice. Maternal age and parity are inextricably confounded to some extent simply because, while a woman is having more children, time is passing, and a concomitant increase in her age is inevitable. Age and parity therefore are not, and cannot reasonably be considered to be, independent variables, and efforts to separate their effects statistically are not entirely satisfactory — women who have had six children before the age of 25 are hard to find and probably not statistically representative of the population as a whole anyway.

These "racial groups" are lately often called "major continental," or "major geographic" "subpopulations" in deference to widespread wishful thinking and misunderstandings, to the effect that ignoring or denying the large and complex realities of biological differences across some of those "racial" groupings will somehow reduce the existence or influence of racist thoughts or behaviors. To whatever extent that the concept of racial grouping might really be meaningless, the uproar over every mention or application thereof is all the more counterproductive willful ignorance. There are profoundly important things to be learned from these differences in human biology.

Quite apart from the variations in twinning frequency among racial groups, there are sound reasons to believe that certain heritable factors can increase the probability of bearing twins for any given mother.

There is also sound evidence for heritable factors which increase the probability of *fathering* dizygotic twins. Against the dimly lit but durable background orthodoxy that only the frequency of dizygotic twins is heritably variable and that it varies entirely as a function of the frequency of double ovulation, these observations of paternal effects on probability of dizygotic twinning generate some seriously inconvenient truths.

It is demonstrably true that fathers almost never ovulate, let alone twice at once, so it is honestly quite difficult to accommodate the notion that something heritable in the father's makeup can influence the probability of double ovulation. This is a severe problem as long as double ovulation is the only acknowledged way to vary the probability of twinning.

Twinning and Fertility

Because of its association with higher parity, twinning was once considered a sign of higher than normal fertility. We have more recently come to understand that natural, spontaneous twinning is in fact a sign of subfertility. For example, when Gordon Allen tested the hypothesized relationship between twinning and high fertility: The twins found in large sibships (with mothers of high parity) were usually late or last (when fertility is declining in older mothers) and seldom early in the birth order. The excess of twins among conceptions following upon return of soldier husbands from war were not concentrated among the quickest conceptions, as might be expected if twinning represents higher fecundability, but were in fact among the slower ones, with greater delays to conception after the reunion than for the single births in the same cohort.

There were reported excesses of twins among illegitimate pregnancies (in the 1950s and 1960s but not anymore), and among couples in which the spouses work in different cities and see each other only on weekends, and among the children of merchant seamen and commercial fishermen.

All of those situations have in common with each of the other situations listed, and with mustering-out from the armed forces, and with increased maternal age and parity, *not* any sign of greater ease of conception or greater efficiency of gestation, but rather circumstances of infrequent and irregular intercourse — which increases the probability of fertilizing an overly mature or otherwise ill-prepared egg cell.

Besides all that, nobody could ever give any interesting answer to my question as to how the commercial fishermen and merchant seamen managed so consistently to choose or be chosen by the most fertile women in town, to explain the excess of twins among the children they fathered. But we do know that they are often away at sea for extended periods …

The prospect that overripe oöcytes might figure prominently in the risk of twinning receives strong support from a multitude of animal experiments in which ovulation was triggered experimentally and insemination was allowed only after varying delays. Not only twinning, but anomalies of chromosome number or structure, malformations, spontaneous abortions, stillbirths — all of the common sorts of birth defects and pregnancy wastage — increased dramatically.

We cannot ethically arrange such experiments in our own species, but the enormous diversity of human reproductive behavior includes some of very nearly . everything, if one can happen to find or recognize what is needed for "experiments in nature." Susan Harlap found this one in Jerusalem: Mosaic Law forbids sexual intercourse for Orthodox Jewish couples during the wife's menstruation and for a period of seven clean days after dabbing the vulva with a white cloth no longer brings back red. This will usually total about 12 days of abstinence. At the end of this time, the Law further requires a mikveh (ritual cleansing bath), after which intercourse is not quite exactly a religious duty, but certainly an act of piety for both partners. This has the generally-excellent-for-their-reproductive-public-health consequence of putting the resumption of intercourse very close to the average optimum cycle day for fertile insemination.

Sometimes, they have a high holy day that also requires a trip to the mikveh. Sometimes somebody's sick, or they're fussing, or he's away on business … and she will put off the resumption of intercourse and wait a day or two to go to the ritual pool. For any number of reasons, the timing may occasionally be imperfect. When resumption of intercourse is delayed one day beyond the couple's usual cycle day of resumption of intercourse, the frequency of twinning *triples* — and the excess is due overwhelmingly to boy-girl pairs! So, is it plausible to suppose that we have somehow summoned a second egg cell by some consequence of abusing the first one to death by allowing it to (over)ripen beyond its optimal state of preparation? Hardly.

Dr. Harlap hesitated for five years to publish those results, which were quite contrary to the common knowledge that dizygotic twinning requires double ovulation. When she did publish them, she caught a load of disparaging commentary, since which time this part of her work has been roundly denigrated or ignored — after all, how could putting the oöcyte through something that ordinarily kills it cause the rescue of the first, dying oöcyte *and* the ovulation of a second one to go with the first, however enfeebled, to make a total of two? and most of the time with the second being different in sex from the first?! (…because …, you see …, everyone knows dizygotic twinning requires double ovulation.) "Those women couldn't possibly have records accurate enough to justify those conclusions!" … and so on — there simply had to be something badly wrong with anything so far out of line with what they had been telling each other for so long!

By the way, Down syndrome[a] was also dramatically more frequent among the progeny of such delayed fertilizations. This is entirely, perfectly consistent with the excesses of chromosomal anomalies in those delayed-fertilization animal experiments I mentioned earlier, and with reported excesses of Down syndrome and Turner syndrome[b] among twins in general populations. There is also a separately reported excess of dizygotic pairs both members of which had Down syndrome.

---

[a] Most cases of Down syndrome involve the presence of a third copy of the chromosome #21. The most common source of the extra chromosome is believed to be an accident in the building of the egg cell in which the separation of the members of the chromosome 21 pair fails, sending an extra copy to the daughter cell which ends up providing the maternal half-genome for the zygote. The risk of Down syndrome increases with maternal age, and increases dramatically beyond a maternal age of about 34. With uncommon exceptions, frequency of intercourse is diminishing under those conditions, and this effect is believed to be due to pre-ovulatory over-ripening of the egg cells.

[b] The majority of cases of Turner syndrome are believed to arise from a similar accident, as a result of which the zygote has only one X chromosome, no second X, and no Y. Because the X chromosome that is present in cases of Turner syndrome who survive to term is most often from the mother, it is generally believed that the chromosome segregation accident that leads to Turner syndrome happens in building the sperm cell. Frequency of Turner syndrome *at delivery* does not depend on the age of either parent.

One of the studies that showed increased twinning after delayed fertilization (this one in rabbits) found the excess twins as "double" embryos inside single zonae pellucidae, and the authors (for no other reason) declared the twins to be monozygotic. Most of the workers in the field were satisfied with that explanation. It was entirely consistent with the fossilized common knowledge, and no one ever tested it further. Of course, no one would ever imagine that twin embryos could have gotten inside single zonae beginning with independent double ovulation. They are absolutely right ... but that idea of independence is a problem, as we will see further along.

There is still no evidence that double ovulation has ever — even once — been the origin of any naturally conceived pair of human twins.

Fathering Dizygotic Twins

If indeed dizygotic twinning were all or even mostly about double ovulation, then only females could bring it down through a pedigree lineage over several generations. John St. Clair and Mikhail Golubovskii have reported a family of Scots which shows paternal transmission of twinning, both monozygotic and dizygotic, through nine generations over the 19th and 20th centuries.

Another study, by Dorit Carmelli and colleagues, found over 400 families in the extensive genealogical records of The Church of Jesus Christ of the Latter-Day Saints (also known as the Mormon Church) with two or more pairs of opposite-sex twins at least as closely related as first cousins (third-degree relatives — includes aunts, uncles, grandparents/children, and great- aunts, uncles, grandparents/ children). In almost exactly half of those families, the twins in the pedigrees were related only through males.

Therefore, either a father can inherit something that increases the likelihood of his fathering dizygotic twins at least as much as anything heritable that increases a woman's chance of giving birth to twins, OR there is in fact nothing to do with double ovulation underlying that variation. The only source of contradiction is the myth and creed of double ovulation as the source of dizygotic twinning. Epigenetic variations must be considered, as must the facts that the male contributes to the circumstances of conception a number of things other than the sperm cells and seminal fluid, and that the sperm cell carries into the egg cell more than just a half-set of the necessary DNA sequences.

Rising Frequency of Twin Births

Over the last three decades, the frequency of twins among live births has increased considerably — by some reports, doubled or more; by every report at least nearly doubled. This increase in the frequency of twins is almost entirely a result of life-style changes involving the postponement of reproductive careers in favor of careers for women outside the home.

Part of the story is not news at all. As long as we have recorded and studied appropriate data, it has been clear that older mothers have more twins, and these career-woman mothers have all (sorry! yes, every one of them) gotten older while they were waiting to begin their families. We have also long known that older women seeking conception for the first time are less fertile than they would have been 10 years earlier and so are more likely to have difficulty conceiving. This brings about the largest single cause of the recent increase in multiple birth frequencies, that being medical treatments for infertility.

Twins in Technologically-assisted Conceptions

It took more than 20 years, but it seems that we have finally convinced at least most practitioners of IVF-ET (*in vitro* fertilization and embryo transfer) to confine their procedures preferably to one and never more than two embryos transferred. As a result, these practitioners have needed to become more proficient at inducing ovulation, and at fertilization, and most especially at choosing for transfer those embryos that are most likely to succeed. Still, when they succeed, it is not as rare as it seems it should be that they get more babies than (they thought) they put in enough embryos to make. This is a seriously interesting result.

By restricting the procedure to only one or two embryos transferred, it is now much less common for an IVF effort to generate four or more fetuses — an unusual untoward result that was once not nearly unusual enough.

The least complex of the common "fertility treatments" is the artificial induction of ovulation by itself (AIO, by "fertility drugs"). AIO is more than sufficiently complex to give us reasons for profound concern, but it is still the least complex because it is (only) a part of all the others. (AIO is the one manipulation common to all of the assisted/artificial reproduction technologies.) AIO remains difficult for the practitioners to get the dosage right, and still frequently results in multiple pregnancies, twins and higher multiples. While the frequency of twinning has roughly doubled, largely as a result of technologically-assisted conceptions, the frequency of triplets has increased as much as 20-fold and more in some observations.

An unfortunate side effect of ART procedures resulting in higher-multiple pregnancies has been the use of "fetal reduction" to reduce crowds of fetuses to safer numbers. Healthy outcomes for all concerned cannot be made certain, but can be made more likely. Imagine if you will, please, having to choose the one/s whose development is to be stopped so that the remaining one/s might have a better chance, at least as far as the multiplicity of pregnancy can be manipulated after the opportunity for a more responsible course of management has passed.

One aspect of all that, seen as very surprising by some observers, is that the multiple pregnancies produced by those artificial ovulation-stimulating procedures are not all polyzygotic — they are not all coming from one separate fetus-per-zygote or vice versa. Whatever embryonic event may be required to initiate monochorionic twinning occurs substantially more frequently among the products of any procedure involving artificially-induced ovulation than it does in spontaneous conceptions. All of the risks associated with spontaneous multiple pregnancies (malformations, chromosome anomalies, etc.) are also associated, and in general more strongly associated, with artificially induced ovulation however it is accomplished.

Beginning some time before we had that realization, we had the reliably repeated observation that the most common configuration of natural, spontaneously conceived triplets is dizygotic (a monozygotic pair plus a third fetus from a second zygote). This always suggested to me that the presumably unrelated mechanisms for monozygotic and dizygotic twinning had a remarkable habit of happening at the same time, to products of the same ovulation. I had also, before the publication of that finding, observed and reported excess frequencies of nonrighthandedness and midline fusion malformations equally in monozygotic and dizygotic twins and the first-degree relatives of both. There were also the observations of reduced asymmetries in the multivariate structural relationships in dental development, in monozygotic and dizygotic twins equally. For over 20 years I have argued that *all of these observations taken together imply/ demand that the monozygotic and dizygotic twinning processes have causes and mechanisms in common.*

ART (artificial reproductive technology) triplets, arising as they do from artificially induced multiple ovulation, would always have been expected to be trizygotic (three fetuses, each from a zygote of its own). Trizygotic triplets do in fact represent a greater fraction of artificially induced triplets than they do of spontaneous triplets (of which they are about one-third). Dizygotic triplets [a monozygotic pair +1 (the MZ pair usually being identified as such by monochorionicity)] are three times as common among ART triplets as monozygotic twins are among *total natural* conceptions. Twinning events resulting in delivery of monozygotic twin pairs are clearly more frequent among ART conceptions than they are in natural pregnancies.

Most of the people to whom I tell any significant part of this story are amazed to hear that there is no evidence that any pair of naturally conceived human twins ever arose from two separate, independently ovulated egg cells. That is true. Not one. Not ever. My focus in all of this has been and will remain primarily on the naturally conceived human twins, but it is informative here to note that even multiple pregnancies arising from induced multiple ovulation in many cases cannot all have arisen from separately ovulated egg cells.

Monozygotic twins are more than three times as frequent in conceptions from induced ovulation as they are among natural conceptions. The same is true of dizygotic twins who are wrongly diagnosed at delivery as monozygotic because they are monochorionic. It has been an erroneous common-knowledge rule of thumb for decades that all monochorionic twin pairs are monozygotic. Over most of that same period of time, there have been occasional reports of dizygotic monochorionic twins. A young physician from Glasgow was shouted down at the International Congress on Twin Studies in Amsterdam in 1986, upon trying to tell us about a few dizygotic monochorionic twin pairs he had discovered in his practice. There was no question that what he was saying was a horrible, heinous error! The only question was where to put the blame for the obvious incompetence those results had to represent!

As per Sir *Winston Churchill*, a recurring theme throughout the background of this work:

> Men occasionally stumble over the truth,
> but most of them pick themselves up
> and hurry off as if nothing ever happened.

A very few papers in the literature acknowledge that the cellular origin of dizygotic twinning has not been soundly demonstrated. With those rare careful exceptions, every paper in the literature of twin studies and twin biology either flatly states that dizygotic twins come from double ovulations, with no offer of evidence or authority ... OR, the author gives a reference to some previous writing as authority for the statement. That previous paper in its turn either flatly states that dizygotic twins arise from separate simultaneous ovulations, with no offer of evidence or authority ... OR, its author gives a reference to a previous paper as authority for the statement. and ... so ... (it goes ...) on.

Follow it as far as you can manage, and you will find no direct evidence, and no better pretense of logical authority than a reference to a previous statement which in its turn offers no evidence.

Invoking Aristotle

One obviously desperate effort to make the assertion of double ovulation seem more like the product of a valid argument from profound consideration is to quote Aristotle. Yes, that Aristotle; not Onassis or Savalas, but that very one ancient (b. 384–d. 322 BCE) Greek philosopher, from back in the days when the concept and the word "philosophy" were evolving to mean "love of knowing," and any philosopher worth a chunk of bread for his breakfast, let alone a little bit of good Greek olive oil to go with it, was expected to know more or less everything about more or less everything.

Among the roles Aristotle played in the dramatic human cultural (r)evolution that sprang forth from his times and culture, we are probably most indebted to him for his teachings about logic and other such uses of human reason. However, to all appearances he was also the premier biologist of his day.

Often not quite right, but somewhere between seldom and never really in doubt, Aristotle had an understanding of the process of learning about the natural world (what we now call "science") that was rather different from ours today. It was, after all, a very new idea — Aristotle has for centuries gotten a lot of credit for spreading the idea that we could use our reasoning abilities to figure out how things happen, instead of giving all the credit or blame for everything to miscellaneous divine whimsies. That idea has spread far and wide, but still has not covered all the gaps — in substantial part due to remnants of the efforts of other ancient Greeks. We are deeply indebted to Aristotle for spreading the idea that the application of human reason and logic to the solving of problems and to learning in general had great advantages over ignorance and superstition and acceptance of gratuitous assertions from false or out-of-their-element authorities.

In the exercise of our current understanding of the "scientific method," we form hypotheses based on our preliminary observations and then test, through further experimental observations, the effects predicted by the hypotheses we previously framed. Aristotle was not much of an experimentalist. His reasonings were generally grounded in "first principles" — things he knew were true by definition and not by derivation from other truths or principles. Perhaps not exactly always the first at-all-plausible thing that popped into his head and was deemed obvious and certain, but maybe not often far from that ... So, he really was not much of an epistemologist, but he was a great logician, and in his day he had no peer in the range and depth of his considerations about what there was to be learned about Life and living things. He was an astute observer, and the absolute grand wizard at reasoning out the meanings of his observations, but experimentation had no part in his conduct of scientific argument. This allowed him, in a number of notable instances, to lead himself subtly but profoundly astray.

With respect to all the ramifications of all these questions about the origins of twins, Aristotle's day for being the greatest of biologists came and went about 2000 years before there were microscopes with which to observe cells and to appreciate and declare the cellular basis and character of Life ... and yet another few hundred years before we knew anything about egg cells and sperm cells and fertilization and syngamy and zygotes and embryogenesis.

In the ancient Greek portion of the meantime, Aristotle had a first-principle kind of knowing that the "life principle" comes from some sort of interaction between the genital fluids of the male and of the female. To Aristotle, that observation had all the necessary qualities of a first principle, because new human life never arose on his watch in the absence of either of the two humans necessary and in such condition as to guarantee the presence of both of those components.

Aristotle knew that twins were an anomaly arising obviously from some untoward deviance in that interaction — and he believed that such deviance must be caused by variation in the temperature and/or pressure acting upon those fluids in that interaction. He further knew that there were some twin pairs with one shared life principle between them, and some in which each had his own.

"Aha! See?!
Alike, one; not so much alike, two!
One egg, two eggs!"

Not so fast … Sorry, but that is not anything Aristotle ever even imagined he knew anything about, first principle or otherwise, and that is not to be found among the thoughts he put into writing. There are several steps missing, that no one would really understand for thousands of years to come.

Albert the Great (1206–1280), Doctor of the Church and teacher of Thomas Aquinas, was called Doctor Universalis, The Universal Teacher, because of the enormous range of his knowledge. According to some accounts, he was accused of sorcery because much of his science was so advanced. Remember Arthur C. Clarke: "Any technology sufficiently advanced cannot be distinguished from magic."

Through his studies and teaching, Albert the Great was Aristotle's primary channel from Greek antiquity through Arabic scholars to Europe's Middle Ages and beyond, via the [European] universality of church Latin and a few other of the then-current vernacular tongues. Expanding upon and explaining Aristotle as was his habit, Albertus Magnus is said to have written that the cause of the deviation-from-process that resulted in twinning was an excess of coital pleasure on the part of the mother on the occasion of the conception of twins. I'll bet you didn't see that coming!

> I felt quite fortunate to have found that particular nugget, but it was not, on reflection, a complete surprise to me. You see, my youngest daughter, when she was only five- or six-years old, told me that she had solved for me the origins-of-twinning problem we had discussed when she had asked me about my work: "You know, whatever it is that you do to make a baby … you just have to do it twice as much!"

Evidence was still not the most important thing about the way what passed for science was done even then, but, unless Universal Doctor Great Albert quantitatively observed and recorded a goodly number of occasions for conception and was able to compare those notes with outcome follow-up to know which subset of those occasions resulted in twins …? Naaaah … I really can't picture that it happened that way. That was probably another one of those first principle things, or a heavenly revelation, or something of that sort.

The double ovulation hypothesis of the origin of dizygotic twins is more or less exactly as soundly scientific as these pronouncements, from Aristotle via miscellaneous Arabic scholars to Albert to Aquinas. To be sure, these pronouncements are as close as anything else available — even today — to being sound evidence in favor of that hypothesis. There is, on the other hand, a substantial body of sound evidence against it.

Dizygotic is the proper, technical term for what most people think they mean when they say "fraternal" twins. Fraternal means "brotherly." Dizygotic means "from two zygotes," and only about one-fourth of dizygotic twin pairs are brothers. Roughly three-quarters of all those "brotherly" twin pairs include at least one sister, and about one-fourth of all dizygotic twin pairs include no brothers at all. One might argue that "fraternal" is okay because everybody knows what it means, but, among other problems, it does not travel well. In French, for example, the non-technical version of "*dizygotique*" translates as "false twins," and nothing to do with brothers.

The idea of dizygotic twinning arising from "double ovulation," the more-or-less simultaneous release of two independently ovulated egg cells — is not much more than a blind assumption, repeated without challenge until it has come to be considered "common knowledge." The idea was not totally the stuff of fantasy — we have known for a while that other mammals that routinely bear litters usually have one corpus luteum[c] for each fetus.

As a first-principle sort of guess, double ovulation as the origin of dizygotic twins makes a good deal of sense. It might reasonably be considered a plausible hypothesis, but some important steps got left out, and it is still an untested hypothesis. The idea can never be more than hypothesis until its necessary predictions are repeatedly confirmed. In the world of what science is really about

---

[c] The corpus luteum (Latin, yellow body) is the ovulation 'scar' on the ovary, formed from the cells of the follicle which just ovulated. During a healthy pregnancy, the corpus luteum serves as a gland secreting progesterone to support maintenance of the pregnancy.

in this 25th century after Aristotle gave us permission and encouragement to use our minds to figure things out with, that is a rule, and we are going to stick to it here.

Dizygotic twinning from two zygotes does not in fact imply and does not require independent double ovulation. There are several normal "kinds" or "forms" or "states" of the egg cell on its way to fertilization. Those forms, or states, of the oöcyte are conceptual "snapshots" from a continuous process involving a number of complex mechanochemical events. We have no really good idea exactly how very many biochemically separate steps there may be, or how many more unusual variations might occur at whatever frequency, but certainly not one of these various states or forms of the oöcyte is the same thing as the zygote.

A zygote is by definition an importantly different thing from an egg cell.

I could have attempted to start this work with zygosity like most previous efforts to tell this story have done. That approach, like all the others, including this one, is imperfect. The writers of all those other versions did that because they saw zygosity as fundamental to all that was to come, and they thought of it as one of the simplest parts of the whole story. I can heartily agree with the "fundamental" part of that assertion, but every effort to give it any appearance of simplicity is false, misleading and counterproductive. It is not simple, and efforts to construct a facade of simplicity around it have generated some serious fundamental errors.

> None of the differences between monozygotic and dizygotic twins is simple. Nothing that is the same about monozygotic and dizygotic twins is simple.

The simplicity of the stories you have been accustomed to hearing about the biology of twinning, and especially all the parts of that having anything to do with zygosity, is false simplicity.

The "zygosity" of twins — whether they are "monozygotic" or "dizygotic" — strictly speaking, means exactly, and only, whether the twins arose from one? or two? zygotes. No more, no less. That definition *does not imply anything about how there came to be two.*

The zygote is a single cell, the single nucleus of which has just been formed by a new syngamy, a new melding of the half-genomes from an egg cell and a sperm cell into a single diploid nucleus containing a double set of chromosomes. One member of each pair of chromosomes in the zygote nucleus comes from the sperm and one from the egg, with a combined set of DNA sequences different in exact molecular detail from any such set that ever was assembled before or will ever be assembled again.

If a zygote is to serve as the cellular origin of a viable embryo, it must, besides that set of genetic material just specified, have a functional system of cell division machinery. It must have the ability to form a mitotic spindle, by the use of which to sort properly and reliably the duplicated chromosomes in every cell division to follow. The human oocyte loses its centrioles during meiosis ... cell division functionality of the zygote requires a pair of centrioles brought in by the sperm.

This cell must also have all the mechanochemical structures necessary for the control of the expression of all the components of that genetic information, to produce each such event very reliably in the right cells at the right time in development. Each of these major subsystems is itself a complex system composed of many molecular parts — all of which are gene products.

Most of those components that are used in the first few cell divisions of embryogenesis were put there in advance by the developmental processes of preparing the egg cell for ovulation and, to a much lesser but clearly essential extent, the sperm cell. It takes some time before the genes of the unique new genome of the zygote are expressed.

Much of the biochemical activity of the zygote is conducted by gene products the production of which was complete or at least already in progress in the egg cell before ovulation, probably even long before ovulation.

Twins may develop from a single cell defined by a single such zygote nucleus, from progeny cells that carry copies of that same one set of DNA sequences. Monozygotic twins develop from subgroups of cells within a group of cells that have all descended from a single original zygote cell. Except for mutations which occur with some finite probability in every cell division, they will continue to carry the same set of genetic information throughout their development.

There is no mechanical "split" that causes monozygotic twinning. Eventually there must be a separation between two groups of cells which will become the two separate bodies, but this is a consequence of the twinning process, not a cause.

To possess or carry the same set of genetic information is not the same as to express all of it in the same ways. We have been for a while, and will be for a while to come, involved in trying to decipher a hugely important additional system of mechanisms of developmental control, which involves variation in the expression of gene DNA sequences rather than variation in those sequences themselves.

There is at least that one other system of controls on gene regulation and cell differentiation, generating individual differences in development even between monozygotic co-twins. We call it "epigenetic" because it involves changes in expression imposed *upon* the "genetic" DNA sequences, rather than genetic changes in the sequences themselves. We do not yet know nearly as much about epigenetic controls as we know about genetics, but it is a current focus of major interest and research, and we know for certain that we have hardly begun to understand all that its systems include.

One other, especially interesting component of human developmental individuality (uniqueness, self, personality) is immunological. Our immune systems need to be able to make highly specific antibodies, to protect our bodies from bacterial, viral, and fungal infections, from parasitic infestations, and from some toxic molecules.

Antibodies are complex protein molecules that can bind to molecular substructures of those invading structures with astounding specificity and avidity. When I was in graduate school, we used to marvel at this, discussing it at great length, most of the time without perceptible progress. We were entirely certain that the human nucleus does not contain nearly enough gene DNA sequences to code for each of the specific structures of all those different antibody protein molecules, especially since some of the substances against which we would need such defenses were only invented last week, and could not, therefore, have highly specific corresponding antibody protein sequences placed in waiting by evolution.

When he tells the story in his Nobel lecture, he seems to have come across the question only after he got to Basel, and I do not remember his participation in those discussions. However, one of my fellow graduate students at UC San Diego, Susumu Tonegawa, won the 1987 Nobel Prize in Physiology or Medicine for showing us how that happens. That also is worth a few books of its own.

Suffice it for here and now to say that the immune response process begins with a small population of cells some few of which can produce antibodies that will fit only approximately to the structure of each new foreign *antigen* (substance which generates an antibody response). From that beginning, a very dynamic *iterative* process of mutation and selection among the genes that code for antibody protein chains results in the evolution of consecutive descendant populations of cells that produce increasingly better-fitting antibody proteins.

Each generation of such cells that can provide a better fit is chemically "rewarded" and stimulated to proliferate and keep mutating to yield an even better product in a later generation. The final result is usually an exquisitely precise fit between an entirely novel newly synthesized antibody protein molecule and an antigen that quite possibly never existed before some living chemist synthesized it last week.

Even if the life histories of monozygotic co-twins were identical with respect to the infections, infestations and chemical exposures they had experienced, the mutations and selections required for their individual immune cell evolution would not, could not, be identical because of the substantial element of chance involved in generating each of the necessary rounds of mutation.

Control of DNA Expression by Modifications of Histone Proteins

I have mentioned earlier the sperm nucleus, having penetrated the egg cell, being tended to by the egg cell cytoplasm in preparation for syngamy with the maternal pronucleus. This is a prominent example of epigenetic processes. (As far as we know to date) DNA base-pair sequences are not mutated in this process, but they must be unpacked and reconfigured from the extremely condensed state of DNA in a normal sperm head, to the variably and controllably looser configurations required for their function within an active cell nucleus.

Basic (as opposed to acidic) histone proteins and protamines and polyamines which have been essential to getting the acidic DNA wound up tightly enough for packaging in the sperm head, now get released, unhooked, swapped around for lighter bonds or none at all in some parts — and this happens differently for different genes and groups of genes, and it is likely to vary individually by chance and epigenetic variation as well.

This process of variation in chromatin condensation by histone modification is fundamental to ongoing cellular function and differentiation. For every function of the DNA, it must be acted upon by enzyme proteins to catalyze the chemical reactions it must undergo. To be duplicated in preparation for cell division, it needs DNA polymerases to read its sequence and build the complementary sequence. For transcription into messenger RNA, it must be labelled by transcription factor proteins and read by RNA polymerases. For all such functions, the DNA base pair sequences must be available to the enzyme proteins, not sequestered in chromatin packages as it is when it is idle (which is the case for most of the DNA most of the time).

Some genes and groups of genes are imprinted such that they will not be read from the paternal copy of a given chromosome in the zygote nucleus or beyond. The locations and strengths of that imprinting are variable.

Some of these epigenetic processes are dependent upon chance variations, such as which of two daughter cells gets a certain combination of imprints or other DNA modifications, and that in turn varies as a function of which cell division in a given embryogenic lineage is happening at the time.

Monozygotic means having developed from a single zygote — a single cell. That original single cell has one and only one and the same nucleus, which has just formed from the egg cell pronucleus and the sperm pronucleus. That one nucleus contains one and only one and the same set of chromosomes. The DNA in the chromosomes in the nucleus generally gets most of the credit for providing a plan for the individual's development. It sets some of the limits, maybe most, but some of the limits are not under control of these nuclear genes. There are several different kinds of exceptions. One exception is the mitochondrial genome.

The Mitochondrial Genome

The great majority of the cell's DNA is in the nucleus, passed down in more-or-less exact replications from each cell to each of its daughter cells. A small fraction is outside the nucleus, in the mitochondria (my'-toe-kon'-dree-ah). Mitochondria is plural; the singular is mitochondrion — another of the many Greek-derived terms in biology and genetics. Mitochondria are the energy factories of the cell. They probably used to be bacteria that came aboard a very long time ago as symbionts (organisms that maintain mutually beneficial commensal relationships; from Latin, "same table" — harmlessly sharing food and other resources).

The mitochondria have their own DNA, which codes the information for some of the mitochondria's own structures and functions independently of the genetic information in the nucleus of the cell. After a long co-evolution, some of the structures and functions of the mitochondria are now coded in the cell nucleus — this has become a deeply interdependent relationship.

These days, most of the people in the developmental genetics business are convinced that we get all of our mitochondria, and therefore all of our mitochondrial DNA, from our mothers. There are a few hundred thousand mitochondria in a human egg cell (it IS the largest and most complex of all human cells, after all), and maybe a dozen in the sperm. The ones in the sperm are all in the collar, there to provide energy to drive the tail the sperm uses as a propeller (it works as a screw, not a sculling oar) to swim to its target.

· Our best pictures of freshly fertilized mammalian egg cells suggest that the fertilizing sperm's collar goes inside the egg cell along with the head. There, along with most components of the sperm outside of its nucleus, the mitochondria are labeled extensively with ubiquitin, a protein that says "this is junk — tear it up, break it down and get it out of here." The ubiquitinylated debris is gathered up by lysosomes, inside of which the debris is cut up in molecular pieces either to be recycled or to be done away with by expulsion from the cell. Lysosomes, when they are "full," fuse with the cell membrane, open, and dump their contents to the outside of the cell. The sperm's mitochondria appear to be destroyed, and appear not to contribute to the mitochondrial DNA of the zygote.

Even if any part of the sperm's mitochondrial DNA should survive to contribute to the zygote's ration thereof, maternal mitochondrial DNA would outnumber the paternal contribution at syngamy by a factor on the order of 20,000 to one.

We know very nearly nothing about exactly how it happens, but we do know there can be more than one version of the mitochondrial DNA sequences in any given cell (this is called heteroplasmy). In theory, those differences might all be due to new mutations, or might be present as an accident in sorting mitochondrial genomes through the cell divisions leading up to the construction of the egg cell or to any point in postnatal development.

Mitochondria are responsible for generating most of the energy needed by cells of the body to sustain life and support growth. It is not inappropriate to think of the mitochondria as the cell's furnace, or its engine room boilers. Mutations in mitochondrial (mt-)DNA are responsible for a sizable and still growing list of human illnesses, which range in severity from inconvenient to catastrophic. Some strike children, even in infancy or prenatally; many of them usually appear only in maturity. The nature and severity of symptoms depend on the organs or tissues involved, and the exact nature of the mutations involved. Most are difficult to diagnose. Some of them can be ameliorated by treatment, but none of them can be cured. The worst of those illnesses involve cells of the brain or nerves or muscles, because the mitochondria are the cell's energy-producing subsystems, and those are the most energy-greedy cells of the body.

That brings to mind another process at work here with major effects on every aspect of the functions of living cells. The primary energy-yielding biochemical pathways of the mitochondria involve oxidation-reduction reactions, which generate some of the most highly energetic molecules in the body. Perhaps the best known of these are the triphosphates of adenine and guanine, ATP and GTP, which function as highly portable molecular "fuel pellets" to provide the chemical energy of activation needed for most of the body's chemical reactions.

The chemical energy of ATP and GTP is harvested from the process of oxidizing (burning) nutrients. "Burning" in this usage is more than a metaphor. The final products of the body's complete combustion of nutrients in the mitochondria are carbon dioxide, water and energy, just the same as from your automobile's internal combustion engine or any other fire burning organic matter. This fire, however, is very stringently controlled, and the chemical energy is harvested in small quantities from several intermediate stages of the gradual oxidation. Some of the activated intermediates and their byproducts are very highly-reactive "free radicals" (compounds with unpaired electrons caused by their high degree of chemical activation), which can cause serious chemical damage whenever they escape the process.

Effects of Mutation

Together with the fact that the mitochondrial genome has no DNA repair capability, this concentration of free radicals leads to the mutation rate in mitochondrial DNA being far higher than that in the nuclear DNA. This fact is also biased against cells in brain, nerves and muscles, because of their greater use of energy.

Our best evidence suggests that it usually takes several generations for a harmful mutation in mitochondrial DNA to become a large enough fraction of the total to produce such harmful effects. Some researchers have reported a huge change (from less than 5% mutant mt-DNA to over 85%) taking place in a single generation, but none of them reported any plausible idea of how the cells on the scene managed to do that. It would seem to require some kind of bottleneck, some stage in building egg cells where only a few mitochondria get passed on to the next generation, so that the mutant proportion can be immediately sharply skewed. To my present knowledge, no one has yet identified any such process.

That will be enough for now about mitochondrial DNA. For purposes of this discussion, the point is this: When one embryo decides it needs to turn into two embryos and grow up to be twins, there is no reason to expect that the cells that are incorporated into each of the embryogenic body symmetries will have the same mixture of mitochondrial DNA sequences.

Another source of variation in the course of development from two subsets of cells from a single original embryo, is mutation in the nuclear DNA. The chromosomal packages of DNA in the nucleus have to replicate at every cell division, so that each of the daughter cells will have a full and complete copy of the DNA. Assuming absolute perfection in every step of that process on every occasion, all the cells in that body will have exactly the same sequences of DNA in all of their nuclei. It never happens quite exactly perfectly that way.

Replication is imperfect; not terribly imperfect, but certainly imperfect; very reliable, but not perfectly reliable. Errors are made in every replication. [With three billion base pairs to be replicated, that is six billion precise base addition reactions to be conducted; even one error in a million generates six thousand errors per half-genome per generation.] Not all of them are removed in proofreading and not all of them are repaired, with the result that the newly replicated copy of a strand of DNA does not have entirely and exactly the same sequence as its parent copy.

Mutational replication errors can be caused by various "chemicals," in wholesome food or drink, or in pollution. They can be caused by radiation. They are often caused by perfectly normal things that happen in the course of entirely normal metabolism and biochemistry.

For example, even without anything extraneous happening or being around, the information-carrying parts of the DNA (the "bases") are just naturally and normally and constantly changing their molecular structures. That is simply the nature of those particular compounds. The pyrimidines thymine (thigh'-mean) and cytosine (sigh'-toe-seen) have six-atom (4 Carbon, 2 Nitrogen) single rings as backbones, and the purines adenine (add'-in-een) and guanine (gwah'-neen) have nine-atom (5 Carbon, 4 Nitrogen) double rings (sharing one 2C side). The "conjugated double" bonds that hold the carbon and nitrogen atoms together to give shape to those ring molecules move, constantly. (That is not perfectly accurate, but it is not far wrong, and it is much easier to visualize than the quantum-mechanical fact that the electrons shared in those bonds exist in all of their possible configurations simultaneously, at all times.) Each of these molecules has at least one alternative chemical structure, an alternative arrangement of the bonds between the atoms, that will cause it to be read as if it were a different base when the base-pairing function of the replication machinery catches it in that alternative shape. Each of these molecules spends a very small fraction of its time in any of those alternative shapes, but that is not zero, and there are billions of opportunities in every replication.

There are very effective and efficient proofreading and editing processes built into the DNA replication machinery, and they detect and repair the great majority of these mutations even before the replication machinery moves on. But those parts of the system are not perfect either, and some mutations get through.

A substantial fraction of them make no difference we know how to identify in the function of the cell in which they occurred.

We have, however, recently learned that some mutations of the kind that we have always called "silent" or "synonymous" (because the mutated DNA sequence codes for the same amino acid sequence in its protein product that it coded for before the mutation) do in fact cause differences in the final protein gene product. These differences show up not in the process of reading the messenger RNA to specify the amino acid sequence of the protein gene product, but in post-translation processing such as folding the protein into its functional three-dimensional structure.

Most of the time, when we talk about mutation, we have in mind those that happen in germ-line cells where they might be passed to offspring. This is not often identified as a source of differences between monozygotic twins, but in fact it can be. Mutations that occur in the last DNA replication before the onset of building the egg cell or the sperm may not be final and fixed at the time of fertilization. One strand of the DNA molecule of one of the chromosomes may be mutated while the other strand remains as it was before the mutation. If the editing and repair mechanisms fail to fix it, to put the mutant strand back the way it was, the next normal replication will yield one daughter DNA molecule, and thus one daughter cell, with, and one without, that mutation. The daughter cells may then establish two genetically different cell lines — differing by that one mutation — within the same embryo. A monozygotic twinning event that might happen a few cell divisions later may then yield two embryos with different fractions of the two resulting cell lines. [This is entirely possible in theory, and we have documented such occurrences in experimental organisms, but no such case in human twins has been documented.]

Here is a bottom line: The probability that every one of those changes would occur in the exact same places all through the lives of a pair of monozygotic twins is zero. No two of us have exactly the same sequences of DNA in every cell of our bodies, including monozygotic twins.

Epigenetic Variations

To bring all of these understandings together, it is also very important to understand that the exact same DNA sequence in a gene does not necessarily have the exact same developmental outcome. This has long caused a substantial amount of confusion in the world of human genetic analysis. For example, it has been clear for a long time that the single best predictor of your personal individual risk of being diagnosed schizophrenic at some point in your life is the closeness of your genetic relationship to someone else who already did. There is without question a major genetic influence. In adoption studies, there is zero evidence for any familial effect that is other than genetic. In spite of all that, the fact that monozygotic twins with schizophrenia do not always have schizophrenic co-twins is still cited as "the best evidence," as if it were actually any evidence at all, for an "environmental" non-genetic contribution.

Misdiagnosis from Misunderstanding

"Monovular" is sometimes used as if it were the same as "monozygotic." It does not mean the same thing, any more than the egg cell and the zygote are the same thing.

I answer many calls and e-mails from twin pairs who go through life "knowing" that they are dizygotic because "the doctor told our mother we were fraternal, because we had separate placentas." But they have come to doubt it; friends, teachers, even close family members frequently have difficulty being sure which twin is which. If co-twins are enough alike that friends and family frequently confuse them, there is an excellent chance they will prove to be monozygotic when properly tested. Several groups of researchers have, over the years, compared the results of various similarity questionnaires with blood typing results. Fairly simple similarity questionnaires have managed to discriminate zygosity with well over 90% agreement with genotyping results.

In general, the literature still insists that a twin pair absolutely must be monozygotic if the twins are monochorionic. There have been published reports of exceptions in the literature for several decades, but we have only recently come to understand the explanatory connection between monochorionic dizygotic twins and chimerism. We know that at least some monochorionic twins are dizygotic. We know how difficult it has been to recognize the few such pairs that have been reported. We know that all such pairs who do not have the kinds of anomalies that have led to the discovery of those few reported chimeric pairs are never investigated for the possibility. For all those reasons, there is every reason to believe that monochorionic dizygotic twin pairs are far more common than we presently believe, and there is no reason to suppose that anyone knows exactly how common or rare they really are.

We also now know for a certain fact that dizygotic twin embryos sometimes arise from a single embryogenic cell mass. We know this just because monochorionic dizygotic twins exist. We know this because there are people whose bodies are built of two or more types of cells which are different in too many ways to be the result of a single mutation. These people are called "chimeras," or "chimeric."

Now suppose that a chimeric embryo, a single embryonic cell mass formed from or otherwise containing dizygotic twin embryos, might routinely undergo the kind of twinning event which is typical of monozygotic twinning — that two sets of body symmetry axes may be established within that single mass of cells. The resulting twins would be monoembryonic, but they would not be monozygotic. They might both, one, or neither be detectably chimeric as adults.

At this point we cannot claim to know what causes, or what allows, or what is required for, any cluster of embryonic cells to become two "embryos" instead of only one.

In later chapters, there are extended explanations of several ways in which sound evidence has made it clear that the embryogenesis of dizygotic twins and that of the monozygotics leave the same subtle anomalous traces in further development. Dizygotic twins differ in their development from singletons at least as much as the monozygotics do, in the same ways and by the same distances.

the Chorion, the Amnion and the Vertebrate Body Plan

A complex system that works is invariably
found to have evolved from a simple system that worked.

John Gall

We can lick gravity, but sometimes the paperwork is overwhelming.

Werner von Braun

Problems worthy of attack prove their worth by hitting back

Piet Hein

The human fetus develops in a fluid-filled membranous sac, sometimes called the "bag of waters," sometimes just the "water," as in "he took me to the hospital after my water broke about two o'clock this morning." The wall of that sac has two layers. The layers are one inside the other, ordinarily very closely adhering over their whole surfaces, and not at all clearly distinct to any sort of observation other than competent microscopy looking at a properly prepared slice of tissue from the right part of the structure. In embryology we usually speak of it as two membranes, because each layer has its own name and a different origin — arising from two different sets of cells, from two different places in the embryo, at two different times in embryogenesis.

The thicker outer layer is the chorion, of which a twin pair may share one or have one for each. Twins may be monochorionic, with both fetuses developing inside a single chorion, or they may be dichorionic, with each having his/her own chorion. In higher multiples, all of the intermediate situations are observed. For example, the most common arrangement for naturally conceived triplets is dizygotic: "a pair and a spare"; a pair of monozygotic twins from one zygote and a third with a different genotype from the second zygote. Often, the monozygotic pair is monochorionic, with the result that the three fetuses occupy two chorionic membranes. Triplets may, of course, be monochorionic, dichorionic, or trichorionic, as well as monozygotic, dizygotic or trizygotic — they may have one, two or three chorions, as well as having developed from one, two or three zygotes and having one, two or three diploid genotypes.

The thinner layer on the inside of the gestational sac is the amnion. Each chorion is lined with at least one amnion. Most twin pairs that develop in a single chorion have separate amnia (proper Greek plural of amnion — amnions is also used). In other words, most monochorionic twin pairs are diamnionic. Monoamnionic twinning, with development of both twins inside a shared amnion, is the most dangerous sort of membrane configuration for twin gestation. Only monoamnionic twins, with no membrane between them, may be conjoined.

Most twin pairs are dichorionic and diamnionic, each member of the pair having his or her own full set of both membranes. The Orthodox Common Knowledge says that this group includes all dizygotic pairs and some fraction of monozygotic pairs.

I still find it amazing, but this comparison apparently was never fully put together from separate publications until I did so in just the last couple of years: The chorionicity fraction among monozygotic twins varies quite distinctly over the major geographic subpopulations ("formerly known as" "racial" groups). Of East Asian monozygotic[a] pairs, ~80% are monochorionic and ~20% are dichorionic; of white European monozygotic[a] pairs, ~2/3 monochorionic, ~1/3 dichorionic; and, of monozygotic[a] pairs of African ancestry ~50% each.

The fraction of monozygotic twin pairs that are dichorionic varies over these same subpopulations in the same rank order as the fraction of all twins that are of unlike sex: Africans and African-Americans have the highest fraction females among all live births, the highest frequency of twinning, the highest fraction of all twins who are in mixed-sex pairs, and the highest frequency of dichorionic pairs among the monozygotic pairs. East Asians (Chinese, Japanese, Koreans) have the lowest fraction of females among all live births, the lowest frequency of twinning, the lowest fraction of all twins who are in boy-girl pairs and the lowest fraction of dichorionic pairs among the monozygotics. The white Europeans

---

[a] Here, "monozygotic" means only "as estimated by Weinberg Difference Method calculations." More later.

are intermediate, but closer to Asians than to Africans, in each and all of those variations and several other variations in reproductive epidemiology (more later, in Chapter Eleven).

Only recently are we becoming unable to ignore the fact, and spreading the news, that dizygotic twins are sometimes monochorionic. The total number of such pairs ever discovered to date has recently roughly doubled. So far, they have, in this decade as well as decades past, been discovered only by the accident of being unignorably discordant for some major substantially heritable anomaly or for sex, or both. According to the orthodox understanding of dizygotic twins arising only from two separate independent ovulations, male and female have no more business being inside the same chorion than they have being parts of one hermaphroditic body. A true hermaphrodite, by definition, has gonadal tissue of both sexes, some ovarian tissue, some testicular tissue, neither necessarily being in a properly formed gonad.

Realizing this drastic restriction on the recognition of dizygotic monochorionic pairs, I must insist that we really have no useful idea of their frequency. Commonly made assertions to the effect that they are extremely rare are totally gratuitous. The monochorionic dizygotic twin pairs recognized to date have all been (and will continue to be) chimeric. The great majority of them have two normal cell lines, of the same sex, and will never give us reason to examine them closely enough to discover their chimerism.

This makes all the sense in the world, if we stop to think about it. In order to develop inside a single chorion, the two genetically distinct groups of cells have to be together, while a single trophoblast precursor to the chorion will differentiate around them. That occurrence IS the formation of the blastocyst/blastula from the morula. That usually occurs on about the 4th or 5th day when the zygote has divided into a couple of dozens of cells still inside the zona pellucida, in a single contiguous mass of cells the outer layer of which, one cell thick, now differentiates in the direction of becoming the chorion.

Like other spontaneous human chimeras, if the twins are normal and of the same sex, we have not been finding cause to investigate closely enough to identify their chimerism. They will be declared monozygotic, without question and without investigation, by virtue of their being of the same sex and being monochorionic. To my knowledge, none of the people concerned with the biology of twinning over the several decades of the development of that science have so far been willing to acknowledge the existence of any doubt. Everybody who is anybody just knows that is the way it is.

Most monochorionic pairs are diamnionic, separated by a double layer of amnionic membrane between them inside the heavier outer layer of their shared single chorion. Only a small fraction of monochorionic pairs are also monoamnionic, developing with nothing to keep them apart in the same volume of amnionic fluid.

Not all monoamnionic twins are conjoined, but all conjoined ("Siamese") twins are monoamnionic (think about it … they cannot be conjoined if they are separated by membranes, whether two amnions, or both amnions and chorions).

> By the way, in case you did not already know it, conjoined twins have been called "Siamese" because of a pair who came to the United States from the land that was then called Siam (now Thailand) and became famous and financially comfortable working in show business. Chang Chun and Eng In (1811–1874) were born near Bangkok of Chinese and Chinese-Malaysian parents. They were joined by a band of tissue below their rib cages. They would probably be easy to separate today, but no surgeon of their time wanted to take the risk or impose the risk on the boys. They worked in show business for 10 years, then retired to North Carolina near Wilkesboro, where they became American citizens and successful farmers, took the surname Bunker, married a pair of sisters from nearby and had a total of 21 children between them (one source says 22). They died within hours of each other, still attached.

Monoamnionic twinning has long been recognized as a developmentally very dangerous situation. Monoamnionicity does not cause conjoinment, but is a necessary condition for it. Monoamnionicity also makes it possible for one twin to be strangled by the umbilical cord of the other. (A singleton fetus can strangle himself with his own cord if it is long enough to make a loop and swim through. It is much easier for a twin to do it with the co-twin's cord if they are both in the same volume of amnionic fluid — mechanically not possible with separating membranes.)

It is generally understood that monoamnionic twins undergo their twinning event and become twins later in embryogenesis, after the differentiation of the amnion had begun, which becomes visible later in embryogenesis than the differentiation of the cells that will become the chorion. Accordingly, they have been considered to be at greater risk for all those disturbances of embryogenic asymmetries that have been imagined to be associated with the "late"ness of the "split."

The number of chorions per fetus in a multiple pregnancy has long been considered an important element in zygosity diagnosis. It has been the general understanding, repeatedly asserted in the associated literature for generations, eventually assuming the status of "common knowledge," that dizygotic twins are ALWAYS dichorionic, and that monochorionic twins are ALWAYS monozygotic. In other words, monochorionicity has been presumed to be peculiar to, and proof positive of, monozygosity.

Monozygotic twins may also be dichorionic, but according to the orthodoxy/common knowledge/received wisdom, dizygotic twins cannot be monochorionic.

If twins are of opposite sex, we know that they are dizygotic … right? About the only way anyone has imagined to get around this is for an XY embryo to undergo a chromosomal mutation in a very early cell division such that one of the daughter cells will have the one X-chromosome and no Y. Absent one or more genes that are normally on the Y-chromosome, male development cannot occur. If development continues, spontaneous abortion is by far (~95%) the most likely outcome. Those surviving to live birth are generally infertile females with a characteristic appearance called "Turner syndrome."

If twins are monochorionic, then — of course, everyone knows — they are surely monozygotic.

If twins are same-sex and dichorionic, we cannot know the zygosity without some sort of genotyping. These days, zygotyping, or zygosity genotyping, generally means DNA analysis, allowing the analysis of many different DNA sequences in each sample simultaneously.

These assertions have been continually repeated for decades, even after published reports about a number of births the very existence of which proves the falsehood of both of these absolute statements. The fact is that, after all, some dizygotic twins are born monochorionic; some monochorionic twins are after all found to be dizygotic. Not all dizygotic twins are dichorionic; not all monochorionic twins are monozygotic. This shows us one more facet of the fact that the story is not as simple as you have been told.

This is astonishing, really. For anyone who understands and believes the Common Knowledge doctrines, this simple little fact Rips the Fabric of the Universe of Twin Biology! This idea generates such cognitive dissonance that the standard response is to swear that one never heard of such a thing.

Everyone knows, of course, that dizygotic twins arise from separate and independent double ovulations. Therefore, according to the common knowledge, they cannot possibly appear at delivery inside a shared single chorion.

But they do, therefore they can.

Do not try anymore to tell us that this is extremely, freakishly, rare. How often it happens that way is in fact quite entirely unknown, because — if monochorionic twins are both normal and of the same sex, no one ever before this understanding has imagined any reason to investigate their zygosity. No one has any useful idea how many monochorionic dizygotic twin pairs have been born but have never been discovered because "everybody knows" they just do not happen, and we almost never find what we know for certain is not there.

This prospect brings with it a very large problem for anyone still clinging to the belief that dizygotic twins arise from double ovulation. Up until the cells that will form the future chorion differentiate, and a bit beyond that, the embryo is still inside the zona pellucida. Before hatching (while still in the zona pellucida), usually followed more or less immediately upon hatching by implantation, there can be no contact between the cells of two embryos that have arisen from two independent ovulations and are involved in independent embryogenesis. Without contact, there will be no fusion.

Researchers in developmental biology have, many thousands of times over the last few decades, fused all or parts of two or more separate embryos of laboratory animals, to form single masses of embryonic cells including cells of more than one genotype. The chimeric animals produced have found many uses, mainly having to do with learning the developmental pathways that lead to particular tissues. The largest reported number of different embryos productively fused, to contribute cells to the development of a single adult, appears to be four, in the mouse. Four was a rare outcome in those experiments; three was much more common. This has led to the proposal that the embryo proper, the core of the embryo as it will continue development, is derived from only three or four cells.

There IS a "catch" here. There is no basis for the assumption that this process of generating artificial chimerism looks at all like the normal, natural, spontaneous process of embryogenesis. In natural embryogenesis, differentiation of the chorion comes before "hatching" and implantation. At the time of the differentiation of the chorion, everything is still inside the zona pellucida. To perform all of these experimental embryo fusions, to make all of these experimental artificial chimeras, the zonae pellucidae (that is the proper Latin plural) have always had to be mechanically or enzymatically removed — without which there could be no contact between the two separate masses of cells, no chance of causing fusion of the two inner cell masses.

Each of the reported cases of monochorionic dizygotic twins has been found in or with some anomaly such that the occurrence can be seen as pathological in one way or another. Scientists in the associated fields of inquiry have generally been able to avoid making themselves try to understand the implications of these twin pairs by dismissing them as pathological and therefore not deserving of being allowed to derail the dogma.

This IS dogma; it IS doctrine. These are the right words, not a bit too strong. Perhaps incompatible with sound, open-minded scientific attitudes, nevertheless, this is the orthodoxy; it is a matter of faith and tradition. It is the Common Knowledge. Everybody who has any sense at all knows — like the difference between right and wrong — that dizygotic twins are always dichorionic and monochorionic twins are always monozygotic. Assuming that dizygotic twins arise from two separate and independent ovulations, they cannot arise inside a single chorion inside a single zona pellucida. It cannot be otherwise. This is fundamental. Also false.

The "catch" is, with no exception published as of this writing, that monochorionic dizygotic twins have been discovered to date only when and if and because they are distinctly more different from each other than monozygotic twins are expected to be, with respect to something arguably pathological but certainly too obvious to be ignored.

Normal, same-sex monochorionic twins are declared to be monozygotic by default. They are not in general examined for any possibility of dizygosity. Everyone knows there is no need — it is a waste of time and reagents in the lab — to genotype such twins to determine their zygosity. "Everyone knows" monochorionic twins are always monozygotic. It is by general agreement quite clear, it is well and widely known, that this is always true ... — ... except, as we are here and now coming to understand, when it is not true.

When we find twins to be boy and girl and unarguably monochorionic (there really is no possibility of confusion given competent examination), the common knowledge understanding of twinning biology has come upon a crisis of faith. Just as monochorionic twins may draw enough curiosity to cause closer examination if they are visibly not of the same sex, the same can sometimes happen when the twins are of the same sex but clearly discordant for some major anomaly. There are several birth defects, malformations and congenital disorders which are known to be largely genetic in cause, but which are reported to be excessively frequent among twins *reported as* monozygotic, *and* frequently discordant in those "monozygotic" twin pairs. Beckwith-Wiedemann syndrome, for example, has been reported several times to be excessively frequent in female monozygotic twins and usually discordant.

In those cases, the twins are being called monozygotic quite confidently, which can only be because they are of the same sex and monochorionic. Although they are discordant for a disorder generally recognized to be genetic in origin, this does not raise a question of their zygosity if they are of the same sex and monochorionic. This is a matter of perspective that needs to change.

There is a problem here, just as there is every time we must face the fact that any profound matter of ancient faith is in error. It is always uncomfortable.

Given that monochorionic twins are not always monozygotic, *given* that dizygotic twins are not always dichorionic, *then and therefore* the common knowledge understanding of the double-ovulation origins of dizygotic twins is ... is ... is ... ... well ... wrong!

This particular dogma, about the relationship between zygosity and chorionicity, is a corollary of another, older and deeper false certainty — the one about how dizygotic twins come from double ovulation. If dizygotic twins can be monochorionic, it has to mean that dizygotic twins can arise from a single mass of cells containing cells of two different genotypes — a chimeric embryo.

At least some dizygotic twin pairs were contained within single cell masses before the trophoblast cells that would become the chorion differentiated from the rest of the cells of the morula, which would become the inner cell mass. At least some dizygotic twin pairs must have arisen from a single mass of cells just exactly in the same way and at the same developmental time and from the same stage of embryogenesis as monozygotic twinning events.

Both monozygotic and dizygotic twinning events occur by way of the onset or assumption of separate destinies by cells within a single cell mass that will become a double embryo and perhaps eventually the bodies of two separate people.

Dizygotic twin pairs arise by the exact same system of cellular events as the monozygotics. Inside one ball of cells, inside one morula, two body symmetry plans are laid out. Two systems of head-tail axis differentiation, two sets of left-right determinants, and two systems of back-vs-belly definitions must establish themselves. Two sets of body-symmetry axes must be established and elaborated. Two subsets of the available cells must be 'recruited' into two separate vertebrate body plans.

It is a fairly simple thing to say "establish a system of body symmetry axes." It is not terribly difficult to understand that the unpacking of your whole body and brain from a single cell … to at least several trillions of cells with at least a few hundred different sets of structures and functions … requires some planning and organization. To understand exactly what all the pieces are and how they must go together to execute such a plan — we have made great strides, but it is going to take a while longer. I can confidently tell you that there must be a plan, preferably right from somewhere very near the beginning, whatever we imagine "beginning" to mean.

We have learned a fair amount about how a fly is built, mostly from the fruit fly *Drosophila melanogaster*. There is also a tiny nematode worm (called *Caenorhabditis elegans*) smaller and simpler than even a fruit fly, about two millimeters long, the adult body of which worm is composed of 959 cells, and we know the developmental lineage of each of those 959 cells — the history of how each cell in the body of the worm came to be the kind of cell it is, working the way it works, where it is.

The house mouse is our favorite experimental mammal for such studies. Chickens and quail, frogs and newts — many animal species from several different "kinds" of living systems have contributed wonderfully to our understanding of developmental biology, "embryology" taken now to cellular, subcellular and molecular levels.

There are many particularly wonderful lessons from those studies. Arguably the greatest of those lessons is about the continuity of Life, the Brotherhood of Every Living Thing, the consistent use of the same cellular mechanisms, driven by gene product molecules the structures of which have been highly conserved — passed down with only a few surviving mutational changes — in many cases for hundreds of millions of years. The Unity of Life and the Consilience of Living Systems never cease to amaze.

Probably my personal favorite example is that of the *eyeless* mutation first discovered in the fruit fly. When that gene is mutated or damaged in such a way that it cannot make its normal gene product, the fly does not, cannot grow eyes. Something that looks rather like a scar occupies each position where an eye would have been expected to be. The product of that gene is necessary for the proper building of eyes in fruit flies.

What We Can See Through the Missing Eyes of Flies

You may already know that insect eyes and mammalian eyes are very different structures. We mammals have "simple" eyes, each a single chamber with a lens out front to focus images on the retina. Insects have "compound" eyes. The compound eyes of insects are composed of multiple tiny eyes, ommatidia (plural), one each behind all of those facets. The number of facets and ommatidia varies among insect species, from dozens to hundreds. Each ommatidium (singular) contains a small cluster of light-sensitive cells surrounded by support cells and pigment cells, in a tube with a transparent cornea on the outside surface and a single nerve fiber on the inner end — by way of which fiber the brain gets one "pixel" from each ommatidium for the composite images it forms.

Spiders are arachnids, not insects. They do not have eyes like insects. They typically have six to eight not-very-beautiful "simple" eyes, but some spiders are reported to have excellent vision. The eyes of such mollusks as the octopus and squid (yes, they are classified as mollusks, the same phylum as snails and cockles and scallops and clams) seem to be nearly as good as ours. They are simple eyes like ours. The simple eyes of vertebrates are single chambers with retinas covered with thousands of light-sensitive cells to capture and forward images.

A colossal squid recently caught in Antarctic waters and studied in New Zealand has simple eyes 11 inches in diameter, three inches wider than a regulation size 5 soccer ball, with a lens the size of a large orange — the better to hunt with, way down deep where photons are feeble and few.

The eyes of crustaceans are compound, like those of their arthropod cousins the insects, but apparently are substantially more complex in function, especially in dim light.

As far as we have tested to date, every organism that has eyes of any sort has a gene the DNA sequence of which is very much like that of the fruit fly's *eyeless* gene. The corresponding gene in mouse is called Pax6; in humans it has an almost identical sequence named PAX6. These are switchable signal molecules which contribute to the control of multiple gene functions.

Some people know how to move genes around fairly easily these days. We can put the Pax6 gene from the mouse into cells from eyeless fruit flies, then put those transgenic cells into fruit fly embryos, and grow fruit flies with mouse Pax6 in place of their missing eyeless gene. Those transgenic flies make perfectly good eyes, normal in structure and function.

Humans with aniridia — a defect in which the eyes lack normal irides (singular iris, proper Latin plural is irides — the colored rings around the black pupils) commonly are found to have mutations in PAX6 that result in haploinsufficiency (half as much of the normal product from only one good copy of the gene is not enough).

Various mutations of PAX6 have been detected in a variety of human eye anomalies, including aniridia, Peters' anomaly, corneal dystrophy, congenital cataract, Axenfeldt anomaly and foveal hypoplasia.

If you have ever been swayed by arguments that the eye is "irreducibly" complex, too complex to be a product of evolution by natural selection (with a resulting current "design" that might easily have been much more intelligently arranged if a competent designer had started from scratch and moved with direction and determination), if you have ever doubted that every form of Life on Earth has descended from a common source … look at this story and reconsider. You and a fly, and animals with every other kind of eye structure ever properly tested, need normal copies of some of the same gene DNA sequences to build eyes properly.

This is a fascinating illustration of the continuity and consistency of the basic animal body plan. There is a tail end and a head end, there is a back side and a belly, and there is the "mirror-image" asymmetry of left vs right. Along the head-to-tail dimension, there is a major segmentation into head, thorax and abdomen, with a tail section that is more or less distinct in various animals. Within each of these major segments, there is a finer segmentation.

Different parts of the brain unfold from three to five segments, followed closely posterior by a set of "gill" arches that may become parts of face and jaws, and/or ears. Finer segments of the thorax become vertebrae and ribs with associated nerves and muscles. The abdomen's subsegments become vertebrae and nerves and muscles. Some species of snakes have as many as 300 vertebrae, vs our 33, and they grow ribs on almost all of theirs. The giraffe's neck and the mouse's neck and yours and mine have the same number of embryonic segments, ultimately making for the same number of cervical vertebrae (neck bones). The difference in size evolved after the number of cervical vertebral segments was set at seven for the mammals. Departures from seven cervical segments in mammals occur only with changes in the functions of Hox genes, which changes are associated with neural development problems and childhood cancers and stillbirths.

There are many more parallels, with variations on these basic themes. Insects are not vertebrates, because they have no vertebrae or ribs or bones or any other parts of the vertebrate **endo**skeleton, but instead have their **exo**skeletons on the outside, composed of plates of chitin, more or less calcified. The segmental approach is the same, a variation on the same basic plan.

We arguably know more about the details of the vertebrate body plan of embryogenesis and its elaboration from studies of the fruit fly (which is not even a vertebrate) than from any one other organism. The fly has a short generation time, for efficient breeding studies. It is small enough that upkeep is almost trivial, big enough for detailed studies and dissection at low magnification. It is an ideal research animal, and many of its genes of developmental importance are highly homologous to genes with similar functions in other organisms, including humans.

Axis Definition: The Armature; Laying Out the Plan

Axis definition — defining the anterior–posterior, dorsal–ventral, and left–right axes — is cardinal and crucial.

Imagine yourself to be a newly synthesized gene product, coming out of the zygote nucleus into a featureless sphere thousands of times your own diameter. In every direction, everything looks the same. Your mission, should you choose to accept it, is to impose upon the structure of the zygote a three-dimensional organization for the embryo that is to be built from it. Pretty much like building a brick wall starting with the middle brick, is it not? Go ahead, feel helpless; you can't help it ... but don't feel too bad. The failing is not yours. It cannot be done. Look at a wall made from bricks or stones. Pick a stone, or a brick, "in the middle" and figure out how you could have built that wall starting from that brick, all by itself, right where it is right now.

Physics gives us the law of conservation of matter and the law of conservation of energy ... neither matter nor energy can be created or destroyed, but only transformed from its form in prior existence to its present form and on to some other form. Similarly, all structure derives from pre-existing structure. Each and every brick wall starts from the ground, never from the middle brick.

If cellular development could not make clear "which way to the head?" and "where is the tail to grow?" — then consider for example the cells that become the limb buds from which to grow out the limbs ... how are they supposed to know where along the length to start growing out? If the back-to-belly axis is not clear, where shall we grow the gut structures and where should we put the spine and nervous system? Where between the dorsal and ventral midlines should the limb buds settle?

A fertilized frog egg has layers of three different colors, and the middle layer is tapered, thicker on one side than the other. That structure is information enough that I can place spots of dye in the cortex of a frog egg in certain positions relative to that banded configuration, and know what parts of the adult frog will carry the color of each spot of dye. From some embryos, I can remove one or another specific cell and cause defects in specific structures of the body that may grow out from what I left of the embryo. From relationships discovered through such experiments, we can create "fate maps" for such embryos. Those are called "mosaic" embryos.

Mammalian embryos are not "mosaic" embryos. Our embryos are called "regulative." After removal or artificial death of an individual cell in the early embryo, the embryo "regulates" itself to the loss. Either the fate of the cell that was removed has not yet been "determined," or other cell/s can assume the function that has gone missing. These days, people are routinely removing a single cell from early human embryos prepared by *in vitro* fertilization, to be tested for sex or for certain genetic disorders (and most recently, for "good" versions of genes found to be the most different between embryos that have survived implantation and those that have not) so that they can choose which of the available embryos they will have transferred to the uterus for further growth.

This is called *preimplantation genetic diagnosis*, and its practitioners claim to have almost as good a "take home baby rate" as they have with unmolested embryos. By the way, these are the same people who told us for all those years that IVF babies did not have significantly more congenital anomalies than naturally conceived babies. That was even actually true for a long time because they had not yet accumulated sample sizes large enough for statistical significance, so no difference observed could be "significant."

> There is sound evidence to be had in the distribution of values among multiple statistically non-significant results. If they are really representing the same distribution of results as in the natural situation, then the results of the various samples should be symmetrically distributed around the average values in the natural samples, instead of always showing (even non-significantly) greater frequency of problems.

These understandings have generally, for decades, been taken to mean there is no "plan" in the mammalian embryo comparable to the predictable structural relationships in "mosaic" embryos like those of amphibians. We are simply to believe that the cells of "regulative" embryos decide what they are supposed to do according to where they are at any given point in developmental time. According to that odd old notion, gene regulation and cell differentiation in regulative embryos are determined entirely by "positional information." In other words, cells know what they are supposed to do next and how to do it because of where they are in the embryo. This does not answer any of the questions, but only moves all the questions back a step or two, and begs the question of what in this world or any other "positional information" might be made of, and how whatever-positional-information-is-made-of might reliably manage to be where it is supposed to be when the time comes for the next developmental change in cellular structure and function. How does the positional information stuff know how to be where it belongs when it will be needed?

The truth is, mammalian "regulative" embryos have at least all the same jobs to do, at least all the same developmental differentiation decisions to make, at least as much need for all the same structural and mechanochemical results of cell differentiation and controls on how things come to be what they must become, at least as much need for a plan.

The organization of the egg cell, which constitutes or contains the plan, is not perhaps as easy to see as that in the frog egg, and not as obviously arranged according to the visible polarity of the egg cell, but there is clearly a need for structure and there are clearly structures to meet those needs, and the substructures of the egg cell's contents remain in the same relative orientations all the way through cleavage (division of the zygote into the multi-cellular morula) in experimental mammals.

The Vertebrate Body Plan

the molecular armature of embryogenesis:
making it possible, and likely, for every cell properly to decide
what it will be when it grows up — and
giving it more-or-less reliable means to do that:

All organisms that have segmented backbones — that is, nearly all organisms that have *any* bones, plus those like sharks with skeletons of cartilage ... the vertebrates, in a word — all do their embryogenesis according to the same basic body plan, and end up with their internal organs in roughly the same relative sizes, shapes and positions.

The establishment of the anterior-posterior (head-to-tail) and dorsoventral (back-to-belly) axes of the overall body plan — and the structural asymmetries that are dependent upon them — are much better understood at molecular levels than is the fixing of the structural asymmetries that depend on the left-right axis.

The asymmetries of the head-tail and back-belly axes are not mirror asymmetries; they are of a much less subtle sort than the mirror asymmetry of left-vs-right. The parts toward the head don't look anything like the parts toward the tail. The back and the belly, and the parts attached more closely to each than to the other, cannot be confused.

An embryo with no head (acephalus) or nothing but head (acardia) must in general fail, and stop and clean up the mess by destroying and digesting itself, very early. These monstrous variations on the theme are never seen among products of human development except when the thing can grow attached to and supported by a more-or-less normally functioning twin within the same amnionic fluid volume.

Coherent global situs inversus, with the normal left-right visceral asymmetries concordantly reversed, is much less disruptive than either of those. Even the worst of the "fusion" midline asymmetry malformations, say anencephaly, or the cyclops version of the holoprosencephaly sequence, for the sake of argument, represent much smaller and subtler departures from the whole plan of embryogenesis than that. If at any point in development beyond implantation the embryo proper is not round, then the very most basic parts of the plan have probably worked.

If you think about the geometry, you should easily appreciate that fixing the direction of two embryogenic asymmetries — labeling two dimensions in any three-dimensional thing — leaves the third with no more unoccupied degrees of freedom and therefore no choice of where to go relative to the other two. When we can see the blastula formed, we know which side of the two-layer-disk embryo will be back and which will be belly. When we see the prochordal plate and the primitive streak forming soon thereafter, we know which end will be head and which will be tail.

If we know where head, tail, back and belly belong, even the least mechanically-inclined among us can think about it a while and come up knowing where left and right are destined to be. But there is still crucial information missing.

The business of putting the remaining pieces in their proper places as they are built remains to be accomplished, and — even though it is clear, from where we sit looking at it, where left and right must be, it is still possible for the pieces and parts to line up wrong, to be built in the wrong shapes because of being based or started on the wrong sides, or for either side of some component to be built as if it should be on the opposite side.

Our logical understanding, through the microscope, of the obvious ultimate three-dimensional layout does not define the left-right differences in the structure and behavior of cells and cell systems and tissues that are just beginning to establish themselves in and grow into the left and right sides. We can see what parts of the overall embryonic structure have no choice but to go on to become the left and the right sides, but this does not include the cellular left-right definition of parts that remain to be put in those places.

What is the difference, after all, between one cell that will develop to occupy a certain position in the skin of your left hand and another cell that will occupy the very same mirror image position on the right? How should the heart tube pinch up into a hump and then "know" which side is the proper one toward which that hump should lean and then roll on over?

How do the cells that will build and become the lungs "know" on which side to build a three-lobed structure and which side gets only two lobes — and to do it the same way very nearly every time, as if that is somehow much more important than it seems from here?

What is it that consistently puts the stomach on the left side of the abdomen and the liver on the right? Not much is wrong with the body in which the whole system of internal organs is concordantly reversed, but things can get really ugly, for example, if only the heart, or only part of the heart, grows up in there backward — as a result of which it gets hooked up wrong to the largest of the body's blood vessels and other parts that are in the correct usual places.

Lessons from a Feather:
Self-Similar Smoke and Fractal Mirrors

If I were to hand to you one of the primary flight feathers from a bird's wing, with at most a little thought you will be able to tell me from which wing it came. If you were to hold the feather I gave you in front of a mirror, the feather you would see behind the mirror would look very much like the feather from the same place on the other wing. But no one can tell us the whole story of how the growth processes of that feather produced the system of differences between that feather and its antimere (the one from the same spot on the other wing).

I have not yet learned the appropriate parts of mathematics nearly as well as I wish I had, but I know we can see in wings and feathers one of the most salient features of things and phenomena that are fractal in character. In this particular context, by which I mean the scale-independent self-similarity of fractals, if we "zoom in" or "zoom out" on a structure like a feather, things tend to look rather similar at various levels of "magnification."

This is how that plays out in the development of a left-right "pair" of primary flight feathers: The limb buds that would be wings grow out from the embryonic torso in opposite directions. Along each wing, feather buds in the skin will sprout feathers at intervals. There are some smaller ones growing out toward the anterior leading edge (to your left as you move from the shoulder along the right wing toward its tip). If you are moving out from the left side of the torso, along the dorsal surface toward the end of the left wing, the leading edge with its smaller feathers is to your right, and the trailing edge to the left. The larger feathers, among which we will find feathers more or less like this primary flight feather, sprout toward the trailing edges of the wings.

Then, moving outward from the wing along the main shaft of each future primary flight feather, the barbs on the shaft of the feather grow shorter to the lateral side (outside, toward the end of the wing) and longer toward the medial side, back toward the torso, the midline of the bird. Along each of the barbs on the feather, the barbules repeat the pattern. Along the barbules on each of the barbs on the feather, the hooklets or hamuli repeat the pattern again.

At every step, the exact counterpart feather growing on the other side has to do exactly the same list of things, in mirror image. The two structures are approximately identical in relative, mirror-image space, and totally different, totally non-superimposable, in absolute three-dimensional space.

> How do the cells that are doing these things know how
> to do these things they are doing?!

A thin surgical glove can become its own antimere, its own exact mirror image, merely by being turned inside out. Where do the cells, the growth processes, get the necessary information and mechanisms to establish all of that for the whole solid arm, hand, leg or foot, or, for a left-right pair of teeth, say the maxillary first premolars, in left vs right jaws?

How can something so very simple come to seem so ridiculously complex! Or is it so absurdly complex, to seem so ridiculously simple?! All this set of paired structures has to do is to do everything! in the opposite direction on the two sides! … how hard can that be for growing cells to arrange? Where does the cell — a cell, each cell, any cell … get the information to know where and when to turn, and which way to turn, to do what it must do next, the same way on both sides but in different directions?

How can we suppose that a cell that is doing such a thing on the right wing of a bird might know how its counterpart on the left wing is doing it, so that they can each be sure to do it the opposite way from the other?!

I find this asymmetric cellular functioning to be reminiscent of, and more or less exactly as weird as, the quantum-mechanical phenomenon of "entanglement" (Schrodinger, 1935). Because every cell in the body at any given point in its development arose from the one same original zygote cell, then, at some point in their shared past, every pair of cells on opposite sides of ongoing embryogenesis descended from one same cell. To set out the parallel to quantum entanglement, from that common origin, they carry forward "memories" that keep them matched, generally in opposition, with respect to a great many dynamical properties that can predict their behavior in interactions with their respective environments. Each of those paired cells has followed a particular complex path forward from the time and place when they were in one same cell. For each to be certain of functioning in this proper equal-but-opposite antisymmetric fashion, opposite to the other in space, then either they have had equal and opposite, mirrored, antimeric, histories or they share one same "entangled" history, to put them in the "same" place, doing the "same" thing at the same time in opposite directions from each other. This shared mirrored path is in fact easier to envision in cellular development than in quanta flying through space. We do know that cells in a developing embryo have an epigenetic switching protocol which must be enacted at each and every cell division over the paths they have taken. Cell divisions in developmental differentiation have been reported to be routinely asymmetric.

Meanwhile, it is easy enough to see that any major feather could not function properly as a working feather if put in the place of "the same" feather on the other wing. Every element of its three-dimensional shape is opposite to the satisfactory performance of its aerodynamic functions in that opposite position. It could not contribute to, but could only detract from, the function of that wing over there, opposite the wing on and for which it was built.

Embryogenic Asymmetries of the Brain

The reliably different structures and functions of left and right halves of a human brain may not be as structurally obvious from across the table as those of a bird's primary flight feathers, but the patterned arrangements of their relative positions in tissue space are at least as important for their interactive functions in cell groupings at various levels, and they have the same kind of need to be reliably — very reliably, not randomly — different. Virtually every major anomaly in the development of human brain or behavior is statistically associated with unusual versions of asymmetries of brain structure and function that are elaborated in embryogenesis.

There is a clear need for every cell to have means to know not only where and when, how fast and how long, to do what, but also to know which way to turn before or while doing it, or from, or on, or out of, which side of itself to do this particular job. It should be clear also that there must be directional spatial referents inside each and every cell, so that the cell's directional functions can be specific.

> Everything that happens between your scalp and your toenails, between your ears, and between your legs, is done by cells.
>
> Everything that cells do, they do with molecules; they can only do what they have (can make or gather) the molecules necessary to do.
>
> Human cells in general do not function alone, but at least nearly always in networks and tissues.
>
> The proper function of each cell within each system of which it is a subsystem, generally depends quite intimately on its proper position within the physical arrangement of the components of that system.

It is, further, a logical and biological necessity that no cell ever does anything it does not "know how" to do. The idea of "knowing how" is more or less obviously an example of anthropomorphic thinking and probably not exactly right in one dimension of the consideration or another. I am not sure how else to say it, but I believe this should be generally understood: Every cell must have the wherewithal — whatever mechanochemical, molecular stuff it takes — to do what it is "supposed to do." In general, this means that each cell has to know when and where and how to synthesize the correct set of RNAs and the correct set of proteins, with those proteins to include the receptors and enzymes necessary to accumulate or build from accumulated precursors the necessary fats and sugars and combinations thereof with proteins, all to put in place the necessary substructures and/or to output the right chemical products at the right time and place and in the right amounts.

We owe much, maybe most, of what understanding we have in these realms, to things we have learned about the embryogenesis of the fruit fly *Drosophila melanogaster*. We have also learned a great deal from the just-barely-big-enough-to-be-seen-without-a-microscope (about two millimeters long, 959 cells in the whole body) nematode worm *Caenorhabditis elegans*. In each of these and other research organisms, we have continued to find, very nearly every time, that the same or closely related gene products do the same job in other organisms up to and including mammals, including humans.

We seldom fail to find that these gene DNA sequences and their products have strong homologies with genes and gene products involved in similar functions in all other organisms including you and me. When mutation and selection conspire to evolve/invent a tool to do a particular biochemical job, from then on that tool shows up very nearly everywhere that same job needs to be done. Occasionally, distantly related organisms find closely related mechanisms by which to perform a particular function, with good reason to believe that the two arrangements were built from unrelated beginnings. We call this "convergent evolution."

Wings of bats and birds (both vertebrates) are both built around arms, but a bat's wing is a hairless webbed hand while the bird's feathers are modified hairs, of which the finger bones hold only a small part of a bird's wing, and the bee's wings (an invertebrate) are quite different from both in origin.

The DNA sequences that carry the code to make each of these RNAs and proteins at the various stages of development have been there all along, but every type of cell must have ways to "know" where, when, how-fast-times-how-long-equals-how-much, to read each bit of that DNA, to express (or not) each bit of the code properly, to the proper extent, at the proper time and place.

The differences among all the different types and functions of cells are not made by differences in the DNA sequences each cell or cell type contains. (With only very few, very special and necessary exceptions, every cell in your body or in the body of any given organism — contains copies of the same set of DNA sequences.) Developmental differentiation depends not on which sequences are present, because that in general does not vary from one cell type or tissue to another, but rather in the complex combinations of which genes' DNA base-pair sequences are expressed, when in developmental time, where in the body's structural space, how fast and for how long.

Beyond the first very few cell divisions of embryogenesis, these gene-regulation decisions are made by interactions among the cells, in many cases involving signals from cells of one type to other cells.

Yes, of course, there has to be a beginning for each lineage of cells and of the corresponding history of signals and differentiation "decisions." That gets more and more important the deeper into the definition of vertebrate body plan asymmetries we probe.

The complexity of the interactions involved can seem overwhelming, but now we have the option to refuse to be, or to stay, overwhelmed:

— now we know the exact DNA base-pair sequences of at least most of the genes so far recognized as functional genes in the human genome [that is more circular than it might appear at first glance].

— we can identify exquisitely specific bits of each of these sequences — bits of base-pair sequence not found anywhere else in the genome except in the expressed parts of this very particular gene sequence.

— we can make abundant copies of each of these bits, and precisely place microdrop puddles of each on a silicon wafer which thereby becomes a "microarray" chip.

— when we extract RNA from cells in a particular state, the patches of the many various sequences on our microarray will, with exquisite specificity, bind to and collect matching bits of the messenger RNA that has been made to express each of the gene sequences that are being expressed in those cells in that state of development or activity.

— we can identify the sequences for which our microarray has collected messenger RNA, and compare the resulting binding patterns between cells doing different things.

— then we can know the differences between the sets of genes that need to be expressed differently for the cells to be doing these different things.

There are thousands of labs running thousands of students and post-docs doing a few such experiments each. Soon we will know EV!erything about how genes and their products work to perform all of the various functions of living cells.

Not a single one of all such experiments ever done so far has come back with a simple one-to-one interaction. One of my colleagues here, for example, decided to try to learn what makes the differences between two different kinds of fat cells in the adult human body. Beginning microarray sampling in their laboratory showed over 250 genes behaving differently between two seemingly just barely different kinds of fat cells. Perhaps no one has yet set up such an experiment to ask a simple enough question for the experiment to come back with a simple answer.

One thing we have recently come to know, is that one crucial determinant of left-right bilateral asymmetry is already permanently localized in the first two or three cell divisions of embryogenesis. One of the first four cells already knows who he is, already expresses gene functions that will be crucial to establishing structural asymmetries in embryogenesis. Only one possible asymmetry signal is available earlier in development than that; namely, the DNA. More about that later.

Each of Us Was Once a Single Cell ... or Two

Meiosis, Gametogenesis, Syngamy, Embryogenesis ...

> The teachings disintegrate when we try to grasp them.
> ... The whole of life is like that.
> This is the truth and the truth is inconvenient.
> For those who want something to hold on to,
> life is even more inconvenient.
> ... We're all addicted to the hope that
> the doubt and mystery will go away.
>
> *Pema Chodron*

Let us consider a bit more closely how we get to the formation of the zygote, about what it takes to get to the point in development where the building of a potential future human individual might be able to get under way, where an embryo can begin to put itself together. Then we can go on to consider the large basic steps in assembling an embryo.

Most multicellular organisms are diploid except in their gametes. That means that we carry our genetic information in two copies of a system of DNA sequences. One copy came from each parent back in the fertilization process that eventually resulted in the life of the individual organism.

Gametogenesis

The gametes are the reproductive cells, for humans the sperm and egg cells. Gametogenesis is the process of making the gametes beginning with the primordial germline cells of the early embryo. This includes spermatogenesis (the making of the sperm cells) and oögenesis (the making of the oöcyte).

"Oöcyte" is the proper, technical name for the egg cell, from Greek roots meaning (oddly enough ...) "egg" (oö-) and "cell" (–cyte). By the way, oö, with the umlaut over the second "o," is the proper spelling to notify newcomers to the word that the double vowel in there is not a diphthong (now that, "diphthong," is an ungainly word, is it not?). Both vowels are pronounced separately, not blended. The proper pronunciation is "oh-oh-site," not "ooh-site," and certainly not "ooh-oh-site," like some people read zoology and say zoo-ology instead of zo-ology, at least in part because hardly anybody ever puts the umlaut over the second "o" anymore. It might be considered a test by which to distinguish between adepts and the uninitiated, but probably has more to do with who learned it from lectures and who learned it only by reading. Back to work! On faut travailler ... as Louis Pasteur is reported to have mumbled like a mantra while pacing.

For normal human embryogenesis, the zygote must have a normal diploid set of DNA around which, and from the genetic developmental "instructions" contained in which, to build an embryo.

Many plants can tolerate doubling of their sets of chromosomes, and many special varieties of plants have been created by humans who have deliberately caused such doubling. The resulting "tetraploid" varieties (having four copies of each whole chromosome in every cell) are usually larger than the normal diploid parent line of plants from which they were derived, and/or have larger flowers or fruits.

But even plants could not tolerate repeated doubling of their chromosome numbers in several successive generations, and no existing animal species is known to live through it to reproduce even once. When it occurs accidentally in human embryogenesis, a tetraploid conceptus almost never even makes it through embryogenesis. If and when such a conceptus does complete embryogenesis, survival to birth of a tetraploid human fetus is extremely rare and none has ever been reported as having survived beyond birth for more than a few hours. Triploid embryos, with three copies of each chromosome, are rather more common than tetraploids among pregnancy failures, but they also survive rarely through embryogenesis, almost never to term, and never significantly beyond.

Normal animal development requires that the egg and sperm cells each bring to fertilization a single copy of each chromosome's DNA strand — half of a normal set, to reconstitute a normal diploid set in the process of fertilization and syngamy.

Normal cell division for the purposes of building growing bodies and replacing dead cells is called mitosis (my-toe'-sis). In mitosis, the whole double set of chromosomes (one copy of each chromosome from each parent) is duplicated, once in each cell division cycle, and each daughter cell carries away from the finished cell division a more-or-less exact duplicate of the whole double set the cell had before it divided.

The genomes of the daughter cells in every cell division are as close to identical (in DNA sequence) as replication can make them, but that is never perfect because it includes inevitable imperfections from mutations (heritable changes in DNA sequence) which arise constantly at low frequency from rare errors in DNA replication and from a number of other sources — radiation, pollution, and so on. Even if uncorrected mutations were as rare as one per hundred million base pairs per replication, that would mean about 32 mutations per cell division in the 3.2 billion base pairs of the human genome.

The process of preparing sperm and eggs for productive fertilization and embryogenesis includes a sequence of two highly specialized cell divisions, which together constitute the process called meiosis (my-oh'-sis) or reduction division. The function of the process of meiosis is to leave the sperm and the egg cell each with only one copy of each chromosome's DNA strand.

In preparation for the first cell division of meiosis, the whole set of chromosomal DNA is duplicated, as it is for any mitotic cell division.

However, in the first cell division of meiosis ["the first meiotic cell division"], the dividing cell does not separate each of the newly replicated strands of each individual chromosome to the daughter cells as it happens in mitosis. Instead, all four copies of each individual chromosome's DNA sequences (both parental homologs,[a] duplicated) zip up together from end to end in sets of four strands. This pairing-up process is called synapsis (from Greek, juncture or point of contact, from synaptein, to fasten together, from syn– together + haptein to fasten).

---

[a] The normal two copies of each chromosome, one from each parent in the normal diploid organism, are called homologs.

The resulting structure, containing all four strands of each chromosome, is sometimes called the bivalent, because it contains both homologs of each chromosome pair. I prefer its other name, the tetrad, because the fact that there are four strands of DNA in there will soon be seen as very important … and in my opinion it sounds much better in my ear, feels much better in my mouth and looks much better on the page … just an all-around much more appealing sort of word.

The tetrads, with all four copies of each chromosome's DNA lined up together from end to end, are each pulled together and held together by a beautiful and complex ladder-like structure made primarily of protein and called the synaptinemal complex (complex for "connecting the threads").

The synaptinemal complex has at least four components that are microscopically distinguishable. There is a lateral element on each side, to which both highly condensed duplicate copies of each parental chromosome homolog are connected. There is a central element (down the center, eh?); and there are rung-like fibers holding each lateral element and its attached condensed duplicated chromosome in some close approximation of point-to-point alignment [no, we really don't know how they do that, either] with each other across the central element.

There are oval structures imbedded at intervals within the complex, which we call "recombination nodules" because they are roughly equal in number with the number of recombination events (explanation soon) that will occur during synapsis.

We know distressingly little about all of the various proteins involved in this process and the associated structures; not even how many different proteins are involved. We generally learn about such processes best by studies of the effects of mutations, but mutations in such basic machineries of cell division have a strong tendency to be lethal and thus rather more difficult to study than mutations in most other processes. We have made much progress in such studies, in microorganisms, with conditional mutations — in which a protein in question might, for example, function properly at one temperature and not at another, such that we can "turn the gene off" by changing the temperature and study what process is changed, and how.

Experimental manipulations such as these obviously cannot be performed on human subjects, and would not work very simply with any mammal anyway, needing as we do to hold body temperature within a narrow range.

Recombination — Engine of Segregation and Diversity

While in this synapsed configuration, the four DNA-strand versions of each chromosome intertwine, break and rejoin, in a complex, difficult but critically-organized process of swapping parts between the maternal and paternal strands. This generates new combinations of DNA sequences unlike any that ever existed before. The complexity of the process makes certain that it is a major source for mechanochemical errors, as well as genetic diversity.

The physical exchange of DNA strands as seen by microscopic observations of the chromosomes is called "crossing over." The genetic manifestation of the resulting new assortments of genetic markers in the progeny is called "recombination" (because the exchanges generate new combinations of the versions of neighboring gene loci on the chromosome). These exchanges happen at least once on each arm of each chromosome, with more on the chromosomes that have longer arms. Recombination events of crossing over occur with greater (~1.5-fold) frequency in females while preparing egg cells than in males building sperm.

The density of probable crossover locations within a given length of DNA sequence varies greatly over various parts of the length of each chromosome, and that pattern of varying crossover probability density also differs between males and females. Some believe that there might be specific "recombinogenic" DNA sequences which attract crossover events to specific places along the chromosome. We have not, however, proven this, and we are more than a little baffled about the mechanisms underlying the different distributions of crossovers in oögenesis vs spermatogenesis. The results are consistent with the possibility that there might be two completely different systems of recombinogenic sites, and enzymes and processes of recombination, for oögenesis and spermatogenesis!

At least as plausible, and quite possibly more so, these variations in crossover density may have nothing to do with genetic DNA sequences themselves. They might instead be determined by epigenetic variations in the chromatin packing and the consequent variation in the accessibility of the DNA sequences to the enzymes involved in the recombination process.

Meiosis is the platform for the operation of all of the rules of Mendelian genetics. Meiosis is where Mendel's Laws of Segregation and Independent Assortment live and work. Recombination during meiosis is the engine of genetic diversity for all sexually-reproducing species. The results are quite sufficiently complex to insure that no two individual human gametes or zygotes have ever had or ever will have exactly the same sets of DNA sequences. Every one of us is one of a kind.

You and I are unique ... just like everybody else.

> I don't get it. Can YOU tell me why that never gets the laugh it deserves? "... unique, just like everybody else ..."
>
> Is it just too subtle, too abstract? How about "one of a kind, just like all the others?" no? still nothing? Oh, well.
>
> Being unique genetically is one thing all humans have in common ... Being different from all the others may even be the only way in which every human is the same as all the others.

Even between full siblings, there are two choices from each of 23 pairs of chromosomes from each parent. Two independent choices at each of 46 opportunities yields two-to-the-46th-power possible variations.

> (2^{46} is about 7.04×10^{13}, which is about 70,400,000,000,000. In words, "about seventy trillion," more than 10,000 times the number of people alive today, and several thousand times the total number of people who have ever lived.)

So ... 2^{46} different possible combinations of parental chromosomes present at fertilization, 2^{46} different possible combinations of whole chromosomes between full siblings. That is way more than quite enough for "unique" already. That is, however, only the beginning. That is the number of arrangements available *assuming all the parental chromosomes are transmitted intact*, which is not ever how it happens in the normal process. That ignores the effects of recombination, which are nowhere near negligible.

Recombination occurs at least once in each arm of each chromosome, and more in the longer chromosome arms. That divides every chromosome into *at least* three segments, with two independent choices from each segment, so that each chromosome may be transmitted in at least 2^3 = eight different versions — at least. This changes all of the above calculations from multiples of two to multiples of eight! That changes the possible outcomes to yield *at least* $(3.5 * 10^{41})$ 350,000,000,000,000,000,000,000,000,000,000,000,000,000 possible combinations even between full siblings. Again, or still, that is a dramatic understatement of the possibilities, because of the additional crossovers in the longer chromosomes and because of the effectively-infinite variability in the exact location of each of the crossovers.

Because there is indeed such a thing as a "family resemblance," we must suppose that there are some constraints somewhere, that put some limits on all this free exchanging of genetic material. Perhaps genes that specify variations in facial structures and body build, for example, move in clusters that fall between common recombination break points. We know that there are a great many such clusters — we call them "haplotypes." More about that a little farther along, maybe.

As far as we can understand so far, genetic diversity is the primary evolutionary/ survival advantage that is responsible for the overwhelming dominance of sexual mechanisms of reproduction in populations of complex organisms. Only the simplest of organisms reproduce asexually — which is generally accomplished by simply dividing the cell into two ... and even a great many of them have mechanisms for the exchange (and subsequent recombination) of genetic information. Most of their reproductive processes are not entirely without a sexual component [the important difference being a mechanism that provides the opportunity to mix genes].

A substantial level of genetic diversity makes it less likely that any "new" disease or other change in the environment can wipe out the species. Genetic diversity strengthens the prospect that at least some individuals will have a combination of genetic variations that will allow their individual survival and reproduction and thus the survival of the species.

The Genome

Therefore, there is, in fact, after all, no such thing as "*the* Human Genome." There are somewhere near seven billion different human genomes, subtly perhaps but definitely, uniquely different full individual sets of human DNA.

Another way to look at it is that there is after all only One Genome, a supersystem comprising all of the DNA base sequence genomes of all species of living organisms on Earth as subsystems. We have abundant evidence, providing very good reason to believe, that Sustainable and Reproducible Life began on Earth only once, and that every living thing on Earth today has descended from that one single common ancestral source. Every species yet examined on Earth today holds some of its gene DNA sequences in common with some of those found in most other species.

With a few so far unexplainable variations affecting a very small fraction of the total system, the Genetic Code is "universal." What that means is: With a very small fraction of partial exceptions that we cannot presently explain in full and exact detail, the same specific three-base sequences in the DNA call for the same amino acid to be put in the corresponding place in the sequence of a protein, universally, in every species of living organism yet examined.

Many of our gene DNA sequences are modular — there are "motif" sequences that mean the same thing, code for the same string of amino acids that will fold into the same bit of protein structure, to do the same job, in genes and proteins of many different species. The evolution of the great many different structural and catalytic proteins in living things has been importantly a matter of shuffling parts.

Some people have been saying for a while that normal humans differ from one another by less than half of 1% of their DNA sequences, and "less than" may actually be "much less than." Well, okay … that comes from one absolutely non-exhaustive way of measuring such things, and it serves social and political purposes far better than it serves any scientific purposes.

Suppose we put that in perspective: Half of 1% of 3.2 billion base pairs would allow for an average of 16 million ! base-pair differences between any two of us. Some fraction of those will not make a detectable difference, and some of them will. Further, more recent evidence, collected at the level of multiple single base pair mutations (known in "the trade" as "single nucleotide polymorphisms" or SNPs "snips"), suggests that the within-species variation is in fact substantially larger than we had been estimating from less sensitive methods of analysis.

Richard Lewontin has famously argued, since about 35 years ago as of this writing, that about 85% of total human genetic variation is due to differences among individuals within subpopulations, and only about 15% is due to differences among subpopulations such as the groups defined historically by "race." That speculation continues to the effect that any two people within any one of the major continental subpopulations are likely to be as different genetically as two people chosen from any two different subpopulations. The resulting famous assertion, that humans cannot be soundly classified into "racial" subsets on the basis of differences in gene frequencies, is tightly wrapped into a profoundly misleading fallacy. The primary problem is the assumption that genotypes and the gene products thereof are independent at each of the loci included for consideration.

Lewontin's assertion ignores or dismisses the fact that most of the variation among people or populations, among individuals or groups, lies in the relationships among the many individually variable DNA sequences. Those relationships among the variations in DNA sequences are far more numerous than the variations of individual gene DNA sequences themselves. If N is the number of gene sequences under consideration, then the two-way relationships alone are $N * (N + 1)/2$ in number, and almost certainly most such relationships will be more complex than two-way, involving more than two single base changes within each of the groupings being studied as haplotypes.

The unit of segregation, the chunks of DNA on which genetic variations are assorted among progeny and moved through large breeding populations, is much smaller than the whole chromosome, but much larger than the individual functional gene. Each of the mutations which caused any of the genetic variations in question occurred within one molecule of DNA in the genome of one individual — embedded in DNA sequences from which it probably will not soon, and may effectively never, be set free by recombination crossing it over into a different background.

On that strand of DNA are other mutations which have accumulated over the ages in the specific population to which this individual belongs. Because of this fact, any given new mutation can, and often will, have an effect on developmental expression that differs from the developmental effect of the identical DNA base-pair change when it has occurred in another genetic background.

We find that a great many clusters of these single base pair changes, too close together in the DNA sequences along the respective chromosome to be sorted quickly to equilibrium by recombination, have travelled together in the human population over thousands of generations. Members of the major continental subpopulations can be readily sorted with a high degree of accuracy, according to these SNP haplotypes.

By mechanisms we have not yet been able to see clearly, the process of recombination does its wonderful job of generating our genetic diversity as a side issue or byproduct of some other function that is essential to proper performance of gametogenesis. Editing the DNA sequences of the chromosomes on the large scale of whole genes and blocks of genes is one suggestion still considered viable, but there seems to be a need for some more basic suggestion, something closer to the core of achieving proper chromosome segregation.

At any rate, an individual whose germ cells cannot properly perform synapsis and recombination cannot make functional gametes. We have recently come to know, for example, that several mutations which cause infertility in human males do so by causing defects somewhere in the system of events leading to synapsis and recombination. They presumably would have the same or similar effects in oögenesis, but — as you can readily imagine — the developing primary oöcytes of first-trimester female fetuses are much harder to sample, ethically and physically, and thus logistically, financially and in every other dimension of practicality.

In the male, spermatogenesis begins at puberty and continues throughout adult life. Each primordial germ cell that will proceed properly through spermatogenesis must perform the same linked set of two highly specialized cell divisions. These two cell divisions separate two functions that take place in every cell division of mitosis.

Spermatogonia (plural of spermatogonium) are sperm stem cells, which undergo successive asymmetric mitotic divisions, in which one daughter cell retains its stem-cell-ness and the other enters the first division of meiosis to become a primary spermatocyte.

In every "ordinary" mitotic cell division, the DNA is duplicated, and the centromeres that hold the duplicate strands (of each copy of each chromosome) together are duplicated and separated, each taking one copy of each chromosome's DNA strand with it into the daughter cell. Thereby each daughter cell gets a copy of each (maternal and paternal) copy of each chromosome.

In meiosis, in the first meiotic cell division, each chromosome strand of DNA is duplicated, but is left attached to its original, unduplicated centromere. In the second meiotic cell division, the DNA is not duplicated again, but now each centromere duplicates and the two copies of each centromere segregate to the daughter cells. Each daughter cell of Meiosis II thereby gets one copy of each centromere, which at that point carries one copy of each chromosomal DNA strand. They are now haploid, having only one copy of each chromosome — half the normal amount of DNA.

In most pictures of condensed chromosomes ("packed up for travel" in preparation for cell division), the centromere appears as a narrowing; it has been called a "waist." A major function of the centromere is to serve as the point of attachment for spindle fibers for their function of separating duplicated chromosome strands ("sister chromatids") in mitosis and meiosis. Some spindle fibers go pole-to-pole, providing a cage-like structure for the process and pushing the poles of the dividing cell apart by elongating, while other fibers go from centromere to pole, becoming shorter and pulling each chromatid to the pole of one of the daughter cells, dragging it by its centromere. It has recently been reported that spindle fibers attach at many places along the chromosome arms as well as at the centromere.

It is the function of the pole-to-centromere fibers to pull the duplicated strands apart into the daughter cells. There are specific kinds of DNA concentrated in the centromeric region of the chromosome strand, which include specific subsequences that serve as labels for the attachment of a complex system of proteins (the kinetochores) that serve, in turn, as spindle-attachment sites.

Each of our 24 different chromosome types (that's 22 paired autosomes[b] plus X and Y) has its own specific centromeric DNA sequences. A collection of several different kinds of proteins gathers on these centromere DNA sequences to form the kinetochore. Spindle fibers attach to the kinetochore to pull the newly duplicated chromosomal DNA strands apart into the daughter cells in mitosis. From each pair of newly duplicated copies of each chromosomal DNA strand, it is imperative that exactly one of each pair of copies should go to each of the two daughter cells.

Think about it mechanically: Suppose you have pieces of yarn of 23 different colors, two pieces of each color. Suppose you have been set the task of dividing them into two sets such that each set has exactly one piece of each color, and you can't see or think because you are only a protein fiber. Maybe you can smell or taste or feel — there might be chemical or mechanical labels? We cannot yet explain it, but somehow the binding of spindle fibers to the centromeres (and at other binding sites along the chromosome arms) is controlled with sufficient reliability that it usually ensures that the two DNA strand copies of each chromosome go in opposite directions, one to each of the daughter cells. There is, further, a subtle but profoundly important alteration of the mechanism in the first division of meiosis, when attached pairs of duplicate strands, and not the separate newly duplicated strands of each chromosome, are to be sorted.

You may be certain, however, that Life "*as we know it*" on Earth would not be possible without these mechanisms for getting the right amounts of all the right kinds of DNA to where it is supposed to go next.

So: meiosis separates, between two consecutive cell divisions, the two major functions that normally happen in every mitotic cell division. In the first cell division of meiosis, the DNA is duplicated, but the centromeres are not; and each daughter cell gets a single member of each pair of chromosomes (duplicated) attached to one haploid set of centromeres, each with two copies of its DNA. In the

---

[b] The *autosomes* are all of the chromosomes other than the *sex chromosomes* X and Y (which are sometimes called the *gonosomes*). The autosomes are paired, and they recombine freely along their lengths. The X-chromosome is much longer than the Y, as a result of which X and Y do not pair and recombine freely along most of their respective lengths. They must, however, pair to some extent for proper conduct of synapsis and meiosis. At each end of each of them, there is a short region of homology over which they do pair for synapsis and inside of which they do recombine. These are called the pseudoautosomal regions.

second meiotic division, DNA is not duplicated, but the centromeres are, so each daughter cell of the second meiotic division gets one copy of each chromosome. At this point, due to the results of recombination, each chromosomal DNA strand is a mixture of parts from the two different copies of each chromosome that the maker of this gamete got from his or her parents.

When all of this is done correctly, in spermatogenesis, all four strands of each duplicated chromosome pair get parceled out, to provide proper half-genomes for four sperm cells. Each strand of chromosome-1 DNA, for example, now consists of some sequences (about half) from the copy of chromosome-1 that came from the father of the man who is making these sperm cells and some sequences (about half) from the copy of chromosome-1 that came from his mother.

Egg cells have an importantly different way of doing these things. The genetic aspects are all the same, but the normal, crucial end product of oögenesis is a single haploid maternal pronucleus in the secondary oöcyte. This differs sharply from spermatogenesis, which puts all four copies of the DNA into separate sperm cells as potential haploid paternal pronuclei. The paternal and maternal haploid pronuclear contributions will fuse in syngamy to form the diploid zygote nucleus.

In oögenesis, after recombination is completed and before the first cell division in meiosis proceeds to completion, the process is suspended, when the female fetus is about 16 weeks along in gestation. The four recombined strands of each chromosome's DNA are still in close association, but the chromosome pairing of synapsis is beginning to relax in the direction of coming undone. At this stage these are the primary oöcytes, which will all remain in this suspended state (called "dictyotene") until time comes to complete the egg cell's developmental preparations for ovulation, somewhere between about 10 years and about 50 years later ...

The chromosomes of the primary oöcyte remain partly decondensed. While the cell division process of meiosis is suspended, the egg cell's chromosomes are at least in part available for RNA transcription and protein synthesis in service of a low level of ongoing metabolic activity. This is necessary for the complex preparation of the oöcyte for its developmental duties in directing embryogenesis in case it is ever ovulated and successfully fertilized.

When the human female approaches sexual maturity and begins menstrual cycling, in preparation for each ovulation a number of these primary oöcytes "wake up" and resume meiosis. The number may vary substantially among women and among cycles in the same woman. There is much to be done, to build and furnish the largest and most complex of all human cells, which must carry and elaborate the basic structural plan for an embryo which might develop if this egg cell should be successfully fertilized. The developing primary oöcyte is surrounded by a cloud of nurse cells, which feed her and install a wonderful variety of pieces and parts and molecular toolkits for specific subsystem processes of embryogenesis.

We will focus here on the DNA genetic material of the egg cell. We left all four strands of each chromosome still clustered, with synapsis and chromatin condensation substantially loosened to allow metabolic activity over the long suspension of meiosis. At the resumption of meiosis preparatory to ovulation, the DNA strands now prepare to complete the chromosome sorting, the segregation, of the first meiotic division. The DNA of each parental copy of each chromosome has been duplicated, and the duplicated strands of both parental copies have synapsed and recombined, but the centromeres have not yet duplicated.

As in spermatogenesis, segregation in this first cell division of meiosis will sort out the unduplicated parental centromeres. Each unduplicated parental centromere carries two chromosomal strands of DNA that were sister strand duplicates before recombination. One such centromere representing each chromosome pair moves toward each daughter cell. The number of chromosomes, as defined by their respective centromeres, is thereby reduced to one (duplicated and re-assorted by recombination) member of each original pair. Meiosis is sometimes called "reduction division" because of the reduction of the number of chromosomes (as counted by their respective centromeres).

In both cell divisions of oögenesis, one end of the cell division spindle is close to the oöcyte membrane. In the first division of meiosis in oögenesis, the daughter cell that will form at the end of the spindle against the membrane is the first polar body. This division is normally as extremely asymmetric as the cell division machinery can arrange it to be, essentially pinching off just enough membrane to wrap and set aside a naked spare nucleus containing two copies of each chromosomal DNA strand, still attached to one copy of each chromosome's unduplicated centromere.

They were sister chromatids, as identical as replication could make them, until they entered recombination. Because of their random and independent participation in recombination, they are no longer anywhere near identical, but they will on the average share very nearly half of their DNA sequence alleles exactly. The DNA in the first polar body includes two copies of each DNA strand, equivalent to the DNA mass of the normal diploid genome.

Some papers have claimed to see first polar bodies outside the oöcytes of some mammals divide again, and some textbooks simply repeat that as if it were a revealed truth, but my interpretation of the photographs I have seen of these processes in the human is that the polar bodies constitute dead storage for DNA genetic material that is extraneous to the further function of this cell and therefore unwanted and dangerous. They may disintegrate. I have seen no reason to suppose that either of them divides or undergoes any other activity. Extra copies of genes in a functioning cell can cause gene dosage problems, in which abnormal amounts of particular gene products can cause severe disruptions of cell functions.

There are some fascinating variations on this story among other animals, in which the contents of the first or the second polar body carry very important developmental information. In some cases, a polar body forms a "polar lobe" the contents of which cycle in and out of the dividing cells of the embryo several times, and development stops and fails if that material is removed. I have not been able to find any evidence that either human polar body, first or second, does anything except to sequester some extraneous chromosomal material, take it out of the egg cell and keep it out of the way.

Still inside the zona pellucida (the translucent, elastic, non-cellular layer surrounding the oöcyte; the "eggshell") with its sister cell (the first polar body), the main product of the first cell division of meiosis in oögenesis is the secondary oöcyte. This is the "egg cell" in the package that erupts from the ovarian follicle at ovulation. This is the rightful objective of the sperm's travels for fertilization. This is the cell that might normally be fertilized, as pictured on the cover.

Immediately upon bursting out of the follicle in ovulation, the secondary oöcyte proceeds with preparation for the second cell division of meiosis.

As in the second meiotic division in spermatogenesis, there will be no DNA duplication, but now the centromeres duplicate and the duplicated centromeres line up on a new spindle prepared to separate a single copy of each chromosome's DNA strand to each of the daughter cells. Again, one end of the new spindle is up against the membrane, most photographs showing it right inside the spot in the membrane where the first polar body sits on the outside (Figs. 1 and 2, p. 220).

The axis of the first cleavage division of embryogenesis is perpendicular to the axis of the two cell divisions of meiosis. The plane of that first cell division of the zygote to begin embryogenesis will pass through the spot on the membrane through which both polar bodies have been extruded. Clearly, there is something special about that spot, but nobody seems to know much about what defines it or how it comes to be the way it is in the human.

Timing of Gamete Encounters

After the completion of their respective maturation processes, the sperm and egg cells must meet.

The egg cell is the largest and most complex of all human cells. It carries most of the plan of its potential development. The sperm is little more than a half-genome in a capsule, with a chemical drill on one end and a propeller on the other. It might be the smallest of all human cells, but it is certainly not the simplest. The human red blood cell gets my vote for being the simplest: It does not even have a nucleus. It is little more than a bag of hemoglobin for sopping up oxygen or carbon dioxide, whichever is more plentiful, in exchange for whichever is less concentrated, thereby causing the transfer of oxygen from the lungs to the rest of the body and of carbon dioxide from the rest of the body to the lungs for release and dispersal.

There is an optimum time and place for the meeting of sperm and egg cell, relative to the ongoing development of the oöcyte — its maturation in the ovarian follicle, its release at ovulation, and other ongoing changes. After its release in ovulation, the secondary oöcyte has a strictly limited useful life span. Within 12 hours, the probability of its continuing to normal development given successful sperm penetration is below 50%. By 24 hours after ovulation, its prospects for participating in normal embryonic development are negligible.

Ideally, sperm and egg meet within a few hours after ovulation and ejaculation. To arrange such timely meetings is the function of estrus ("heat") in nearly every mammal but the human. In the interestrus or anestrous part of the estrus cycle, the vulva and vagina are dry and insemination is difficult if not impossible (and reproductively a waste of all the effort that would be necessary). Evidence for human counterparts to estrus is subtle and has been difficult to obtain, but there are some variations in sexual desire and behavior of human females who are not using hormonal chemical contraception, that correlate strongly with the timing of ovulation. There is also some evidence — none of this is very robust — that at least some men sometimes recognize some subtle clues of a woman's ovulatory status and pay more attention (it has been reported, for example, that lap dancers get more and bigger tips on their fertile nights).

It has been hypothesized that human evolution gave up the protections of estrus timing, in exchange for constant sexual availability, in service of the pair bond which keeps both parents available for the prolonged dependency of our slow-maturing, neotenous[c] young.

The "reflex ovulation" characteristic of cat and rabbit, as examples, is especially interesting. As the ripening follicles closely approach readiness for ovulation, the female produces a pheromonal scent signal which notifies the male of her fertile condition, and the final stage of ovulation is triggered by some component/s of the act of copulation (it may be the mechanical stimulation of intromission, a pheromonal stimulus from the proximity of the male, perhaps a hormonal substance absorbed from the semen through the vaginal mucosa), so that sperm and egg begin, together, their travel to their optimal meeting place in the ampulla of the oviduct.

There is reason to believe that the human, like the pig, is a facultative reflex ovulator. Women are not nearly as amenable as sows to experimentation appropriate to prove this, but: The sow's ovulation can be advanced by insemination to occur a few hours before spontaneous ovulation is expected.

---

[c] Neoteny is a concept of developmental biology reflecting the retention by adults of traits previously in evolution seen only in juveniles. In the broadest context, it refers primarily to the appearance of sexually mature individuals who are otherwise juvenile in development. It is applied to the human context primarily with reference to the relatively large ratio of head to body size of human infants, and the maintenance thereof into adulthood as a contribution to, or a manifestation of, our evolution of larger brain to body ratio. This also is related to the long infancy of the human young: In order to deliver those big-brained babies before their heads are too big to pass through a normal birth canal, we have to deliver them when everything else is less well-developed.

If ovulation is not reflexively triggered by insemination at the optimal time, it occurs spontaneously in response to hormonal cycles shortly thereafter. Strictly speaking, it is the spontaneous ovulation which is facultative. The system can facultatively perform the "spontaneous" mechanism of ovulation if ovulation has not been triggered in time by the reflex response to insemination as the primary driver of optimal timing of sperm-egg contact for fertilization.

Peak Fecundability

The peak of human fecundability, when insemination is most likely to result in clinically recognized pregnancy, is the day before ovulation, after which it declines very rapidly. The data reported in the literature to date have not been presented in such a way that I can readily learn from them whether or not fertile human ovulations occurring the day after insemination would have occurred spontaneously at that same time in the absence of timely insemination. Probability of viable conception falls from its peak to effectively zero in the 24 hours after ovulation. That drop from the peak is sharper than I would expect for coincidence, but that is an impression I have been unable so far to translate into a sound statistical argument or testable predictions.

At ovulation, the egg cell is poised at metaphase of the second division of meiosis. The chromosomes are aligned on a plane between the poles of the spindle, ready for the copies to be segregated (Fig. 1). Penetration by the sperm normally triggers the completion of that division. That division proceeds to segregate one half genome out into the second polar body (Fig. 2). The other half genome in the egg cell will be prepared and gathered into the maternal pronucleus and begin its migration to the center of the egg cell to meet the paternal pronucleus (Fig. 3, p. 221).

The first polar body contains two copies of the nuclear DNA, still attached to one half-set of chromosome centromeres. The second polar body is now just outside the membrane next to the first, and contains one single-copy half-set. One single-copy half-set of nuclear DNA remains inside the oöcyte as the maternal pronucleus, the maternal contribution to the zygote nuclear genome.

Meanwhile, the single-copy half-set of nuclear DNA brought in by the sperm is being ministered to by activities in the cytoplasm of the egg cell. There the sperm DNA is being unpacked, partially undressed and decondensed, on its way to becoming functional as the paternal pronucleus. DNA-binding proteins and polyamines responsible for the extreme condensation of DNA in the sperm head are being removed, rearranged, or replaced with different ones in varying degree over many component DNA sequence elements.

No, we are not there yet. There is much more to fertilization than the sperm meeting and penetrating an egg cell. It is a process, not an event. It is not finished at this point, and this is not yet even a zygote.

The two pronuclei must move together, to meet in the middle of the cell and mingle and meld into a single zygote nucleus (Fig. 4, p. 221) in a process we can name ("syngamy"), but our naming of which does not indicate significant progress in understanding.

We have no useful idea about how the two pronuclei "know" which way to go to come together, what forces of propulsion and steering they use to move in the right directions. We know very little about how the melding of the pronuclei progresses toward the normal structure of the nucleus or the establishment of the new system of nuclear membrane and chromatin organization.

We can be fairly certain that these movements are performed or guided by protein fiber components of the cytoskeleton, assisted by the mechanisms involved in the re-formation of the nuclear membranes, and probably also involving the protein fiber components of the spindle being re-formed from/by the centrioles brought in by the sperm.

The Centrioles

Centrioles are worth a book or two of their own. When visible, they are usually the most recognizable components of the cell division machinery, inside the pericentriolar bodies "anchoring" the ends of the spindle. There remains much to be learned about the entire spindle apparatus of which the centrioles may be the most complex components. The cell division apparatus has a number of functions, the underlying structures for which we have yet to visualize microscopically. The individual centriole is a cylinder composed of multiple proteins arranged in a pinwheel pattern of nine triple tubes. The structure is very similar (some say identical) to that of the basal bodies. Basal bodies are basal to (found at the base of) the flagella which sperm and other motile animal cells use to propel themselves, and to the cilia which are used, for example, by cells in the brain to move cerebrospinal fluid, in the oviduct to move the egg cell from the ovary to the uterus, and in the respiratory tract to sweep out germs and dirt wrapped in mucus.

The centrioles of cell division occur in pairs, always at right angles to one another, most often in an "L" configuration, near the nuclear membrane in most cells most of the time. The paired centrioles act together as organizer and nucleus of the centrosome. As part of the cell division cycle, the centrioles must replicate, in phase with duplication of the nuclear DNA. Just like the DNA inside the nucleus, they must somehow generate exact copies of themselves to be segregated between the daughter cells in every cell division. They appear to be composed totally of proteins, and not to contain any nucleic acid. Their replication occurs via mechanisms that for now remain unknown.

Also like the strands of the DNA double helix, each member of the "old" pair generates its "new" counterpart, by assembling, on and perpendicular to each member of the pair, a daughter centriole otherwise "identical" to itself in size and structure. During duplication, a daughter centriole lacking discrete microtubular organization first appears near the wall of the mother centriole near the end away from the other member of the "old" pair, and in the direction opposite that of the angle between the two "old" centrioles. It grows in length at right angles away from the mother centriole and terminates growth when it matches the length of the mother centriole.

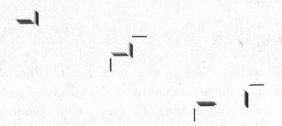

How this process is conducted and controlled, in molecular detail, to assemble daughter centrioles of the correct size and structure, appears in the literature to be no better understood than what I have just told you.

The new pairs move apart to opposite sides of the nucleus, from which positions they appear to generate the new spindle that will conduct the segregation of the duplicated chromatids into the daughter cells of this current cell division. The poles of the spindle appear to be anchored in structures surrounding each pair of centrioles, called the pericentriolar bodies (a falsely specific-sounding term that specifically represents a condensation of the vague notion of "things that look like some sort of structures around the centrioles" — which is very nearly all we really know about them). The whole assemblage of centrioles plus pericentriolar bodies is called a microtubule organizing center (MTOC), because it apparently has a great deal to do with the organization of the spindle fibers, which are made of microtubules, which are in turn made of the protein tubulin.

Syngamy

The path from the oöcyte to the zygote is far more complex than anyone can fully explain today, and far less certain and reliable than seems to be widely assumed. Contact between sperm and egg cells is a very long way even from successful fertilization, let alone from anything that might reasonably be called "conception" or the "beginning of human life." Absolutely nothing has become inevitable at that point, as of the meeting of sperm and egg.

When the genetic material of the maternal pronucleus from the oöcyte and of the paternal pronucleus from the sperm have melded to become a functional nucleus for the zygote cell, and the protein structures and accessories of the spindle that are necessary to mechanically conduct a proper cell division are in place and ready to begin a cell division ... this is *syngamy*; now, the process of fertilization is successfully concluded and that is a zygote. Now we are "there yet."

The Zygote

The zygote is this single cell that has resulted from successful fertilization by way of the fusion and melding of the sperm's and the egg's separate half-sets of chromosomes (in the pronuclei) and genes and cell division machinery, from the contents of which single cell an embryo might proceed to build itself if all continues to go well.

It is still inside the zona pellucida. It will not increase in volume or mass until after several cell divisions, so it is in fact running all of its processes by way of chemical and structural transformations of the egg cell's original contents. It is not a zygote until all those fusion and melding things are done correctly. It occupies that transitional state of being a zygote for no more than a few hours, ceasing to be the zygote when it performs the first cell division on the way to building an embryo (Figs. 4, 5 and 6, pp. 221, 222).

Many would, and we could, set the story of human development in terms of "beginning" with a single cell … but it would be silly to say something that far from true or correct on purpose. In fact, there is no such beginning. That IS what I meant to say: There is. no. such. "beginning." The difference is much more than semantic, and far from trivial. Enormous amounts of time, money and energy have been and continue to be expended in grand arguments about what "begin" means in the context of the development of (what may or may not become) the life of a fully realized human individual.

Any "single cell" involved in the processes at issue here (say, a zygote) is itself already a product of a complex developmental process that brought it to its position in this story from **The Beginning of All Life** (whatever we may imagine that to have been — and not necessarily even on this planet). To the extent that any of us is a member of the current version of the human species, it is a hereditary distinction. You hold your membership in the human species entirely because your parents were members. So were their parents, and their grandparents, and so on back to time before knowing, before any form of life could be called human, back to the time when the first thing that might have been called alive succeeded in reproducing itself and leaving behind some "progeny" entities that could survive and grow to do that again. We have no reason whatever to believe that any life, or any living thing, has begun from nothing or from any other source besides that one grand sweeping continuity, since that time.

We have been able to find no reason to imagine, let alone believe, that Life succeeded in beginning more than once on Earth. There may have been any number of failed attempts or false starts, but all evidence says that it succeeded and continued one time, in one place — one event. From several viewpoints, it will be more productive to think of this too as a horizon rather than an event.

Every living thing we know about on this planet is genetically related to every other living thing. Every living thing we know about on this planet has acid gene sequences in common. On this point, there remains no room for doubt or argument. There is zero wiggle room. Every living thing we have yet found on this planet and tested appropriately is descended from a single common ancestral source.

You and I and that tree over there and the caterpillar that is eating the leaves of that tree and the microbes that feed on the caterpillar's droppings when they fall to the ground and the bacterial cells in our bodies (several times more of them in fact than of our own cells), and the extremophile organisms that live in boiling or even pressurized and superheated water, of extreme acidity, alkalinity or salinity … all of them and all of us have gene sequences in common. These prominently include, but are not limited to, fundamental "housekeeping" genes, without the proper functioning of which life does not go on in any of its forms — things like DNA polymerases and RNA polymerases and ribosomes and the enzyme systems of energy-yielding metabolism.

You have almost certainly heard of "The Human Genome Project," the plan to map and sequence all of the genetic information of the human organism. It became a great, epic race; one government-funded consortium of labs vs one privately-funded maverick lab, with different ways of proceeding, one with a tried and true method and one with a novel, potentially faster and less costly approach the experts said could never work. It has been all over the scientific literature and the news media for several years now. It has caused a revolution in biotechnology. We expect further revolutions in drug design and in most other research approaches. We are being swept along in a revolution of our ability to ask and answer questions about how our developmental biology unpacks and unfolds single cells to build humans.

We have understood well for several decades that the developmental differentiation of complex organisms is achieved more or less entirely by differences in the expression of genes. The many different types of cells in our bodies all have the same unique set of DNA sequences (except for the results of the inevitable occasional somatic mutations), but each reads and uses a different subset. That is our cover story, we have already stuck with it for a long time, and we know it is not the full story, but we do not yet know what we are missing.

In development, we see changes in sizes and shapes and numbers of cells. We see differentiation in cell structure and function among the many different cell types. These arise from changes in the systems of structural and biochemical activities conducted in various kinds of cells. The final outcomes of development, the great variety of functions performed by the many different types of cells and tissues and organs in our bodies, are produced by differences in expression of many subsystems of the information content of the genome. Successful development of the normal body of a complex organism requires synthesis of the right gene products, by expression of the right genes, in the right cells, in the right amounts, at the right time.

The proper unfolding of that protocol depends largely on control of which genes are transcribed into messenger RNA, in which cells, at what stage in the development or function of those cells, how fast, for how long, to make how many copies, to determine how strong or how fast a particular reaction will be done.

Messenger RNA in general is very labile, rapidly degraded by a large variety of omnipresent ribonuclease enzymes. This is entirely appropriate for the primary control point in a complex system — any given gene product can only be made while the corresponding messenger RNA is present; the more message, the more gene product. For some gene products that are needed in large quantities, we have multiple copies of the genes that code for the production of those products.

For embryonic induction, for the actions of pharmaceuticals, for all of the controls on variations in cell functions, most of the cell-to-cell signals responsible for orchestrating the changes are *transcription factors*. Most transcription factors are small proteins which bind specifically to small subsequences of complex control sequences in the DNA near functional genes. The binding of these transcription factors acts as a combinatorial code to attract RNA polymerase to specific gene sequences to initiate transcription and control its rate.

Alternative Splicing of Messenger RNA

The DNA sequence is *transcribed* over to a corresponding sequence of messenger RNA (same language — nucleic acid base sequence). Most of our functional gene DNA sequences include sequences that will be expressed in the final product (called *exons*) and intervening sequences (called *introns*) that will not participate in the expressed information. As the messenger RNA is being made, before it can go on to be *translated* into protein (different language — amino acid sequence), the intron sequences must be removed and the exons *spliced* together to form the mature message.

Some are still astonished at the idea that our genome has no more coding gene DNA sequences identifiable as such than are to be found in the genome of a mouse or even a rice plant. Considering the organism as a whole, human and mouse, both mammals, are quite similar in the organization and complexity of their bodies, but ... a rice plant?! We have no more genes than a rice plant has?! The first and easiest (but probably not final) answer to that particular puzzlement is the one about how so many of *our* genes are transcribed into messenger-RNAs that can be arranged to form dozens, even hundreds, of different sequence combinations from the same DNA base pair sequence. These can then be translated into the same numbers of different protein products, with different structures and different functions, all from that one same gene DNA sequence. "The Central Dogma" from early molecular biology "one gene, one messenger RNA, one protein gene product" is not quite blown all to pieces by alternative splicing, but only a slender but still robust framework remains. We have come to understand that one gene's messenger RNA can be rearranged into many different sequence combinations for different protein products or for situations in which RNA sequences are the final functional product. This includes even the possibility of swapping sequences with messenger RNAs from other "genes" DNA sequences entirely.

The way that splicing happens is variable. Each particular exon may be included, to be part of the mature message, or left out, as if it were continuous with its adjacent introns and equally useless for this present function. The splicing process has at least these two choices (in or out?) for every exon, with the result that a single gene DNA sequence may end up coding for as many different final gene products as $2^{N(\text{exons})}$ — two choices to a power equal to the number of exons in that gene DNA sequence. A gene with five exons can code for 32 different products; with seven, for 128; and every additional intron-exon pair doubles the number of possible products of each given gene DNA sequence. The sequence always begins and ends with an exon, so the number of exons is always one more than the number of introns. There are also some additional variations in the form of accessory splice sites which can be used either to include or omit *portions* of an intron or an exon from the sequence of the mature message.

To make alternative splicing even remotely as developmentally useful as it obviously is, the splicing mechanisms in the nucleus have to "know," have to "be told" by some sort of signal, where, when, how much, etc. to do each of the possible alternative splicings. Alternative splicing has to be under reliable developmental control, so that each possible alternative product of splicing is synthesized in the right cells, in the right amount, at the right developmental time. The gene DNA sequence is not changed, but its expression is dramatically variable by way of this variation in the final spliced sequence of the messenger

RNA. The control of gene expression and developmental variation by way of alternative splicing is epigenetic by definition. And that is not at all the only other (beyond those previously understood) layer of control on gene expression.

. Again, there is no such thing as "THE" human genome. There are now nearly seven billion human genomes. No two of them are identical; each and every one is "one of a kind." Even full siblings have effectively zero probability of sorting out two sets that include all of the same choices of gene sequences from their parents.

Even monozygotic twins are never identical ... there is zero probability that there has been no unshared mutation during all the cell divisions of embryogenesis, zero probability that the mutations that refine the specificities of the antibody-producing cells of their immune systems will happen the very same way, zero probability that female monozygotic twins will inactivate the same X-chromosome in all the same cells, zero probability that all the cell-to-cell epigenetic variations in the expression of the same gene sequences will happen the same way.

No two human genomes have all the same DNA sequences. Even when monozygotic twins share what we understand as a single genome, and have *almost* exactly all the very same sequences, there are variations in expression. As far as we know there is not a single human DNA gene sequence that is read exactly the same way, to make exactly the same product, in exactly the same times and places and amounts, every time it is present.

The Zygote and Embryogenesis

As you might imagine, if human developmental biology is indeed the greatest story ever told, or even just the most complex, there are any number of places one could begin the telling thereof. There are also any number of paths that one might take through the telling, because all the parts are interdependent, and all the paths that are complete must pass through and intersect at all of the same points.

The path we are following now has taken us through the meeting of sperm and egg, to syngamy and the formation of a new diploid zygote genome unlike any other genome ever before or since. The zygote ceases to be the zygote when it divides to begin embryogenesis.

Cleavage Forms the Morula

The next stage of embryogenesis beyond the zygote is called cleavage. The zygote divides, all the way through (human cleavage is "holoblastic"), and it divides again, and again. Human cleavage divisions do not proceed synchronously. We do not see one, two, four, eight, sixteen cells in successive cleavages, with daughter cells of each division close to equal in size. Many people (including some textbook writers) picture it that way, extrapolating from the "synchronous holoblastic" cleavage style of the starfish embryos they may have seen in a well-supplied high-school or early college biology course.

Cell division in human embryogenesis is asynchronous. For at least the first few divisions, only one cell will divide at any given time. We can look at any picture of normal human early cleavage and know that the largest cell remaining will divide next. Division of the larger of the first two leaves the smaller of the first two now the largest of the three ... and so on. In general, cell divisions in human embryogenesis are asymmetrical, and the daughter cells of each of at least the first several divisions are unequal in size. After those first few divisions, it becomes difficult to follow. Far more important, but much more subtle to our limited observations to date, the daughter cells are unequal in functionalities, with different "settings" of the epigenetic controls on differential expression of genes. Cleavage continues through Figs. 6–10 (see pp. 222–225). One cell divides into two, unequally. The larger of the first two cells divides to make three. The smaller of the first two is now the largest of the three, and it divides to make four.

This soon leads to a solid ball of cells called the *morula* (Latin, mulberry). So far, all of this is occurring inside the *zona pellucida*, a translucent, tough, elastic "eggshell" layer surrounding the egg cell from before ovulation through "hatching" when the time comes for implantation. The substance of the zygote is divided among the blastomeres (cells of the morula), with no change in total volume until after hatching.

... Twinning Happens in the Inner Cells of the Morula, Inside the Zona

Twinning is a matter of defining two sets of embryogenic axes in the single contiguous mass of cells within the morula, inside the zona pellucida, within the substance of the secondary oöcyte as ovulated. [For those of you who thought you already knew how twinning happens, this will seem quite strange. The development of the understandings through which we may now be certain that this story is true and that old one is false will become clear in later chapters.]

Monochorionic monozygotic twinning is the simplest in concept. The outer layer of cells of the morula must differentiate and zip together with tight junctions to form the trophoblast layer. If two embryogenic axial patterns have been established before that happens, the two embryos that may result will be together inside the single differentiated trophoblast, and may become monochorionic twins – or even monoamnionic. Differentiation of the trophoblast layer toward becoming a single chorion for monochorionic twins occurs after the establishment of two sets of embryonic axes.

Dichorionic twinning requires that each embryo-to-be should have its own separate trophoblast layer. The differentiation of two trophoblasts within a single morula seems to be the more complex process, because trophoblast cells do not differentiate only from the outer layer, but instead must divide the morula into two masses of cells separated by tight junction trophoblast layers, all still within the morula within the zona pellucida, before hatching and any increase in size of the conceptus.

We have long known that monozygotic twins can be either monochorionic or dichorionic. We now know that the same is true of dizygotic twins (more later). Figures 11 through 15 show what appears to be the primary mechanism for dizygotic twinning. The twinning event is the same; to recruit cells from a contiguous mass into two sets of embryogenic axes. The difference is whether those cells are all of one genotype, from a single zygote or of two genotypes, from two zygotes.

Illustrations of Embryogenesis

The picture on the cover is a photomicrograph of a human secondary oöcyte, taken by Professor Louis Zamboni and published by W. B. Saunders and Company in 1975 as Figure 33-18 in the 10th edition of *A Textbook of Histology* by Professors William Bloom and Don W. Fawcett. It is an old picture, but it has not to my knowledge been improved upon in the intervening 35 years.

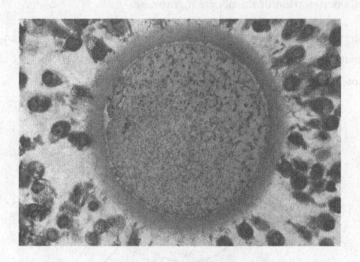

This shows the "egg cell" as we best understand it, ready for fertilization. This is what we believe to be its most normal configuration at completion of the first meiotic division, just after ovulation, ready to be fertilized. It is enclosed in the zona pellucida, surrounded in turn by cells of the cumulus oöphorus ("egg-bearing cloud") released from inside the ovarian follicle with it at ovulation. The first polar body has been pushed out and pinched off between the oöcyte membrane and the zona pellucida, in the most exquisitely asymmetric cell division known. The chromosomes and spindle of the second meiotic metaphase are visible just inside the membrane, in preparation for putting the second polar body out through the membrane right next to the first when the second meiotic division is completed.

Notice the variation in fine structure across the oöplasm (appearing here as varying granularity in the egg cell cytoplasm). Technology now exists for mapping the various effector molecules the respective distributions of which in the oöplasm are believed to constitute the "plan" for embryogenesis, but that work has not been done.

In what we believe is the most "normal" path through fertilization, the second meiotic division waits to be triggered by cortical and oöplasmic reactions to the sperm cell's penetration of the oöcyte membrane.

The drawings that follow, prepared for this work by Alan Branigan, display diagrammatically how things happen after that: The first 10 images show the progress of "ordinary" single fertilization.

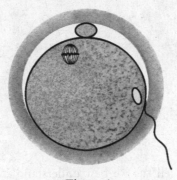

Figure 1.

Penetration by the sperm in Fig. 1 triggers completion of the second meiotic division, resulting in the extrusion of the second polar body beside the first, and the formation of the maternal pronucleus in Fig. 2.

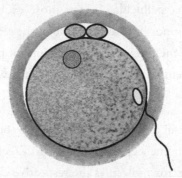

Figure 2.

Enzymes in the oöplasm remodel the sperm nucleus to form the paternal pronucleus and the two pronuclei move together in Fig. 3,

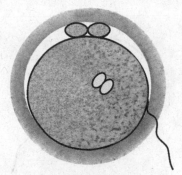

Figure 3.

and meld to form the zygote nucleus in Fig. 4.

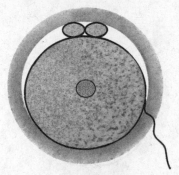

Figure 4.

In Fig. 5, the zygote nucleus begins its first mitosis,

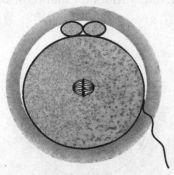

Figure 5.

and the first mitosis is almost completed in Fig. 6.

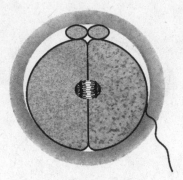

Figure 6.

In Fig. 7, the larger of the first two blastomeres divides, leaving three cells.

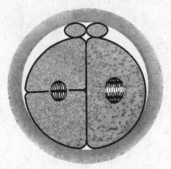

Figure 7.

Then the largest of the three divides for four cells, and so on; Figs. 8–10.

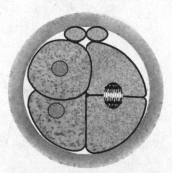

Figure 8.

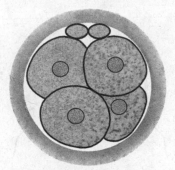

Figure 9.

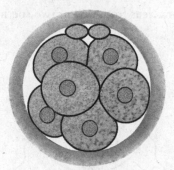

Figure 10.

Figures 11 through 15 model double fertilization. In Figs. 11 and 12, the second meiotic spindle of the oöcyte divides internally, with no second polar body produced (often called "retention of the second polar body).

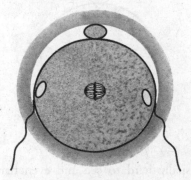

Figure 11.

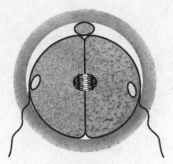

Figure 12.

This generates two maternal pronuclei (a "dipronuclear oöcyte"), each of which melds with a paternal pronucleus from one of two sperm nuclei, in Fig. 13.

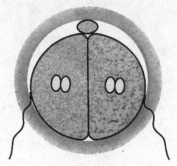

Figure 13.

Division continues in Fig. 14 in exact equivalence with single fertilization after the first zygote nuclear division, except that there is only one polar body.

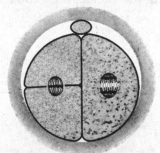

Figure 14.

[Polar bodies are usually hard to see and even harder to count, making nondestructive detection of the difference between single- and double-fertilization unlikely. This goes a long way toward explaining how IVF can generate more babies than embryos transferred, why ART pregnancies generate "monozygotic" twins three times as often as spontaneous pregnancies, and monochorionic dizygotic twins are found more often.]

Continuing cleavage in Fig. 15 and on to the morula stage are indistinguishable in single- vs double-fertilization, although the cells of the morula arising from double fertilization are of two different genotypes.

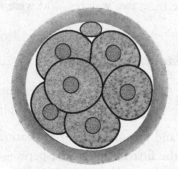

Figure 15.

Twinning of either zygosity takes place here, by way of the same mechanism. In either situation, the next step is to begin differentiation of the embryonic axes and recruitment of cells from the morula into either one or two body-plan armatures.

This may happen before or after differentiation of the outer layer of cells of the morula into the trophoblast layer that will become the placenta and chorion. That step is called "compaction." Compaction is microscopically apparent as the outer layer of cells remain loosely attached to cells inside by way of gap junctions, while coming to adhere tightly to other outer layer cells by way of tight junctions. This is followed by the pumping of fluid through the outer trophoblast layer, separating most of the trophoblast (the "mural" trophoblast) from the inner cell mass and "hatching" out of the zona pellucida.

If two anteroposterior body axes form *before* the differentiation of trophoblast, then two sets of trophoblast cells may divide the cells of the morula into two domains and form two blastocyst "embryos," which may go on to form dichorionic twins. If a single trophoblast layer differentiates before the beginning of anteroposterior differentiation, then twinning will be monochorionic. Despite decades of blind assumptions to the contrary, chorionicity has not been shown to be a strict function of zygosity.

If cells of the morula are of two genotypes (dizygotic), the likelihood that delineation of two body axes will each recruit cells of only one genotype would appear to be negligible — most dizygotic twin embryos will be chimeric, with some cells of each zygote's genotype.

[Perhaps most of us were once single cells, as in the chapter title, but many of us have cells that came from two zygotes, and were never entirely contained within a single cell.]

The Blastula, or Blastocyst:

future embryo proper (inner cell mass) separates from
future extraembryonic support structures (trophoblast)

On about the fourth and fifth days after fertilization, cells in the outermost layer of the morula are the first to undergo a permanent differentiation. In a process that has been called "compaction" of the morula, the outermost cells flatten and bind more tightly to each other side-by-side with tight junctions. They then become separated from the rounder interior cells (of the "*inner cell mass*") by a layer of fluid, forming the fluid-filled cavity called the blastocoel, surrounded by the *trophoblast*. A few cells on one end become the *polar trophoblast*, and the rest form the wall, the *mural trophoblast*. The mural trophoblast, separated from the inner cell mass by a layer of fluid, is the precursor to the *chorion*, the heavier outer layer of the gestational membrane sac.

The *inner cell mass* remains attached to the *polar trophoblast*.

This lump-inside-a-ball stage of embryogenesis is called the *blastula* or *blastocyst*. The process of forming it is called *blastulation*.

The polar trophoblast will soon lead the way to imbedding in the wall of the uterus, interacting with uterine cells to create the placenta from an interactive combination of maternal and embryonic contributions.

Hatching; headed for implantation
shedding the zona

At roughly the same time that these steps in the formation of the blastula have reached completion, the zona pellucida starts to break down, and the blastocyst swells, stretching and dissolving the zona until it is all outside and the zona is left empty ("hatching"). The zona is no longer in the way for the blastula to implant into the wall of the uterus.

The majority of the available published photographs are of egg cells collected after they have been released by artificially induced ovulation. I cannot state with any certainty whether these behaviors are the same as those that would occur in naturally ovulated oöcytes undergoing the same processes; I believe they are not. There have been reports to the effect that the zonae of oöcytes produced by artificial induction of ovulation are "tougher" and less amenable to naturally splitting open. The mechanism by which hatching occurs — what it looks like while it is doing that, whether there is a wholesale collapse or a hole through which cells are extruded — is actually somewhat important. The possibility of pinching the embryo into two pieces on its way out of the zona has been proposed as an explanation for the excess of monozygotic twins among artificial conceptions.

In *implantation*, the cells of the polar trophoblast invade and interact with the cells of the uterine wall, eventually forming the placenta as a joint venture between the fetus and the womb. Hatching, and the resultant freedom from the hindrances and restrictions of the zona pellucida, allows the blastocyst to begin expanding rapidly in size.

The Inner Cell Mass Becomes the Bilaminar Disk "Embryo"

The inner cell mass next separates itself also from the polar trophoblast with another pocket of fluid and spreads out into the bilaminar disc stage of embryogenesis. For a short while it will be a flat disc two cell layers thick between two pockets of fluid inside the trophoblast. The smaller fluid-filled chamber between the *bilaminar disc* and the polar trophoblast becomes lined with a layer of cells that will become the *amnion*, the thinner, inner layer of the gestational membrane sac.

[Microscopically Visible] Definition of the Embryonic Axes of A/Symmetry

The *prochordal plate* (sometimes called the "prechordal plate") appears near one edge of the bilaminar disc. Its position there defines the future *anterior* (head) end of the embryo. The *primitive streak* appears in the midline near the *posterior* (tail end) edge. The prochordal plate will serve as a target for the migration of the *notochord* that will form in the midline from the mesodermal cells from the primitive streak migrating in between the layers of the bilaminar disk, the *ectoderm* and *endoderm*. The notochord is the evolutionary and embryogenic precursor of the spinal cord in all the chordate animals, which include all the vertebrates. In the completed human, the cartilaginous inter-vertebral discs (those banes of the aging back) are the final developmental products of the notochord.

The side facing the smaller fluid layer toward the polar trophoblast will be *dorsal* (back side); the side facing the larger fluid-filled chamber toward the mural trophoblast is *ventral* (belly).

At this point, by these cellular events, the three dimensions of the armature upon which the embryo will construct itself are irreversibly differentiated. There is little room to doubt that the cells have made all their appropriate molecular decisions necessary to produce these differentiations before we were able to see them microscopically, and that we will someday know what they are, but we cannot make that claim today.

Gastrulation: Forming the Three-Layer (Trilaminar Disc) Stage

A particular patch of cells on the endodermal surface of the bilaminar disk near what will be the anterior margin become taller in the ventral (belly) direction. (The endoderm is primarily the precursor to the gut and its derivatives.) This will be the prochordal plate. It will serve as a target for the migration of the notochord and the mesoderm from the primitive streak forward between the ectoderm and endoderm layers.

Another particular patch of cells in the dorsal layer (ectoderm, future skin and nervous system) grow taller on the dorsal (back) surface of the bilaminar disk near the opposite, posterior margin, to become the primitive streak.

The cells of rhe primitive streak divide rapidly, creating a flow of new cells down between the ectoderm and endoderm layers, moving forward toward the prochordal plate, spreading to both sides as it moves forward. This process creates the third layer of the *trilaminar* disc embryo, the layer known as the *mesoderm* (precursor primarily to muscle and bone).

Neurulation: Beginning the Nervous System

Cells around the perimeter of the newly generated middle (mesoderm) layer signal the overlying ectodermal layer to rise from the dorsal, ectodermal surface to form a rim for the *neural plate*. The edges of the neural plate continue to rise and move up and in toward the midline, where they will fuse. The fusion begins between the middle and the head end. Then the fusion moves, zipper-like, toward both anterior (head) and posterior (tail) ends — to form the *neural tube* (from which will develop the future brain and central nervous system).

While the neural tube is closing on the dorsal surface, the perimeter of the whole trilaminar disc is being pulled together toward the belly side like a purse, around where the *umbilicus* (the root of the umbilical cord, the attachment to the placenta, of which the navel is the marker after birth) will be. Near the anterior edge of the trilaminar disk, "forward" of the prochordal plate, two tubes of differentiated cells fold down and under and come to the ventral midline and fuse. With later remodeling under guidance from cells that will come from the neural crest to join in, this fusion of the *cardiac tubes* will form the heart.

Building the Face Around the Mouth

The branchial, or pharyngeal, arches of human and all other vertebrate (animals with backbones) embryos are developmental homologs of, and (when they first become visible) look like, the gill arches in fish embryos.

The *first branchial arch* (most anterior) will form the jaws except for the middle of the upper jaw — which instead comes down, as the *frontonasal process*, from the forehead to form the *philtrum* (the crease in the middle of the upper lip). As it executes that movement, the frontonasal process splits the *eye field* in two and forms the nose except for the rims of the nostrils. The ears come from between the first and second branchial arches. The parathyroid glands — fundamental to the management of calcium in the human body — also come from the arches, right where their evolutionary precursors function in the gills to manage calcium concentration and calcium transfer for fish.

The Neural Crest Cells

Along the middle of the back, where the lateral edges of the neural plate have zipped up to form the neural tube, the cells of the *neural crest* form what may well be the most awesome posse of cells in all of embryogenesis. From the dorsal midline "seam" of the neural tube, they migrate out to the sides and down and around toward the belly, to many different locations, where they differentiate into many different cell types within the embryo. They drive themselves and other cells through many different forms of differentiation to become pigment cells, the sympathetic, parasympathetic, the enteric and the peripheral nervous (sub)systems, skin, hair, and teeth, to shape the face and the heart, to form crucial reactive cell types in the adrenal glands and the gut …

Along the spinal cord neural crest cells form the sensory dorsal root ganglia, which will "direct traffic" between the nerves of sensation and the spinal cord and on to the brain. In the head region, neural crest cells migrate into the pharyngeal arches and form many different structures. There is good reason to suppose that we have barely begun to know all of the things neural crest cells do. Every time I check that literature, they have been found going places I never heard of them going before, and are being given credit for a few more fundamental, crucial jobs.

The Germ Line Cells

Primordial germ cells (future eggs or sperm) move anteriorly (toward the head) and dorsally (toward the spine) from the *primary yolk sac* into the *urogenital ridges* (two long oval piles of cells on either side of the gut and spinal cord) that will become the kidneys and *gonads*. If these cells carry X *and* Y chromosomes, they start changing each undifferentiated gonad (from part of the cells of the urogenital ridge) into a testis. When testicular tissue has begun to form, signals from the testes will change the embryonic phallus to a penis and build a scrotum and the other accessories of maleness.

If the primordial germ cells moving into the urogenital ridges have two X-chromosomes, they start building ovarian tissues, which will send signals to build a clitoris instead of a penis and labia instead of a scrotum, and so on.

(neuro)Musculoskeletal Development

Limb buds sprout to become legs and arms. Mesodermal cells (future bone and muscle) divide up in blocks along the spine (the *somites*) and start turning into vertebrae and ribs and the matching muscles. Nerves grow into those patches, where they interact in crucial ways with the muscles. Without the proper input of signals from nerve cells, the muscles will not grow properly, if at all … and vice versa.

Building the Brain

Up at the head end, wave after wave of future brain cells divide off from the bottom of the neural tube and move out to form the multiple layers of future brain tissue. It matters very much how many of which kinds of cells go how far out and settle where and in what relative positions with other cells to form which layers. Some of this layered migration surely is probabilistic, but apparently all of it needs very reliably to get pretty close to some special pattern if it is going to work well when everybody takes his/her place and the music starts.

Somewhere near here, these migrating cells gain what they will need, to know whether and how to build a left brain or a right brain when they get to where they are supposed to go. That may seem a subtle difference but it certainly has huge and fundamental implications. It constantly amazes me to think about how cells "know how" to do these many different things, and "know" when and where and how fast and how long and how many times — AND do it all, in general, very reliably.

A little patch of gut tissue, where the roof of the mouth will be, moves to meet a patch of brain cells coming the other way. They meet and fuse to become the *pituitary*. We used to call the pituitary the "master gland" until we learned how much its partner-boss, the brain, controls and masters it. You can feel with the tip of your tongue, in the (left-right) middle of your hard palate (~2/3 of the way from just behind the teeth toward the soft palate), a little "hole" where Rathke's Pouch left the dorsal midline of the gut and "went up" to meet the corresponding bubble coming "down" from the ventral midline of the neural tube to form the pituitary.

There are many more pieces than I have been able to include in this quick thumbnail sketch of embryogenesis. Have I told you already, or have you gotten the idea on your own, that there is nothing in the universe more complex and compelling than building a new human — by way of unpacking the programmed "potential" in a single-cell zygote?

Many of the gene products involved in these activities are signals, whose functions are to cause changes in the functions of other cells and other genes and gene products, to keep it all coordinated and continuing. Failure of any one of them at any time may bring the whole system to a halt. This can cause whatever has been built up to that point to take itself apart, pack up, move out and disappear. The dismantling occurs by processes much like the removal of the webbing between the embryonic toes and fingers by programmed cell death.

This very complex system of processes that adds up to embryogenesis occupies the first six to eight weeks of pregnancy. Embryogenesis is complete, and crosses the horizon of the fetal period about the time that the mother is missing her second consecutive menstrual period. She is then coming to recognize that she may be pregnant and thinking about seeing her reproductive health care physician who may pronounce her clinically pregnant.

At the point of completing embryogenesis, about when the second consecutive menstrual period goes missing, the product is about the size of a "baby lima" bean or an average adult fingernail — but it has its own fingernails! All the hardest parts of developmental biology are complete. All organs and organ systems are in place and well on their way to completing the differentiation of the specialized cell types necessary for their functions. By this time in pregnancy, most of what can go wrong already has, and — in the best of circumstances — over two-thirds of even the most privileged conceptuses have already failed.

Embryogenesis Ends; The Fetal Period Begins

As the embryo is finished building itself, it becomes by definition a fetus and no longer an embryo at all. Over two-thirds of all the ovulations and inseminations that resulted in successful fertilization at the same time as this one have failed. From this point forward, in the privileged conditions of well-cared-for pregnancies, yet another 10–20% of those remaining alive at this point will fail before term and delivery. Failures of recognized pregnancies are called miscarriages or spontaneous abortions. Each day that the pregnancy survives brings a higher probability of surviving another day.

Chapter Ten

Getting to the Bottom of
Creation's Best Trick:
How Human Lives Begin

> Nothing is more usual than for philosophers to engage in disputes of words,
> while they imagine they are handling controversies
> of the deepest importance and concern.
>
> *David Hume*

A human Self, a human Person, is — throughout every instance and every aspect of its observable reality — of no real interest to us except insofar as it is a living thing, a member of living humankind. A human person is a living system, a supersystem of many different tissue subsystems which are themselves supersystems of many different cell systems, and a subsystem of numerous levels of organization up to and beyond its species — the many layers of family and other groupings, neighborhood, state, nation, etc., and on into considerations of the species' place among species.

You may imagine a human self or person to be also, or instead, some sort of psychological entity, or a philosophical entity, or a moral entity, or even (a container for?) a sacred and immortal something or other. None of those ways of seeing a human person can have any demonstrable reality before or after, or independent of in any other dimension, the biological, cellular entity. The business of being or becoming human can never be understood at any of those levels or from any other viewpoint without an understanding of the structures and functions of the cellular substrates of Life and of Mind.

Everything about human life is cellular in origin and in execution, in the same ways as all other kinds of life we have ever been able to know about. Each and all of its instances have beginnings and middles and ends that are like those of all other kinds of life. That is sacred enough.

I have thought for some time that it should long since have died of suffocation under its own huge breathless weight, but it lingers. The question "When does human life begin?" has accumulated a great deal of hard mileage over recent years, over many a strange path, through some very rough terrain. Around the question swirls a clinging, fetid smog — a stifling cloak of fuzzy thinking smoldering, never quite to burst into flame, lacking as it does the oxygen of honest inquiry. The question, as it is usually meant and heard, you see, is not intended, nor in general will any answer to the question be allowed, to replace ignorance and superstition with understanding, as should be the purpose of an honest question.

This question seems intended to appear as one about a biological process, to bring the weight of science to bear on answers to questions of kinds other than those purely of science. This one is soon enough, however, recognizable as nothing of the kind. All reasonable, biological answers that may be offered have long been and apparently will continue to be dismissed one after the other. The answers given have varied over centuries and more quickly in recent decades, as understanding of the biological components of the processes at issue has changed. The perennial failure of any factual, biological answer to satisfy tells more about the question and the questioners than it does about the answers.

This question is a subterfuge, intended to give scientific cover to an argument that is never really scientific and is more often than not anti-scientific. The question comes from a place that is passionately protected from argument and evidence, so that any answer that might be given hardly matters. I have not yet abandoned the hope that reason must sooner or later prevail.

There are any number of excellent choices of origin stories and creation myths from which comforting mythical answers to this question might be derived without severe disruption. Problems arise only when anyone pretends that their personal chosen truths derived from faith or myth are supposed to be considered rationally equivalent to truths of history or fact or science or of any other act of observation or of reason.

We are each and all free to choose what we want from either of these systems of concepts from which to build our respective personal viewpoints. We are not, however, logically or ethically free to try to make either one into, or out of, the other. We are not ethically or morally free to make the choice on behalf of, or to attempt to force our own choice on, any other person.

It is not a trivial thing to understand, and it may be worth some effort to consider ... what exactly does the question mean?: "When does human life begin?"

— Does that mean: "What point in Time marks the onset of the existence of all Human Life, from a prior situation in which nothing that could be called human life existed"?
— Or is it about "**a**" human life, an individual human life?
— Can the case be truly made that the above two concepts are actually different? Does there exist any such thing as "**a**" human life apart from all other human life?
— Should, or even can, human-ness and life be considered separately, on an individual basis?

Are we to suppose that there ever exists any entity that is always human but for which there must be some instant before which it is not alive and after which it is?

Can anything be human throughout becoming alive from a state of not being alive? We know there are things which can safely be called human that are no longer alive, but is there, has there ever been, such a thing as being human but not yet alive, while "beginning" to become alive?

Perhaps we should expect each other to believe that there is some instant in the process of a human-life-coming-into-being before which the entity in question is living but not human and after which it is both? Can any living thing become human from a state of being other than human?

I find any of these ways of considering the question awkward at best; actually, absurd. The question has no reasonable form or substance.

To be truthful about it, the question represents a desire to draw a bright line by which such matters, when considered to be matters of right, wrong and conscience instead of matters of observable biological fact and rational analysis, might be set in order, and to claim that it was all along a question about the real natural world and that the answer comes from science.

... why ... at conception ... of course!
(whatever THAT means...)

The answer we are all expected to know, the answer that comes from and belongs to those who believe that everyone else really should believe in the same things they believe in, is that "Human Life Begins at (The Moment Of) Conception."

"Conception" is routinely offered as that bright line on one side of which there is not, and then on the other side of which there is, human life — the life of a unique human person with a unique, sacred and immortal human soul and a personal Self, a personality. That's simple enough, is it not? Why can't everybody just learn that, and hang on to it?! "Conception" is when human life begins.

"Human life begins at "conception"." It is only one word. It should not be hard to remember ... except for the fact that, in this circumstance, it is empty noise. It may be just one word, but no one knows what it means!

That conceptual territory contains quite enough logical landmines to keep us busy for a long, long time, all of which are somewhere over on the other side of the bright line we are trying to put in place. Right here and now, on this side of the act of trying to draw the line, there is the obviously necessary assumption that someone somewhere knows what "conception" means. This implies a trusting that "conception" can be perceived or defined sharply enough to make a credible pretense that it is, or that it might in some way help to establish, the desired bright line in time and cellular space between the absence and the presence of human life, which transition thenceforth to call "The Beginning of Human Life."

Suppose, for the sake of clarity and forward progress, we just step around that particular little pile of mess, and put "becoming pregnant" in the place of any concern about a definition of "conception"? That IS what "conception" means, is it not, by any definition germane to our present considerations? That is exactly what the medical dictionaries say that it means, and that is all of a biological nature that any source I have been able to find says that it means. That is the most fundamental and the simplest part of this problem. If we cannot get that part sorted out, there is no hope for the rest.

Does this different approach give us any advantage, or leverage, or any better grip on the question? Perhaps we should be able to mark an instant in time when a woman becomes pregnant? Can we point to an instant in time when a woman is pregnant which follows immediately upon an instant when she was not pregnant, with an infinitesimal instant of becoming in between? {Ffffwoompf! ... with or without a ding!} If indeed there is any such thing, then what is its instantaneous biological equivalent? What is the instantaneous cellular component or counterpart of "becoming pregnant"?

There is one widely touted notion out there to the effect that the "moment of conception" is to be defined as/by the "instant of fertilization", which is to be defined, in turn, as the instant of the meeting of the sperm and the egg cell.

To declare that the instant of sperm-egg contact is the instant of "conception" immediately raises a whole system of problems based in two quite simple facts:

1) the fact that the word "conception", as representing the biological process of becoming pregnant, has had and can now have and will yet have any of several definitions all of which are entirely reasonable from one viewpoint or another, and,
2) the fact that not a single one of those possible definitions realistically defines <u>both</u> a reality of cellular biology <u>and</u> an instant in time.

Given that the intention here is to establish the biological reality of this by stating it in terms of things that certain specific kinds of cells can and actually might sometimes do, then some people are disheartened to learn that that particular answer affords no progress at all.

We may, in fact, reasonably consider that collision-of-sperm-and-egg to constitute an instant, an event. It has just that much going for it as a possible answer, that it can in fact reasonably be considered instantaneous. To be sure, that collision is the **only** part of the entire process that **can** be considered instantaneous, because ... there is in fact an instant before which that collision has not happened and after which it has.

Swim ... swim ... wiggle ... swim ... wiggle ... swim, swim ... wiggle...
huff ... puff ... boink! ta-daaaaaa!

or — from the oöcyte's other-end-of-the-trip point of view...

hum-da-dum ... waiting ...
... relaxing here, floating along down the ampulla
basting in warm juices seasoned just right ...
awaiting my destiny ... whatever! ... waiting, waiting ...
waiting for it ... boink! attaboy! ta-daa**aaaa!** ......

aaaah! yessir! hoo..rah! yessirree! Oh, happy day! "boink! ta-daa**aaaa!**"
... and there you are! ...
... never a phalanx of trumpeters around when you need them!

... *le moment ou je parle est déjà loin de moi* ...
... are you finished already?
I am certainly glad we got that behind us.

After looking at it thus closely, I hope there can be no need to spell it out for you that nothing of any real interest to these questions at issue here is changed in that instant by that happening.

Pre-boink! vs Post-boink! big deal!
what's the difference?! what am I supposed to see?!

There is nothing there that was not already there.
Nothing that was there has vanished.

Nothing is or becomes set,
nothing becomes fixed in place or set in motion for once and forever...

nothing has become inevitable,
nothing has become as it forever henceforth shall be.

Nothing.
Nothing sacred.
Nothing profound.
Nothing significant.
Nothing interesting.
Nothing useful.
Nothing.

Not one single thing is instantaneously changed in any interesting way by the collision of a sperm cell with an egg cell.

Nevertheless, that collision is the only component part of the whole process that might be called instantaneous, might be called an event, might be considered the <u>instant</u> of fertilization, or the <u>moment</u> of conception … and … it … is … nothing … it is an absolute non-event. This is clearly and simply not the magic great divide, not the ridgeline at the rim of a new watershed, not the point where the beginning has ended nor the point where the ending has begun.

Suppose we skip ahead just past this, pretend it IS at least a definitive step in the direction of what we wanted it to be, and say "Okay … given: let us just suppose that a sperm cell has encountered an egg cell. Given. No sense in arguing about that. We know it happens."

For reasons I have never heard anyone explain (therefore my natural tendency is to think no one really understands), one sperm, actually anything less than several millions of sperm cells per milliliter of semen, is — in general — the functional equivalent of no sperm at all, zero. We can let that slide for now — I don't think there exists a very useful answer anyway, but the majority of men who have sperm counts lower than about ten million per milliliter of semen (50 million per teaspoonful, which is about the volume, and about the minimum functional total sperm count, of a normal ejaculate) are clinically defined as infertile because an insemination with sperm cells at a concentration that low almost never manages to cause a pregnancy to occur. That may be a correlation and not a cause. It may be the case that men who cannot make large numbers of sperm cells also usually cannot make any really good ones? Clearly, it would seem that this little fact should cloud the issue, but I have never heard it mentioned in the context of the "When does Human Life begin? …why, at conception, of course!" conversation.

Anyway. Given. Let us suppose there has been a "moment," an instant, of sperm-egg encounter. Given. Let us suppose that we have gotten to the other side of that and that everything up to that point has happened as it should. Now what? There is no certainty that the sperm — any one given particular individual sperm, or even any member of the crowd in any one particular given ejaculate — will be able, or will be allowed, first to penetrate the cumulus oöphorus ("egg-bearing cloud" of small cells surrounding the oöcyte upon its release from the ovarian follicle).

Assuming penetration of the cumulus, there is no certainty that penetration of the cumulus will be followed by penetration of the zona pellucida ("very clear area" surrounding the egg cell inside the cumulus; a cell-free, fibrous, elastic jacket, the "eggshell").

There is no cause for confidence that penetration of the zona will lead certainly, or immediately, or ever in fact, to penetration of the egg cell membrane.

And then, just inside the membrane there is the cortex of the egg cell — much thicker and more complex than the corresponding space just inside the membrane of any other kind of cell.

All of those barriers must be penetrated if anything important is to happen. Each penetration will use up some space and some time. None of those penetrations is instantaneous. We have traveled already well beyond any reasonable concept of the "instant of conception," and have encountered nothing even probably transformative.

Even in a normal, healthy, high-quality, high-motility ejaculate, a sizable fraction of sperm cells cannot motor themselves up through the fluids in the female"s reproductive plumbing to reach the vicinity of the egg cell. Of those that can do that, a substantial fraction cannot manage all that is required to achieve all of those necessary penetrations of consecutive layers, to reach the cytoplasm of the egg cell, to get all the way in, to reach a place where the depositing and unpacking of its half-genome-plus-centrioles cargo might do some good, where there might be any chance of productive consequences for all the effort it had to expend to get this far.

Even if we should suppose that sperm penetration does, will, must happen at every unprotected insemination with a viable egg cell, the sperm's successful penetration of the cumulus, the zona, and the egg cell membrane and cortex does not constitute fertilization.

Successful penetration is only a threshold, opening onto a complex system of processes that must, from that "beginning," unfold properly once the sperm nucleus is inside the cytoplasm of the egg cell, if successful fertilization is to be achieved. There is no certainty that it will lead to a successful fusion of the pronuclear genomes in syngamy, which might … finally … be called success in fertilization.

If syngamy is successful in completing a process of fertilization, there is still no certainty that the process of forming an embryo will begin, to say nothing of the prospects for successful completion of pregnancy, such that there might eventually be a new human growing from any such "beginning" into some prospect of taking an observable, fully qualified place here among the rest of us.

"Conception," becoming pregnant, does not happen in an instant. It is not an event, but is instead a complex biological process occupying a considerable span of time and space. The beginning of pregnancy, conception by any name or definition, is a <u>developmental horizon,</u> perhaps clear and distinct at a considerable distance, but not amenable to definition at close range by any meaning of the word. It cannot serve the purposes of this inquiry.

> <u>"Conception" is not a scientific answer to the question</u>
> <u>"When does human life begin?."</u>

Fate of the Average Human Conceptus

In all of the hubbub and stew (I have no idea where it came from, but that is what my mother used to call such multivariate situations) surrounding questions about human embryos and stem cells and cloning and abortion and such, there is one great compelling chunk of the facts of the matter that I have yet to see addressed in any of the discussions I have encountered in the various relevant literatures. This particular observation cannot fail to make a great deal of difference, regardless of the perspective from which anyone may be approaching any such question:

> **In the most privileged of circumstances [that is, with healthy parents of proven fertility and in the optimal age range for reproduction, under research-level medical attention], over three-fourths of all human "conceptions" fail before they achieve any realistic eligibility for survival up to and beyond live birth, at which point any one of them might actually begin growing into a fully realized version of what we normally think of as a human person's life.**

> **That is the way things are in the best of circumstances: Fewer than one-fourth of even the most circumstantially privileged of human conceptions will survive to term birth.**

> Over the population of the world at large, where circumstances of pregnancy are seldom as good as those in the samples I used to find that number, one surely must suppose that the odds are rather worse. It would be quite reasonable to imagine that the worldwide survival of human conceptuses to full term live birth may be as low as 10% of all "successful fertilizations" over the whole human population. There is no evidence against any hypothesis that it would be even less.

The fact that — even in the best of circumstances — most human "conceptions" fail, including over two-thirds of them before the pregnancy is even recognized, must put certain limits on our thinking. One quite certain such limit is this: no point in human prenatal development, from building the sperm and egg cells to term birth, over however many generations back one cares to look, does anything become inevitable about the existence or structure or content of any potential future individual human life. Every day survived brings better odds of surviving another day, but none of its possible outcomes ever becomes certain before it becomes observable fact.

A large majority of the natural aggregations-of-cells that might be considered to be "human embryos" prior to completion of embryogenesis and clinical recognition of pregnancy (both occurring about six to eight weeks after fertilization) have in fact no potential at all for live birth and a subsequent fully realized human life. <u>A large majority of natural human "conceptions" will fail.</u>

The same is true — and more so — for each and every aggregation of undifferentiated cells arising from any source other than natural fertilization (any kind of "artificial" embryo). No artificially contrived aggregation of cells has ever demonstrated the potential — shown itself capable of continuing development — eventually to result in the birth of a live human. In fact, we will be heroically delighted if we can ever reliably maneuver any such cell or group of cells into a situation of performing any useful function of human cells.

The <u>Unique</u> New "Human" Genome

If sperm penetration and "fertilization" succeed and progress through a successful syngamy to the formation of a viable zygote, then indeed an absolutely unique new individual human genome will be formed. This uniqueness is the foundation of some of the arguments about this question. The new genome that might be formed if syngamy succeeds will, indeed and necessarily, be different from any and every other human genome, past or future. It will not necessarily, however, be supported by functional cell division machinery. It will not necessarily be a functional genome. A newly constituted genome means precious little by itself.

> **Your own particular unique set of DNA sequences <u>is not You</u>.**
> **Your DNA is surely necessary, but at least as surely is not sufficient,**
> **to define You as a realized human being, as a person.**

Until and unless that genome can function in a workable zygote nucleus, and on down through many generations of the progeny of that zygote nucleus, to guide the building of a proper, functional new vehicle for that genome's transit to another next new generation, nothing is inevitable about whether or how that new genome may be constituted or expressed. Nothing. Period.

> **That unique new set of genes is not, at that time of its formation,**
> **and odds are that it never will be,**
> **a new human person,**
> **nor even any part of a new human being.**

The "Potential"

> The aim and the end of all becoming
> is the development of potentiality to actuality,
> the incorporation of form in matter.
>
> *Aristotle*

One may imagine the presence, within a newly constituted diploid genome, of a "potential" new human being, but potential (noun) is, of its essence, by definition, always potential (adjective) and exists only in the imagination of the beholder.

Potential is always entirely contained within imagination and supposition. It can never be certainly known or shown to exist in any given instance; it can only come to be known to have existed (in the past perfect tense) after it has become actual and is therefore no longer potential (noun or adjective).

We look at this thing, a cell or group of cells, and perhaps we see that it looks like some other thing we have seen before, before that other thing we saw changed into some still other kind of thing. From there we go on to suppose that this (thing that looks like that other thing looked before it changed) might change the same way that other thing it looks like changed. It might become that for which our imagination has assigned it the totally imaginary prospective outcome for what we are accustomed to calling a human embryo. It might do that. Until it has finished doing that, we cannot know whether or not it <u>can</u> do that, and all talk about any "potential" it may have is flapdoodle, with no concrete reality and no means of gathering any.

> Any definition of a "human embryo" that does not include some concept
> of its potential is empty. Any definition of a "human embryo" that
> <u>requires</u> the concept of potential has no present meaning in observable
> reality; only potential, and imaginary. I can write "human embryo," and
> expect you to know what I mean, for the sake of discussing the processes
> of human reproduction. However, in any situation requiring moral
> certainty, or any other sort of certainty, it is a meaningless construct.

If and only if any given particular product of conception survives to gain, grow into, accept, acknowledge and promulgate its unique human individuality will its unique human genetic individuality have any meaning ever, any significance at all. If and only if this particular entity is one of those fortunate few "destined by chance" (a bewildering concept, is it not?) to be born and live a unique personal human life, only then does anything anyone ever said about personhood, from whatever imagined beginning, have any chance whatever of becoming true.

Do not, please, let the irony get away from you, that each and every unique individual "human" "destiny," in all of its glories or shames in each of its many dimensions, depends absolutely on genetic and epigenetic and developmental and environmental chance. That is besides, apart from, in addition to, the many influences of plain old dumb luck, in the form of so very many things that cannot be foreseen, foretold or controlled, but which certainly can instantaneously go wrong and ruin the whole process without warning and without recourse.

Except perhaps from perspectives of quantum theory, it is usual and customary — some would say necessary — to believe that <u>nothing happens without cause</u>. Several different philosophers have said so, and one or more of them has called that concept "the law of sufficient reason"; nothing happens without sufficient reason. It is also quite common to suppose that chance does not quite really qualify as a cause. This does not, however, of course, mean that we will be able to know the causes in any particular situation, let alone always, and chance has a great deal to do with that. Chance is routinely used to stand in place of knowing all of the causes of any particular situation, event or outcome.

For example: In explaining the basics of probability theory, I have taught for decades that the exact physical results — exact to the level of the atoms and electrons — the exact physical results of smashing a rock with a hammer, the exact sizes and shapes of each and every one of the resulting pieces, might in principle be known before the blow is struck. We can know <u>all</u> of that exactly as well as we can know in advance <u>all</u> of the forces that will come into play, atom-to-atom and within each atom"s structure, at the instant when the hammer hits the rock.

Probability is by definition unknown. Probability is always all about, and only about, what we do not know. It always has a value between zero and one; never one, never zero. Zeros and ones are facts, measurements, observations; they are not probabilities. Probability is only about what we do not know. It is easy enough to suppose that there is no realistic prospect of any normal human knowing exactly the path of every atom in the hammer and the rock, into, through and beyond the instant of impact. For such reasons, Science can and generally does agree to let Chance carry the load, "averaging over" all the causes that we will in fact never know …

The building of a new human appears to be a process of rather greater complexity than the smashing of a rock with a hammer. Every step, every "event" in that process has molecular and atomic elements of perceived randomness. These events have probabilistic unknown causes, and their outcomes are components of causal chains leading to later events. The final outcome is a multidimensional vector sum of all those intermediate outcomes, and the choice of paths through all of that which can lead to a "normal" result is in fact an exquisitely narrow fraction of all the possibilities.

> … without perception, there can be no reality.
> There is neither time nor motion without life.
> Reality is not "there" with definite properties waiting

to be discovered, but actually comes into being
depending upon the actions of the observer.

Robert Lanza & Bob Berman, Biocentrism

We are what we think.
All that we are arises with our thoughts.
With our thoughts we make the world.

Buddha

It is true that every oak tree was once an acorn. It is equally true that acorns
are not oak trees. It is also true that very nearly all acorns never will be oak trees.
It is really okay for me, or a pig, or a squirrel, or a deer to take nourishment from
an acorn without worrying about how the future of the universe will be changed
by the absence of the oak tree that acorn might ever have, or might never have,
but now certainly will not, become.

Human Destinies

Should you choose to believe otherwise, and should you wish that your belief
might make sense, then you must find a way to incorporate all the many different
kinds and outcomes of "destiny" or "fate." Whatever you imagine that to mean,
you must understand and specifically include the fact that the great majority of
human "lives" end before they might usually and reasonably be said to begin at
birth.

The great majority of so-called potential new human beings are "destined,"
"fated," in fact and indeed "intended" never to see the light of day. (If
one believes that the existence of any new human arises from any conscious
intention — especially as imagined in unique individual "special creation" —
then this conclusion is inescapable.)

More than three-quarters of even the most biologically privileged of all
human conceptions end before live birth. (Healthy parents, of proven fertility
and of reasonable reproductive age, under research-level medical attention ...
conditions better than those of the great majority of pregnancies in the world ...
can fairly be called "privileged.") To insist that each of those doomed conceptions
receives and carries away with it a uniquely and personally human sacred and
immortal soul is to require of the Creator, at best, an outrageous waste of the

great majority of all performances of His Very Best Most Special Trick. Every plausible idea of God I have ever heard of, by whatever holy name, is smarter than that.

So much has falsely been made of that idea of unique genetic individuality, in arguments concerning personhood and sanctity among unborn "human" "lives." This is the basis for the claim that a new human person begins when a unique new diploid genome is formed at syngamy, that human life begins at conception defined as syngamy. It is certainly true that every one of such entities is absolutely uniquely one-of-a-kind, just like every one of us who is already here, and definitely also in that regard just like every snowflake in every flurry and every avalanche, and like every grain of sand on all of the beaches of the world.

Uniqueness in and of itself, human or otherwise, has no real utility or value of any sort, human or otherwise

Every one of those entities of aggregated cells, just like every one of us representing the fraction of past occurrences of such entities that have succeeded, and just like every snowflake or grain of sand, is of no significance whatever in the solitary splendor of its uniqueness. I have not checked it out in detail, but it has been said that the languages of the indigenous peoples of the Arctic have many words for (presumably, many subtly different kinds of) snow. Whether by one of those words, or by any word in any other language, "snow" has meanings that no snowflake all by itself can ever have.

Every one of those unique entities gains any meaning it will ever have entirely from its participation in the crowd of all the others of its kind. Every one of them is for all practical purposes interchangeable among all of its kind, however all the while absolutely unique, not quite exactly like any other of its kind. Even if it could be built to exactly the same structure, it would yet be built from a uniquely different collection of atoms.

Every existing thing, right down to the molecular or atomic level, is <u>that</u> thing, <u>that</u> one and <u>not any other one</u> of <u>that kind</u> of thing. Uniqueness is overwhelmingly abundant in any existence as complex as this universe. It is therefore extremely, ridiculously cheap.

My moral upbringing included a goodly number of different lessons to the effect that "there is no irreplaceable man." This was clearly intended to support my prospects for building into my personality a virtuous level of humility. If my uniqueness in and of itself is of any value, it must have something intimate to do with the definition of my Self and would therefore be of value uniquely to me.

If the individual conceptus, at any stage of its development, is not one of those few "destined by chance" to become a fully realized human individual, then it is debris, planktonic detritus, and it will surely find no greater possible ultimate value than as food for the microorganisms that populate the reproductive tract. Such is the future of a very solid majority of the <u>products of "conception"</u>.

> ... to be an Error and be Cast out is a part of God"s Design.
>
> *William Blake*

The possibility that any such cellular material might find some valid utility, either therapeutic or in basic scientific efforts targeted on potential therapeutic outcomes, gives it far greater real value than any it might have in and of itself in its present circumstances.

If this thing were in fact Created and intended by an omniscient, omnipotent Creator to become a real, live, complete in-the-image-of-God, human person, then ... can you really imagine that you or I, puny mortals, should find ourselves able even to slow its progress toward that Purpose?

> ^ That question is intimately related to the core questions of theodicy, which has been considered by some a whole separate field of study — a "scientific counterpart of theology." Professor Ehrman's recent analysis is worth your time to read, if any question in this neighborhood interests or concerns you.

Other definitions of "Conception"

Among the numerous alternative definitions for conception, it has been considered reasonable by some to define conception (becoming pregnant) as taking place at <u>implantation</u>. The blastocyst "hatches" from the zona pellucida and the polar trophoblast interacts with the endometrium (the lining of the uterus) to bury itself in the wall of the uterus and initiate the development of the placenta. [This corresponds with the onset of production of gonadotropin from the chorion — which is the target signal of most of the take-home pregnancy tests.]

Another horizon that has been widely used as a definition for conception is to say that a woman is pregnant (has conceived, has experienced a "conception") only when and after her body/mind tells itself, recognizes, that she is pregnant. Because of hormonal changes, some women seem to recognize some of their pregnancies by way of certain ways of feeling. The most common and best understood signal is the absence of the second consecutive menstrual period. As soon after that signal as an appointment can be arranged, the reproductive health care physician can usually (no, not always, actually) make a definitive diagnosis that a pregnancy is in progress. By the time when maternal, and then clinical, recognition of pregnancy defines its "real" onset for obstetrical purposes, about two-thirds of all such processes that were initiated at the same time have failed.

Still, in the USA for at least the most recent several decades, 10% to 20% of all of those <u>recognized</u> pregnancies will yet fail by spontaneous abortion, commonly known as miscarriage. Those are pregnancies which have gone far enough along in development that the mothers have become aware of their presence, and then are lost.

Take-home biochemical pregnancy tests in technological societies notwithstanding, most instances of maternal, and then clinical, recognition of pregnancy occur in that old-fashioned way, sometime after the mother has missed her second menstrual period in a row, no sooner than about six weeks after fertilization or about eight weeks since the last normal menses.

In that dark quiet time before anyone knows for certain it is there, a viable conceptus progresses from the single zygote cell to about the size of a "baby" lima bean or an adult fingernail, with all organs and organ systems more or less finally situated and formed — including its own fingernails.

By the time the existence of a pregnancy is recognized by the mother and then her physician, the hard work of developmental biology has been done. There is little left for it to do besides grow, and still nothing about its future is inevitable. Still, each day survived brings an increased probability of surviving for one more next day.

It seems easy enough to understand that this period before maternal recognition, which is technically called <u>embryogenesis</u> (= the building of the embryo), is where very nearly everything that must be done right gets done if it ever will get done at all, and therefore also where very nearly everything that can go wrong will go wrong.

Not so current today, now that ultrasound has become so powerful and widespread in its use, but it was once and for a number of centuries considered that "quickening" (the mother's first awareness of fetal movement within the uterus) was the first real sign of life, the first credible evidence of pregnancy, representing the threshold beyond which one could know that "conception" had occurred. The fact that quickening tends to occur close in time to "showing" (when the swelling of the pregnant belly becomes more or less readily visible) reinforced such considerations.

Mole, or molar, pregnancies, for a contrary example, include only a shapeless mass of tumor-like tissue, but they give every other preliminary indication of being ongoing pregnancies — hormones change, periods stop, bellies swell. For want of nerves and muscles and bones to move limbs with which to make their presence felt against the uterine muscle, such false conceptions never "quicken".

These are the available biological answers, and I am poignantly aware that none of them will satisfy those who ask the question "when does human life begin?" and who believe that everyone's immediate answer should be that "human life begins at conception", and who believe that conception in that context should mean "at the instant of sperm-egg contact."

No biological definition of conception will suffice as an answer, because in the context that follows that question around, this is not a biological question. It is rather about something considered sacred and the circumstances of its changing from something-less-than-or-other-than-sacred to become something sacred.

Not all religious groups have pronounced on this issue, and there is considerable variety among the pronouncements of those that have. Not all religious traditions involve any concept of an individual soul or self, and those that do so have various definitions and assign it various properties, not one of which readily corresponds to any biological reality. The observable facts of biology cannot be cut or twisted or hammered or pressed into any shape that resembles an answer to this question in this context. It should be obvious that the people with the problem here are those who are certain that their way is the only right way.

Ideas like these are behind my consistent teaching use of "becoming human" where many would say "<u>being</u> human." Development continues throughout life. Change never stops, even at and beyond what we usually recognize as the end of life. Not one of us is today the same person as yesterday, nor are you now entirely the same as the person who will wear what you think of as your face tomorrow. A very large number of the cells in each of our bodies will die today and be replaced by others, and the same is true of the thoughts and emotions of which my Self and my Soul and my Person are composed.

There is no place in ongoing unique individual human development to draw the kind of line the desire for which is represented by the question we started with here. Each unique individual human life is a constant beginning, with a constant and forever incomplete, imperfect and unknowable unfolding.

bearing no relationship to, and never meant to be part of,
any ordinary physical Earthly reality.

"at" is the only word in that assertion
with a credible definition.

Believe those who claim to be seeking the truth;
doubt those who claim to have found it.
Andre Gide

Insight, I believe, refers to the depth of understanding that comes
by setting experiences, yours and mine, familiar and exotic, new and old,
side by side, and learning by letting them speak to one another.
Mary Catherine Bateson

The beginning of wisdom is found in doubting;
by doubting we come to the question,
and by seeking we may come upon the truth.
Pierre Abelard

Any and every intermediate structure in the process of building and becoming a new human can be called a miracle, with the true and simple and exact meaning that we cannot explain it within our present understanding of the way things work.

Chapter Eleven

Timing Is Everything

Somewhere on this globe, every ten seconds,
there is a woman giving birth to a child.
She must be found and stopped!

Sam Levenson

Time is nature's way to keep everything from happening at once.

John A. Wheeler

This grand show is eternal. It is always sunrise somewhere;
the dew is never all dried at once; a shower is forever falling;
vapor is ever rising. Eternal sunrise, eternal sunset, eternal
dawn and glowing, on sea and continents and islands,
each in its turn, as the round earth rolls.

John Muir

Like all physical realities in this universe-as-we-know-it, every scrap of observable physical reality about human reproduction occupies space and time. Differences made by the passage of time in biological processes of human reproduction have been mentioned here before, in several different contexts:

- reflex ovulation;
- the decay of the oöcyte over time after ovulation;
- the timing of fertilization relative to the ongoing development of the oöcyte;
- the advantages for those mammals that enjoy the protections of estrus against errors in the timing of fertilization, and implications of the fact that we humans do not;
- the effects of age and parity on a woman's risk of twinning, and the mechanisms thereof;
- the difference between male and female embryos in the timing of events of embryogenesis;

- how women of different racial groups progress through the events of their reproductive careers at different rates; and
- the timing of twinning events relative to the differentiation of the trophoblast cells precursory to the development of the chorion ...

... to name those I recall at the moment — this list is certainly not complete.

I want to discuss with you what those various phenomena have in common, how they may tie together, and then back off to a reasonable distance for a broader perspective.

It should be clear from some of my previous statements that the preparations of the egg and sperm cells for fertilization and embryogenesis involve complex systems of developmental processes none of which can reasonably be considered instantaneous, but rather take place over a period of time.

It is from our efforts to understand the biology of twinning and chimerism, and of all of the variations and departures from usual developmental pathways that are associated with twinning and chimerism, that we get our best opportunity to dissect those processes. It appears to me that at least very nearly all of those variations have something to do with timing, independent or relative, of one or more of those sub-processes.

The egg cell is the largest and most complex of all human cells (I think I told you that already ... It is still true, and still important). Its preparation for fertilization and embryogenesis is a correspondingly complex developmental process. A great many pieces and parts must be made (not only in and by the egg cell itself, but also in some of the cells of the follicle and the nurse cells of the cumulus oöphorus) and put in the right places relative to one another and relative to the whole structure of the egg cell. Any change in the overall speed of that process carries a risk of changing the relative timing of the sub-processes, and thereby changing their respective likelihoods of success. This is the most obvious candidate as explanation for the many anomalies observed among the products of every assisted reproductive technology that involves artificially induced ovulations, for example.

We know that the egg cell is in its optimum condition for fertilization and successful embryogenesis for a very short time after its release from the follicle by ovulation, and that it deteriorates quite rapidly after reaching that optimal state. Within 12 hours after ovulation, less than 50% of mammalian egg cells, including those of humans, still have any reasonable probability of successful development should they be fertilized. By 24 hours after ovulation, they are almost all useless, dead or dying and falling apart — busy dismantling and digesting themselves.

We know from experimental manipulations of the timing of insemination in laboratory animals that increasing the time between ovulation and insemination by delaying insemination results in dramatic increases in dizygotic twinning and in chromosomal anomalies. We know from the experiments-in-nature in the work of Susan Harlap and direct observations by Ursula Eichenlaub–Ritter that all of those same changes occur as well in human egg cells. Because spontaneous embryonic chimerism is a common variation on the process of dizygotic twinning, chimerism clearly may be expected to increase as well in all of those same situations. Chimerism (notably in the form of, and discoverable because of, monochorionic boy–girl twins) has recently been reported to occur at much greater rates among the products of artificially induced ovulations.

The timing of fertilization relative to the ongoing developmental maturation of the oöcyte is of critical importance.

The Human Is a Seasonal Breeder

We are not as sharply seasonal as cats, dogs, rats, rabbits and so on, but in USA data our annual high and low conception rates differ from our monthly average conception rate by over 20% of that average. This variation is very clearly real but also clearly not absolute or universal. It may be that some women are not subject to annual seasonal rhythmic changes in their ability to conceive, so that we see the rhythm only among and because of those who are. It is also possible that indeed all women are subject to such rhythmic variations, but not all of them are in phase, or in synchrony. They are not all at the same place in their annual rhythm at the same time of the year. Perhaps some women do have individual annual fertility peaks, but have them at different times of the year from that of the majority. Except for the fact that a sizable fraction of all American women of childbearing age are on roughly the same annual schedule, I would not have been able to see the rhythms that I have reported.

The "basic" annual rhythm is consistent over all sizable subsets of my database. Ask any delivery-room nurse in any industrialized society north of the tropics. At their places of business, August and September are different from the other months. By virtue of having such a large populational database (over 18 million birth records from 12 American states, from the late 1970s to the mid-1990s), I have been able to report some of these variations that just would not be recognizable or statistically credible if recognized in smaller samples.

I would be willing to bet that all of you have already noticed that summer days have longer periods of daylight as well as having higher temperatures on the average, and that winter days get smaller doses of (also weaker) sunlight as well as being cooler on average. In most of the world throughout most of the history of collecting and recording information about conceptions and births in populations, humans have shown a peak conception rate in the longest days of the year, at the beginning of summer, with the resulting peak in birth rate coming in spring.

Beginning in the US roughly in the late-1940s to mid-1950s, the pattern gradually changed in all of the major industrialized societies. First, there came to be a small peak in the bottom of the winter trough on the conception rate curve. Over the next few decades, that second peak grew at the expense of the main peak, until the phase of the rhythm had more or less completely reversed — the peak and trough had traded places. In much of the data I collected for those analyses there is still a bump in the conception rate curve in what otherwise looks as if it should be the bottom of the trough that now comes in summer.

In North America and most other major industrialized societies north of the tropics, conception rates now peak in the shortest days of the year — in December and January, and delivery rooms are most active in August and September.

A woman in Minnesota sees it slightly differently from the way a woman in Florida or the southern part of Texas sees it because of variation in the variation of the length and structure of daylight with latitude. (Yes, that is what I meant to say; the structure of the annual variation varies, with latitude, with distance from the Equator.) Every subpopulation is at their own average conception rate at the spring and fall equinoxes, when everyone everywhere has the same 12 hours of daylight.

The Human Ovary Knows What Day of the Year It Is

There are clear and statistically significant annual seasonal rhythms in sperm count and in the frequency of intercourse. These rhythms do not correlate well enough to be considered as possibly driving the rhythm in conception rate. Frequency of pregnancies initiated by artificial insemination with controlled-quality donor sperm is equally seasonal. In short, the ovary clearly appears to be in charge. We do not, however, yet know where she gets her guiding information or how she enforces her decisions. Some conclusions have been jumped to, and some corresponding more-or-less-plausible opinions have been reported, but clearly and certainly no one truly understands the mechanism by which this seasonal rhythm is produced for humans — or, indeed, for any other organism. We are even farther from understanding the mechanism behind the observed phase-reversal that has occurred over the last several decades in industrialized societies.

The best information we have about the determination of the human annual seasonal periodicity is from one study done in Chile. Chile stretches about two-thirds of the length of the continent of South America, covering about half of the whole possible range of variation in latitude — half the distance between the equator and the pole. Altitude varying from sea level to the ridge of the Andes along the whole North–South length of the country offers considerable variation in temperature on any given day of the year.

Most of the variation in conception rate in that Chilean study was shown to correlate with some combination of the effects of day length and temperature. This is in good general agreement with results from most such studies in non-human animals. We have used that understanding, for example, along with artificial lighting, to exert artificial control over seasonal time of breeding in sheep, so that we might enjoy fresh young "spring" lamb throughout the year.

We can readily agree that annual seasonal variation in animal reproductive function and/or behavior — including that of the human — is strongly correlated with, and arguably somehow regulated by, some function of, or correlate of, the annual seasonal variation in day length and temperature. However, until recently, no one had offered any significant insight into the mechanisms by which such changes in the geophysical environment are used by which organs of the body as signals that can be transduced into biochemical signals which can reliably cause corresponding changes in reproductive physiology or behavior. There is no reason to suppose that all seasonally rhythmic variations will be driven by the same geophysical variables. With respect to some annually seasonal variations in brain function asymmetry development, Marzullo and Fraser and I have recently gathered evidence to the effect that the "real" trigger that correlates with day length is a system of variations in the intensity and wavelength composition of the sunlight. There are two receptors involved, with two different wavelength specificities, both involving modulation of the concentrations of reactive oxygen species in the blood.

It is good to remember here again that correlation is not cause. There are a number of geophysical processes that vary with an annual seasonal rhythm and that must sooner or later be considered and examined as candidates for triggering various such rhythms. These are, for the most part, more subtle than day length and temperature, but some of them have been shown to correlate strongly with annual seasonal rhythms in the reproductive physiology or behavior of other organisms. For example, organisms moved from their Atlantic Ocean intertidal homes to the American Midwest have been reported to maintain behavioral rhythms normally entrained to the tides. They are no longer in rhythm with the tidal changes where they used to live, but with gravitational tidal forcing as if the tides were rolling right on across the American great prairies. They are not in time with the movement of the water, but with the Earth–Moon–Sun gravitational forces that cause the waters to move where there are waters to move.

There are many biological processes, in both plants and animals, variations in which are strongly correlated with annually rhythmic variations in light and temperature. In those situations, it is typical, probably even universal, for biological responses to light to be responses to specific wavelengths, in other words specific colors, of light. A little thought suggests that this makes all the sense in the world because nothing works in biology except by way of chemistry and physics — there simply must be, somewhere in the process, one or more molecules that can be changed by the arrival and impact of a photon. In every such situation we understand at any depth, that interaction depends on the wavelength (color) of the light, which is functionally related to the energy of the photon, and on the resonance properties of a specific chemical bond at issue.

The phase-reversal change in the annual seasonality of conception rates began in the US and has since then spread throughout the major industrialized societies. Two things changed in the US about the time this change in seasonal reproductive rhythms began which might be imagined to have had something to do with the change: (i) We began suppressing the sale of raw milk in favor of pasteurization and began the regular addition of Vitamin D to the milk supply. The Vitamin-D-deficiency bone disease rickets was then a rampant public health problem in cities where tall buildings and child labor kept children from collecting enough sunlight on their skin; (ii) Daylight Saving Time, by fits and starts, became widespread and consistent. Both of these took several decades to go from the original idea to more or less nationwide usage, both taking place over about the same time period during which the phase-reversal of the annual conception rate curve was occurring.

Natural biosynthesis of endogenous Vitamin D requires ultraviolet sunlight to strike the skin to catalyze the necessary chemical reactions. Vitamin D is indeed a vitamin, as a cofactor in several variables of calcium metabolism. It is also a corticosteroid hormone that arises from a branch of the same biochemistry that generates the corticosteroid androgens and estrogens, and, like those molecules, it has a variety of effects on reproductive functions in a variety of mammals. All of the several chemical versions of Vitamin D are chemically closely related to the androgens and estrogens, and like those molecules are also synthesized from cholesterol.

In many industrialized nations outside of North America, raw milk is still readily available and Vitamin D fortification is not universal. A great many people are chronically deficient in Vitamin D worldwide but apparently worse in Northern Europe, where natural exposure to sunlight is least for half of the year (and least for UV even in high summer) and perversely where fortification has been most vigorously resisted.

On the other hand, Daylight Saving Time is now pervasive in industrialized societies. The changes between standard and daylight time, in both directions, accelerate the perceived change of illumination early in the day relative to the ongoing natural change in day length. The rate-of-change of the seasonal variation in day length abruptly changes on those two occasions.

I am unaware of the existence of datasets suitable for analyses to determine which if either of these factors correlates well enough with the change in conception rate rhythms to be considered for further research as a plausible cause of the change.

You may or may not remember that the reason for the variation in day length and temperature over the Earth's seasons is that the Earth rotates around an axis which is not parallel to the axis of its orbit around the Sun, not perpendicular to the plane of that orbit. The rotation of the Earth is tilted. At one extreme position in our orbit around the Sun, at the northern hemisphere's winter solstice, about December 21st, the North Pole is tilted 23.5 degrees away from the Sun and the South Pole is tilted 23.5 degrees closer. At the six-month, 180-degree opposite point in our solar orbit, at the time of our northern hemisphere summer solstice and the southern hemisphere's winter solstice, the differences are reversed, because the tilt does not change except for a wobble that is only a very small fraction of the tilt.

I know it is easy to misunderstand this. I misunderstood it for years, thinking the wobble (technically, the "precession" of the spinning Earth's axis of rotation, like the wobble of a spinning top) was the reason for the seasons. There are some changes in the planet's orientation to the sun due to the precession but they constitute a very small fraction of the total variation, which is due primarily to the Earth's tilted axis of rotation. When the North Pole is closer to the sun, the southern hemisphere has shorter days and cooler temperatures, and vice versa. The summer solstices correspond to the year's longest daylight periods and shortest nights, and the winter solstices bring the shortest days and longest nights of the year.

Many of you probably already know that the annual seasonal differences in the day length cycle are greater in Alaska or Finland or Tierra del Fuego, places far from the Equator. It is greatest at the poles, and least — actually negligible — at the Equator. Few people realize right offhand — I never thought about it until I found myself studying seasonal rhythms in human reproduction — how much farther north most of Europe is than most of the USA. Rome and New York City are at about the same latitude. Most of the USA is south of New York, and most of Europe is north of Rome.

The closer one lives to the Equator, the smaller is the annual seasonal change of day length and average temperature over the course of the year. So, in most of Europe, the annual differences are greater than in most of the USA. In the USA, the differences are somewhat greater in Minnesota than they are in south Florida or south Texas.

In northern European data, twinning and malformations and chromosome anomalies are more frequent in conceptions that take place on the shoulders of the peak (a few weeks before, or after, the peak) of the annual-variation-in-conception-rate curve. Those special variations on the theme of human reproduction are also more frequent in European families who have used "the rhythm method" to manage the size and age distribution of their families. Both are obviously questions of timing.

The oddity associated with rhythm method family planning probably has to do with non-optimal time-in-cycle for insemination. This would be especially the case for older women, who ovulate about the 10th day of the cycle much more frequently than do the younger women. 10th-day-of-cycle ovulations "should be in the safe period" when they are thinking in terms of cycles with fertility peaks closer to the 14th day, which is normal for younger women. How non-optimal annual timing generates virtually the same list of problems as those associated with non-optimal menstrual timing is an obviously worthwhile question to pursue.

With particular reference to the release of the egg cell at ovulation, timing is critical around and on either side of ovulation. The oöcyte to be ovulated in the current cycle is involved in a very rapid and complex system of developmental changes in preparation for ovulation, fertilization, syngamy and embryogenesis. There are plans at issue for the organization of both time (relative timing of a great many structural and biochemical processes) and space within the oöcyte, and embryogenesis.

Chapter Twelve

Twins and Handedness and [Illusions of] Mirror-Imaging

> Our life is an apprenticeship to the truth that
> around every circle another can be drawn;
> that there is no end in nature, but every end is a beginning,
> and under every deep a lower deep opens.
>
> *Ralph Waldo Emerson*

I have briefly mentioned "mirror-imaging" in earlier pages here, but have not explained its involvement with the myths and histories of twin biology. There is a true and fascinating substance to the idea of "mirror-imaging" in twins and what it means for the particulars of human embryology, and those are subjects that should not be left out. It is, however, another one of those things that simply is not what the simplest version of the story may have left you believing. Not all true things are simple. Not all simple things are true.

"Mirror-imaging," "mirror-twinning," or the state of being "mirror twins," is widely considered a fascinating and meaningful distinction among twins, and is always high on the list of topics of conversation whenever twins or parents of twins gather, whether or not they know or suppose there may be a scientist within earshot. Mirror-imaging between co-twins is real, and it almost certainly means something important about embryogenesis, but — after looking at it long and hard, as far as I can tell — any such relationship is entirely accidental and meaningless in this context, the context in which it is most often discussed.

No one really seems to know anything much about what it really means, but I am calmly and entirely certain that motor hand preference (handedness) has nothing specific to do with monozygosity, and that its relationship with the timing of twinning events during embryogenesis is of a very different kind than has been generally imagined.

There is no evidence and no other rational basis for the idea that the human monozygotic twinning event requires the tearing of any product of conception and embryogenesis into two pieces that might, from that "new beginning," become two embryos and eventually perhaps two people derived from the cells of one single mass of embryonic cells. There is no evidence or other rational basis for the notion that the determination of handedness in twins has anything to do with the timing of the twinning event, or even anything to do with whether the twinning event in question is monozygotic or dizygotic.

There is no evidence that he himself had any intention of causing all of this nonsense, but the idea of human twin embryogenic mirror-imaging has long been attached to the memory of the Nobel-laureate Austrian developmental biologist Hans Spemann (1869–1941) and some work he did with Hilde Mangold's very important help, mostly in the 1920s, published mostly in German. The idea that discordant handedness in twins was related to their work came later.

Spemann won the 1935 Nobel Prize in Physiology or Medicine for his discovery of embryonic induction, the phenomenon whereby certain cells in the embryo induce specific developmental differentiations in other cells or networks of cells. One subsystem of the work that brought him that honor, and the one that installed him permanently in the lore of twin biology, involved some astonishing consequences of tying newt embryos in half with a child's hair.

For the sake of perspective for those however many or few of you who do not already know it: all newts are salamanders, but not all salamanders are newts. Newts and other salamanders are amphibians, with smooth, moist skin through which they do a substantial part of their oxygen–carbon dioxide gas exchange (I just don't feel right calling it "breathing"). They hatch as tadpoles from clear jelly-coated eggs laid in water, like frogs and other amphibians. (For that matter, "lays jelly-coated eggs in water" pretty much defines amphibians.) Unlike frogs and toads they do not lose their tails when they grow legs.

Some folks, mostly living a few hundred miles to the west of the Atlantic Coastal Plain where I live and work, call salamanders "lizzards" (definitely with two *z*s), catch them by the dozens and sell them for fish bait. Salamanders are not to be confused with actual lizards (one *z*). Lizards are reptiles, which are more closely related to snakes than to frogs, and which have dry, scaly skin, that does not contribute to their "breathing."

Largemouth bass will eagerly eat a salamander, even when preoccupied on the nest and in general not taking any nourishment, because the bass seems to know or believe that the salamander loves to eat fish eggs and will keep coming back to get some if not distracted first by whatever alarm he might feel upon finding himself being eaten.

Newts have also been excellent bait for embryologists because newt embryos develop a very typical and representative full vertebrate body plan [all their organs have about the same shapes and are situated in about the same relative positions in the body as yours and mine] from eggs about an eighth of an inch in diameter that develop in plain water. These features make them excellent subjects for developmental biology because their embryonic development can be studied closely in a dish of ordinary water on a lab bench at room temperature. There is no need for costly incubation equipment or for finicky and costly growth media. Drugs that can change their developmental progress for experimental purposes can in many cases be administered by simply putting the drug in the water.

Spemann's work was a Nobel-worthy watershed in developmental biology. It was a major revelation that is still inspiring new research efforts at many levels of cell biology. In brief: Many of the changes embryonic cells make in structure, function or position in the course of development depend on signals from other cells. Much of embryonic development depends on events and processes in which some cells induce changes in other cells.

The embryonic structure known as "Spemann's Organizer" is a patch of cells on the future "upper lip" of amphibian embryos ("dorsal lip of the blastopore") that move from where they first acquire the molecular properties of being Organizer cells to induce widespread developmental changes throughout the embryo. Our (human) counterpart of the Organizer begins in the node of the primitive streak and eventually becomes in large part the cells of the neural crest, which figure prominently in the story of human embryogenesis. We now know a little bit about which gene product molecules need to be produced in those cells to give them the properties they need to do their jobs as components of the Organizer. Spemann opened up the process of discovering the behaviors and properties of these cells beginning with some experiments in which he tied newt embryos "in half" with a child's hair.

When the embryos were pinched apart all the way through, always at least one of the "half" pieces died. WHEN all the cells of the embryo were left in a contiguous mass, with the "halves" being left connected by only a small bridge of cells, AND the tie was along the head–tail midline, more or less around the circumference of the sagittal plane, AND each half part got a piece of the "Organizer," THEN both "halves" could sometimes survive and continue development to become two embryos that would separate later in development to yield two newt tadpoles. These experiments were central to Spemann's developing understanding of the pivotal role of the cells of the Organizer.

The production of two embryos from a single embryo that grew from a single zygote clearly satisfies the definition of monozygotic twins. The tadpoles they became were sometimes "mirror-imaged" in the locations of their internal organs: About half the time, the twin that developed from the "right-half" embryo had situs inversus — its internal organs were on the wrong sides of the body and turned the wrong way! — heart looping right instead of left, stomach and liver right and left instead of left and right respectively, and so on.

This seems to be the origin of the whole idea of "mirror-image twins." There is no reason outside of a lively imagination, no physical evidence, to suppose that any of the causes of natural human twinning has anything to do with any process at all like mechanically "splitting" an "embryo" into two parts. The cells of one "embryo" eventually must separate to become two separate people, but there is every reason to believe that is a consequence of the cellular event of twinning and not its cause. The separation of human twins from a single embryo happens after the cellular event that decides twinning, not before.

For one thing, the natural human twinning event happens while the conceptus is still inside the zona pellucida. It is tight in there. There is no spare room in there into which, or within which, the cells of the embryo might move, to enact or enforce or acknowledge by any behavior any kind of separation that might be generated by any mechanical means. We also know, from tens of thousands of experimental manipulations of pieces of mouse embryos, when part or all of one embryogenic inner cell mass is placed inside a blastula with part or all of another inner cell mass, twinning is never the result. If they live, the pieces always fuse to form a single embryo. Conclusion: Mechanical separation of groups of cells within an embryo (as is usually visualized from the use of the word "split"), if indeed it ever happens, has nothing to do with the causes of spontaneous human twinning.

Spemann's mechanical production of monozygotic twinning in newt embryos, and the situs inversus it seemed to cause, had an immediate appeal to some American researchers who were trying to understand the developmental genetics of twinning and handedness, and the apparent relationships between the two phenomena. Here was a glaringly obvious smack-yourself-in-the-forehead-with-the-palm-of-your-hand kind of revelation. Here was a single, simple and apparently plausible mechanism for the origin of monozygotic twinning ["splitting the embryo"], and simultaneously for an apparent deficiency of matching for handedness in "identical" twins. Many twin pairs who appeared to be "identical," but who differed in handedness, could now be explained by "mirror-imaging" imagined to result from disturbances of asymmetric embryonic development by the "splitting."

HH Newman and some other researchers, who were contemporaries of Spemann, took the newt results to represent a mechanism of twinning that could take the place of their then-current version of our enduring ignorance of the exact cellular mechanisms of human twinning.

This work provided something people could visualize, to take the place of prevailing complete superstitious ignorance about the mechanisms of twinning. Beginning there, it came to be, from that time forward, that "everybody knows" that monozygotic twins are produced by the "splitting" of an embryo. The corresponding image produced in the collective mind of the twin-conscious masses has been some sort of mechanical splitting that must somehow resemble tearing the embryo into two parts with a fine hair or some other such force or implement — the way Fraulein Mangold did with Herr Doktor Spemann's newts, of course.

Now that they could picture it, should it not have been obvious all along that monozygotic twinning must occur by way of "splitting" an embryo? How else could a single embryo become two people? Under that assumption, it is almost exactly as obvious that the excess of nonrighthandedness among twins must be due to the effects of "mirror-imaging" disturbances of embryonic symmetry determinations, caused by the "splitting." The "splitting" must be disturbing asymmetric decisions already made or in the process of being made by the cells of the embryo before, or at the instant of, the imagined "split."

As the story goes, when this imaginary split happens late enough in development, it might result in disturbances of ongoing development that could cause mirror-image reversal of normal developmental asymmetries. If the establishment of the body plan, with its various a/symmetries, should be significantly underway before the twinning "split" generates a doubling of the body plan, then cells that were already prepared to become the left side of Bill might have to change their plans and use their remaining resources to build instead the right side of Bob. Imagination, the same imagination that brought us those concepts and without which in general we would get nowhere, is the only constraint on thinking about what the consequences might be for developmental asymmetries more generally.

All of this was happening a decade or two before blood-typing was readily available for any non-clinical scientific use. Blood typing would become the first reasonably sound and objective means of determining whether a pair of twins is monozygotic or dizygotic. Absent any sound objective source of zygosity diagnosis, for a generation or so, "mirror-imaged" twin pairs (with one righthander and one lefthander) were declared to be "late-splitting identical" twins — not only certainly monozygotic, but specifically "late-splitting." They were discordant in handedness because the "splitting" had redirected embryogenesis after some asymmetry decisions had already been made. It was certainly no wonder that a great many additional events and processes in embryogenesis might be rearranged!

If one member of a pair of "mirror" twins had red hair and one was blond, and/or one had blue eyes and one had brown ones, well … this was just further evidence of the way that "late splitting" could disturb the later events in ongoing development. Some reports suggested that even boy–girl pairs who were "mirror-imaged" in their handedness might have been counted as monozygotic in some studies, such was the disruptive nature and power imagined for the "splitting" of the monozygotic twin embryo.

So … if every twin pair with only one lefthander must be declared monozygotic … then, well, of course … it came to be reported that the excess lefthandedness in twins was due to the monozygotic twins. Pairs that were both lefthanded missed that net unless they otherwise appeared to be "identical." Counted that way, adding all of the left–right dizygotic pairs to the monozygotic total, there will obviously, necessarily, be scored an excess of lefthandedness among monozygotic twins. This follows that as night follows day, and that is exactly the way I found the situation when I became enthralled with the issue of handedness in twins in the early 1970s.

In all of the records of medicine and science, the entire number of twins (or any other people) ever found to have fully inverted positions of the internal organs, like Spemann's twinned salamanders sometimes had, is small. I have seen frequency estimates for complete situs inversus as high as 1/8000, but most estimates are more like 1/25,000, and not enough evidence is supplied to justify believing even those. Partial inversion, as in dextrocardia (anomalous placement of the heart rightward of its normal position left of midline, caused by anomalous rotation of the embryonic heart primordium), seems to be rather more frequent than complete situs inversus, and, taken together with other associated anomalies, may explain the discrepancy in those numbers.

"Isolated" dextrocardia (when the asymmetry of the heart is out of line with the asymmetries of the other internal organs) can cause severe cardiac anomalies, such as double outlet right ventricle (both major arteries, the aorta and the pulmonary artery, are attached to the right ventricle), single ventricle, stenosis (severe narrowing) of the pulmonary artery to the lungs, and large defects in the wall between the left side and right side chambers of the heart (atrial and ventricular septal defects), often in complex combinations requiring complicated corrective surgeries that often do not provide fully normal function.

"Mirror-image dextrocardia" is another name for complete situs inversus, wherein unusual heart rotation is consistent with the reversed asymmetry of all the other internal organs. In the situation of complete situs inversus, there are often, probably even usually, no structural defects of any of the internal organs, and the heart is normal except for appearing to have developed on the opposite side of a mirror. A frequency estimate I find more credible for complete situs inversus is given as one in 130,000, but the basis for that estimation is also not well explained. It clearly does not happen often, or at least does not reach live birth often — perhaps because the incomplete versions can be so disruptive.

Twins are several times as likely as singleton newborns to have a congenital cardiovascular anomaly. In this relationship there is no zygosity difference — monozygotic and dizygotic twins are involved in the same proportions as they are present among the sample of uninvolved control pairs. Among all the twins in that sample with congenital cardiovascular anomalies, fewer than 10% of all the anomalies were of the class of anomalies ("cardiac looping") most clearly and closely related to the sidedness problems in Spemann's newts.

Among that small sub-sample, there was reported to be an excess of monozygotic twins, but there was no explanation of protocol or criteria, and no evidence given for the zygosity diagnoses. It is often the case that an unexplained diagnosis of monozygosity of newborn twins is from monochorionicity alone, or even from same sex alone, especially if one of the twins has one of these malformations that are known to be excessively frequent in twins and assumed to belong more or less exclusively to monozygotic twins. I have discussed in more detail elsewhere in these pages how monochorionicity is not in fact a reliable indicator of monozygosity as it has long been thought, perhaps especially in twins with developmental abnormalities.

There may be an association of this particular group of anomalies (cardiac looping) to monozygotic twinning in particular, but the evidence accumulated to date is inadequate to provide any confidence in drawing that conclusion. Assertions to that same effect have been proven false so many times before, that there really appears to be no anomaly soundly attributable to monozygosity either exclusively or predominantly.

Liveborn twins do visceral asymmetry differently rather more often than singletons do, but even among twins situs inversus is not common among liveborn individuals.

On the other hand, nonrighthandedness is not rare in the general population. It is substantially more frequent among twins than among singletons, and a great many pairs of twins differ in motor hand preference. That includes many pairs who otherwise resemble each other closely in most visible traits. So, drawing from Spemann's results, the logic goes, sometimes the "identical" twinning process — the "splitting" — must occur late enough in embryogenesis to result in "mirror-image" handedness, and thus discordance in handedness came to be considered emblematic of monozygosity and of developmental anomalies due to "splitting," especially "late" splitting.

When I found myself enrapt in these curiosities in the mid-1970s, the literature said that twins were more often lefthanded than singletons, and that this was primarily due to the monozygotics because of their "mirror-imaging" when the embryo from which they arose "split" "late." Every paper that claimed a statistically significant excess of monozygotic twins among lefthanders had included data gathered in the 1920s and 1930s, when there was no reasonably accessible and reliable method for genotyping twins to determine zygosity.

Blood typing as an approach to genotyping and individual genetic identification was just barely becoming understood when most of those papers were written. It was not yet very sophisticated and it was not readily available for use in such clinically frivolous pursuits as the zygosity typing of twins. The ABO blood antigen group was discovered by Landsteiner in 1901, Rh also by Landsteiner but not until 1940. Fifteen to 20 others have gradually since then been added to the roster of blood typing antigen genes (coding for variations in structures of proteins carried on the surface of red blood cells) with sufficient variability in the population to make them useful for making genetic distinctions among people. Effective and reasonably powerful genotyping by "blood typing" did not penetrate medical science beyond its utility in transfusion for a few decades after the work of Spemann and was still considered too expensive to be routinely used for twin zygosity determination, or to be used for research samples large enough to be statistically meaningful, well into the 1970s and beyond.

In most of the papers that were more or less current in the 1970s, zygosity was determined primarily by similarity questionnaires which had been tested against blood typing in smaller samples and found to be better than 90% consistent with results from blood typing. Additional handedness data had been collected and included with the data from the earlier papers.

The newer data considered separately showed the same sort of excess lefthandedness among twins as the earlier papers, but the concentration of the excess among the monozygotics was not there. For a while, however, the data continued to be reported in the aggregate, so that the mixing-in of the old data perpetuated the reporting of the excess nonrighthandedness among twins as being due primarily to the monozygotics.

When I removed from the data in those papers the numbers that had been mixed in from other papers written in the 1920s and 1930s, and recalculated using only the newer data, from all samples with any sort or degree of similarity diagnosis of zygosity, there was no longer any consistent, let alone statistically significant, zygosity difference.

In fact, in most samples collected and studied to date, dizygotic co-twins are, clearly and consistently, considerably more likely to differ in handedness (that is, to "be mirror-image twins") than monozygotic co-twins are. So, if a difference in handedness means an embryonic "splitting" that may have happened fairly late in embryonic development, then we simply must conclude that dizygotic twins are more likely than the monozygotics to have experienced an embryonic "splitting"! Another thing about it: no one, ever, no twin, no singleton, no one, ever, has been soundly documented as having the normal asymmetries of brain anatomy and function completely reversed the way internal organs are reversed in complete situs inversus. Not one. Not ever.

The normal left–right differences in brain structure and function are very basic and important facts of normal development. That provides no reason to suppose that they are never done wrong in development, but it may indicate that errors in those processes are poorly survivable. The deeper levels of asymmetry of brain anatomy might be among the many processes in which, if development is not being done correctly, with all the appropriate signals being sent and received, the whole process stops and what has gotten done so far disconnects, digests and dissolves itself.

I had hopes of shedding some light on these questions about the determinants of brain function asymmetry (as represented by handedness) in twins (and the rest of us). With the help of member clubs in the National Organization of Mothers of Twins Clubs, I collected questionnaires from over 800 twin families [in response to a distribution of 500 questionnaires at their 1975 convention]. I got usable handedness data on 773 three-generation twin families, over 10,000 people, classified as lefthanded, righthanded, or ambidextrous.

Inclusion of the choice of "ambidextrous" represents the fact that motor hand preference — "handedness" — is seldom absolute. For the purposes of the questionnaire, this avoided forcing an ambiguous choice in the responses, and left it possible to determine during the subsequent analyses whether the ambidexters belonged to the right-handed or left-handed fractions, or if they should really be considered separately. As it turned out, in every comparison, the results make much better sense when the ambidexters are pooled with the lefthanders (together constituting the nonrighthanders). Results from the ambidexters do not fit with results from the strictly righthanded individuals.

My handedness-related questions were included as if they were incidental among a few dozen questions on other aspects of family history of twinning, twin pregnancy, twin parenting, and membership in twin families. I diagnosed the zygosity of the twin pairs by embedding a similarity questionnaire previously shown to be over 90% consistent with blood-antigen genotyping in pilot samples.

Results of that questionnaire occasionally disagreed with the parents' direct statement of the zygosity of the children, almost always because "the doctor said they're fraternal because they came in separate sacs," or "the doctor said they are identical because they only had one placenta." In most cases, the responses included three generations: the twins and their siblings, their parents, aunts and uncles, and their grandparents.

The results are straightforward: Twins do in fact have a higher frequency of nonrighthandedness than the general (primarily single-born) population. Twins and their parents and siblings are clearly more often lefthanded or ambidextrous (nonrighthanded) than members of the general population.

The second-degree relatives of the twins (aunts, uncles, grandparents) showed frequencies no different from those in general population samples.

Monozygotic twins and dizygotic twins have the same frequency of nonrighthandedness. Monozygotic twins do not have a higher frequency than dizygotic twins. There is no difference in frequency of lefthandedness between monozygotic and dizygotic twins. No difference. There, I seem to have said it three different ways, which is certainly a number rather smaller than the number of papers in the literature that have reported a false result because of counting dizygotic "mirror-image" opposite-handed pairs as monozygotic.

The monozygotic twinning process is not in any demonstrable way related to what Spemann and Mangold did to their newt embryos. Nor is there any reason to suppose that the monozygotic twinning process differs from the dizygotic process in any of the causes of nonrighthandedness.

We believe that monochorionic monozygotic twinning happens later in embryogenesis than dichorionic monozygotic twinning, because it just makes sense that — since they share the chorion — their twinning event happened within the inner cell mass *after* the mural trophoblast, precursor of the chorion, had already begun to differentiate around them. Lefthandedness is not more frequent among the members of monochorionic twin pairs than it is among dichorionic twins of either zygosity. There is no evidence to suggest that a twinning event happening "later" in embryogenesis is more likely to disrupt ongoing determination of embryogenic asymmetries.

The single-born siblings of the twins in my sample families also showed a substantially higher frequency of lefthandedness than the general population. There was no difference in handedness distribution between the siblings of monozygotic twins and the siblings of the dizygotic twins.

Twins have a slightly higher frequency of lefthandedness than their single-born siblings. The difference is not statistically significant (that means the difference is not large enough to be reasonably certain that it is not a chance error in sampling). So, nothing about *being* twins causes lefthandedness. Twin embryogenesis does not increase the frequency of lefthandedness over that of the single-born siblings of twins.

Nothing about twin pregnancy causes nonrighthandedness. Nothing about who spent more time on top or on whichever side in the uterus. Nothing about any of the events in the process of twinning, twin gestation, causes nonrighthandedness. There is no statistically significant difference between twins and their siblings in frequency of lefthandedness.

There is, however, an interesting interaction at hand. Among the twins in my sample, if only one member of the pair was lefthanded, two-thirds of the time the single lefthander was the second-born twin. The excess of nonrighthandedness in second-born twins is more or less exactly enough to explain all of the difference there is between twins and their single-born siblings. So, why is there more nonrighthandedness in second-born twins? When those twin-family handedness data were collected in 1975–1978, ultrasound examination of pregnancies was still a university medical center research curiosity. More than half of all twin pregnancies were discovered to be twins only after one baby had been delivered and somebody realized that the woman on the table with her knees up in the air was still more or less as pregnant as she was when they helped her climb up there.

Because of that delay in discovery of the multiple nature of some pregnancies, there was in those days quite commonly a delay between the delivery of the first twin and the delivery of the second. Entirely too frequently, the delay sometimes amounted to several minutes. During that delay, especially if placental separation had progressed significantly, the second twin was subject to a greatly reduced flow of maternal and placental blood. The undelivered twin was subjected to reduced supplies of food and oxygen, reduced waste and CO_2 removal — asphyxiation, in a word, more or less exactly.

The right side of the fetal brain normally has a substantially larger blood flow than the left, so the left hemisphere is more vulnerable to the effects of such temporary asphyxiation. Therefore, a small excess of lefthandedness in twins relative to their single-born siblings, although statistically minimal, might be a biologically real and relevant consequence of some degree of damage to left-hemisphere function due to temporary, incomplete asphyxiation of second-born twins.

The excess of nonrighthandedness among second-born twins was thought at that time to be pathological. Since the 1980s, serial ultrasound examination of ongoing pregnancies has become routine. Most twin pregnancies are known to be such well in advance, delivery teams (usually separate teams per fetus) are well prepared, and it seems that most of those pregnancies are delivered by Caesarean dissection. Twins born since the 1980s have not been found to show such a birth-order difference in frequency of nonrighthandedness. Dutch twins and Norwegian twins sampled later appear not to follow the same birth-order patterns.

So where do twins and their first-degree relatives get their consistently observed excess of nonrighthandedness?

The best answer from the data I have collected seems to be that twins and their siblings inherit nonrighthandedness from their parents, with about 70% heritability. That means that about 70% of the variation in handedness in this sample population can be explained by genetic relationships. Twins "catch" lefthandedness just like singletons do, by genetic and/or epigenetic mechanisms that are not yet altogether clear in molecular detail.

In those 773 twin families, each nonrighthanded parent increased the probability of lefthandedness among his or her offspring by about 50%. Children with one lefthanded parent show about half-again the frequency of lefthandedness that we see when both parents are righthanded. When both parents are nonrighthanded, children have a little more than double the background frequency of nonrighthandedness. There is no difference in this pattern of transmission between (the families of) monozygotic twins and (the families of) dizygotic twins.

Mothers of twins are almost twice as often nonrighthanded as their own sisters, the maternal aunts of the twins. Fathers of twins are almost twice as often nonrighthanded as their brothers, the paternal uncles of the twins. I compared the sexes separately because of well-known sex differences in frequency of lefthandedness — in general population samples, more males than females are nonrighthanded.

The mechanism of this apparent inheritance is not a simple Mendelian pattern. There are families in my sample whose handedness distribution fits the expectations of a single recessively transmitted genotype which is expressed about half the time. There are other families in which the distribution fits that of a simple dominant genotype, again with about 50% penetrance. (Penetrance is computed as the fraction of people in a pedigree in direct line of descent — having both ancestors and descendants in the pedigree showing the presence of the gene — who express the phenotype corresponding to the presence of the genotype.) With penetrance at or below 50%, that sort of dominant genotype pattern is virtually impossible to distinguish from a multifactorial model. Analyzing the sample as a whole, the best fit is a multifactorial model with about 70% heritability.

Most twin conceptions — just like single ones — result in no births at all, and for every pair of live twins born, at least 10–12 people who grew from twin embryos are born single. It is actually possible that every lefthander in the world is a twin, developed from a twin embryo. I am not sure yet whether I believe that or not, but the observed relationships make that a reasonable suggestion. The numbers are possible and plausible, there is no evidence against it, and it would tell us enough about developmental mechanisms to make it worth our trouble to track it down. Ever since I first said that, in 1977 at a neuropsychology symposium in San Francisco, I have heard or seen it quoted on fairly frequent occasions. There is at least one problem with that conjecture from a scientific viewpoint; namely, it is worthless until and unless we figure out a way to test its predictions. The core of the problem would be to develop a reliable, straightforward and inexpensive way to identify the sole survivors of twin conceptions.

There may be more here than you ever thought you might want to know or think about handedness. Simplicity is great. It really helps to understand and to explain things. But false simplicity is … well … false! Most of the simple stories about the biology of twinning are false. Not all true things are simple, and not all simple things are true.

Mirror-imaging is real. It does happen, and when we talk about teeth later on, we will see at least a cellular and molecular context, if not perhaps a clear mechanism, for its reality. Handedness is not representative of most instances of the phenomenon and is therefore very much the wrong place to look for it. There are other asymmetries which are sometimes mirrored, maybe even a great many of them that have never been noticed. There may be a wrinkle in the right ear of one twin and the left ear of the other. Twins sometimes "mirror" in tooth emergence sequence (the order in which the various teeth appear as they grow in). For example, while Judy grows the left maxillary (upper) canine ["eye tooth"], Trudy sprouts the same tooth on the right side. There may be a difference in size of the palpebral fissures (the opening between the eyelids). Play with those. Look at your twin/s in a mirror — that reverses the appearance of any asymmetry, such that a difference that you are so accustomed to that you never saw it before can really show up when it is reversed.

I am convinced that mirror asymmetries of these sorts are minor manifestations, accidental side effects, of a very real and much deeper phenomenon (which we will see in better light when we get to the teeth in twins, soon).

Twinning and nonrighthandedness are somehow genetically related. The exact mechanisms involved are not clear. But it is a good example of something very important that most people do not understand about human developmental genetics. Many of the most important genes do not show up every time they are present. We call that *incomplete penetrance*. It is usually a finagle factor; a cover-up for some degree of perfectly ordinary ignorance, but it does allow some situations to be explained by two or more versions of a single gene which could not be explained without the assumptions represented by "incomplete penetrance." There is a great deal remaining to be learned about the relationship between the genotype and the phenotype.

Several web sites that offer to answer questions for their visitors are especially eager to expound upon mirror-imaging in twins. Among those, several include a statement to the effect that mirror-imaging occurs in 25% of all monozygotic pairs. I can find no source for that 25% number in the scientific literature, and I believe it has to be a wild arm-waving guess that has been passed along because it is nice to have a number and nobody can readily provide a reference to any evidence that can show that this suggested number is wrong.

To this day, even at levels of cellular and molecular analysis undreamed of in Spemann's day, developmental biology has not yet fully explained the meaning of his results in terms of the cellular behaviors necessary to produce them, what it means for the embryo, or how it comes to be that way.

The thought that dizygotic twinning must also be associated with disorders of embryogenic asymmetry was unthinkable [because everybody knows dizygotic twins come from independent double ovulation and therefore have perfectly normal embryogenesis]. Therefore, the unthinkable had not been thought, and the facts that would require one to think about thinking about the unthinkable remained invisible, under heavy cover.

There is, however, a fundamental truth manifesting itself here: All of these observations about twins and handedness that are founded on sound evidence are the same among monozygotic and dizygotic twins. All of the observations about relatives of twins are the same among the families of monozygotic twins and the families of dizygotic twins. The excess of nonrighthandedness among twins and among their first-degree relatives has nothing to do with monozygosity; nothing to do with "splitting"; nothing to do with "mirror-imaging" resulting from splitting.

A difference in handedness between co-twins has nothing to do with zygosity; dizygotic twins are in fact more likely to differ in handedness, as is to be expected for any trait with significant genetic elements among its causes. These results have been replicated in several samples, notably including a large sample from the East Flanders Prospective Twin Study in Belgium, where chorionicity data (shared vs separate chorions) were also available.

Monochorionic twinning is generally believed to happen later in embryogenesis than dichorionic twinning. It might therefore be associated with the "late-splitting" version of monozygotic twinning thought to favor "mirror-imaging." However, neither the excess of nonrighthandedness in twins nor its distribution among them differs as a function of chorionicity. There is no difference in any aspect of handedness between twin pairs born in a single gestational membrane vs those delivered in separate membrane packages.

Most infants in my twin-family sample were reported as being of unknown handedness. Frequencies of "righthanded," "lefthanded," and "ambidextrous" all rose rapidly at the expense of the "unknown" category over the second and third years. "Righthanded" and "lefthanded" rose steadily at the expense of "ambidextrous," after the frequency of "ambidextrous" peaked at about two years of age and then fell as rapidly as it had risen. Beyond about four years of age, the fractions were stable. Children under four were not included in the numbers on the basis of which I made the above statements.

It had been generally considered to make sense that the excess of nonrighthanders in twins should be due entirely to the monozygotic twins. After all, monozygotic twinning is itself obviously and directly an anomaly of body-plan asymmetry determinations in embryogenesis. For monozygotic twinning to occur, two full body symmetry plans must somehow be established in a ball of cells that would normally build a single body on the armature of a single body symmetry plan. Surely that deviation from the normal flow of embryogenesis might be expected to be associated with other anomalies of the complex process of establishing the normal asymmetries. Assuming its origin from independent double (ovulation plus fertilization plus embryogenesis), dizygotic twinning should have no reason or excuse for any such anomalies of a/symmetry.

However, according to clear results from several independent studies, dizygotic twins are indeed different from the singleton majority in the development of brain function asymmetry as represented by asymmetries of motor performance or motor hand preference (handedness). Dizygotic twins are as different from singletons as the monozygotic twins are, in the same ways and to the same degree. The differences are not due to being carried in the womb as twins or being delivered as twins, but give every appearance of being inherited with the same distribution in the families of monozygotic and dizygotic twins.

This was, and remains, a quandary that is still considered unresolved in the world outside my office and laboratory, but which has been confirmed by several other competent studies. If dizygotic twins come from two independent ovulations and proceed through independent fertilizations and independent embryogeneses, then how can it be true that dizygotic twins appear exactly equal to the monozygotics with regard to the profound and subtle anomaly of embryogenic a/symmetry determination that is represented by minority hand-use preference? And that's not all ...

Chapter Thirteen

Twins and Malformations

> Our report identifies for the first time the severe and previously
> hidden global toll of birth defects. This is a serious, vastly
> unappreciated and under-funded public health problem. This
> report demonstrates the importance of state birth defects
> surveillance programs and also the need for more research to
> identify the causes of many birth defects.
>
> *Dr. Jennifer L. Howse*
> President, March of Dimes Birth Defects Foundation

Birth defects are the leading cause of infant mortality. We have been able, by concerted effort over recent decades, to reduce the overall frequency of infant deaths. During 1980–1995 overall infant mortality in the USA declined by nearly 40%. However, infant mortality attributable to birth defects (IMBD) has not declined as has overall infant mortality. From 1968 to 1995, the proportion of IMBD increased from 14.5% to 22.2%.

Malformations, chief among the phenomena known as "birth defects," are anomalies of development in which one or more body structures are not properly formed. The most common major malformations include neural tube defects (anencephaly, spina bifida, and other developmental defects of the brain and spinal cord resulting from incomplete or improper embryonic closure of the neural tube), clefts of the mouth and face (cleft lip, cleft palate, and some manifestations of the holoprosencephaly sequence), and congenital heart defects. All of these are "midline," or "fusion," malformations, involving midline structures formed by fusion of bilateral half-structures followed by remodeling of the fused, composite structure under the influence of neural crest mesenchyme.

Twins have excess frequencies of all of those most common major malformations. There is strong evidence that each and all of these malformations are strongly genetic in origin, with the best-fitting genetic model generally being the multifactorial model.

Like the twin excess of lefthandedness, the literature in general attributes the twin excess of malformations more or less exclusively to the monozygotic twins. Affected twin pairs are, however, <u>usually</u> discordant — seldom both afflicted with any of these disorders, in spite of high heritability with respect to each of those disorders calculated from family studies that do not include twins. This has always given me cognitive indigestion. The more I think about it, the worse it gets. This has been especially true since I have come to understand that monochorionicity *is not* proof of monozygosity. The discovery of that wonderful little truth has largely depended upon the results of investigating discordant anomalies in same-sex monochorionic twins (who have been gratuitously assumed to be monozygotic).

At least much of the discrepancy lies in the way that excess malformations have been attributed to monozygotic twins. To begin with, there seems always to have been an initial assumption that monozygotic twins will be found responsible for the lion's share of all the anomalies that are in excess among twins. If we should suppose instead that we should wait for the data to show us what their distribution is, it *will* be the case that the majority of these malformations will occur in same-sex twin pairs simply because the majority of twin pairs are same-sex. However, there is much more to it than that: These malformations are in fact under-represented in liveborn boy-girl twins, and the resulting apparent "concentration in same-sex twins" gets translated inevitably into "monozygotic" by the circular logic of the Weinberg Difference Method. There is no other way it could work, no other answer it could give. The Weinberg Exercise is a foregone flimflam.

Neural Tube Defects

The neural tube is an embryonic structure from which the central nervous system (the brain and spinal cord) develops. The most common major defects in neural tube formation involve failure of closure when the edges of the neural plate roll up from the left and right sides of the embryo to meet in the midline and "zip up" to form the neural tube.

Best results suggest that the closure most commonly begins at about the thoracic level (of the future rib cage) and spreads zipper-like in both directions toward the head and the tail ends. Multiple layers of embryonic cells are involved in the proper final product. When the tube closes properly, the spinal cord develops surrounded by the meninges (membranous tissues that surround the brain and spinal cord to contain the protective cerebrospinal fluid), bone and muscle, and the whole package gets covered by the skin. Defects may occur at, and usually all the way out to the skin surface from, any of those levels.

Some very special cells will distinguish themselves at the peaks of the infolding edges of the neural plate, along the narrow surfaces where the roll will close to form the neural tube, not unlike the foam on the crest of a breaking wave. That edge rolling up toward the midline is called the neural crest. Some of the cells in the crest are the neural crest cells, and they are certainly among the most wonderful of all the wonderful cells of the embryo. They will migrate away from the seam in the closed neural tube, down around the tube and out to the sides and into many structures of the body, where they will perform a still-not-fully-known-and-still-growing number of inductive functions.

Some will become the pigment-producing cells of the skin and the eyes and the hair. Some will be fundamentally involved in the placement and growth and shaping of the jaws and teeth, of the face as a whole, of the structural and functional components of the ears, and of the heart. Some of them will form most of the peripheral nervous system (and the enteric nervous system — not always considered separate from the "peripheral").

These neural crest cells will, very importantly, in all of their functions, "know" left from right and participate fundamentally in building accordingly. They are fundamentally involved in all of the common major midline fusion malformations, in ways we have barely begun to understand. I believe they are the primary "messengers" and "executors" of all our fundamental left-right developmental asymmetries, but I do not yet know the molecular mechanisms of the whole pathway, or how to prove it in detail.

Spina bifida ("open spine") and anencephaly are the most common and best known of the neural tube defects. Spina bifida results from failure of the spinal neural tube to close, sometimes in the thoracic (rib cage) area on the head side of the two-way "zipper," but more often in the lumbar (lower back, where the vertebrae have no ribs attached) region where closure of the neural tube is moving toward the tail end.

Once upon a time, most cases of spina bifida ended with early death from infection invading the opening in the spine. With modern medical attention, including surgical closure of the inappropriate openings, most cases are survivable, usually with paralysis posterior to the level of the defect. In some cases, spina bifida is hidden (spina bifida occulta) by closure of the skin over defects in deeper interior layers of the tube closure.

A pilonidal cyst is a hair-lined dimple just anterior of (toward the head from) the cleavage between the buttocks. This apparently is thought by some to be merely a very specifically vulnerable spot for infection around ingrown hairs. I must side not with those but with the others, who see it as a very specifically placed minimal defect in closure of the posterior neuropore, the hindmost end of the neural tube.

In anencephaly, usually all of the parts of the face are present and in almost proper relative positions, but they appear compressed longitudinally — shortened from forehead to chin and stretched wide. Ears are present, but very low-set. The forehead is low and there is no cranium; no brain, skull or scalp, no head above and behind the face and ears. The anterior (front) end of the neural tube has not closed, and the cerebral hemispheres of the brain have not formed. Survival beyond birth is seldom as much as a day.

Several other varieties of neural tube defect concern other places along the length of the neural plate and neural tube, each with its specific associated difficulties. Iniencephaly is a neural tube defect in which the failure of neural tube closure occurs around the junction of the skull and spine. A cascade of other major malformations is associated with this defect, due to the frequent absence of the neck and of the posterior cranial neural crest cells, which are responsible for driving the development of numerous musculoskeletal structures of that region of the body.

Some variations within the various types of neural tube defect result from weak or partial closures.

Most infants with neural tube defects in or near the head survive no longer than a few hours or days. Modern medicine surrounding neural tube defects is based primarily on prevention by supplementation of the mother's diet with folic acid. Numerous studies have demonstrated that mothers can reduce the risk of neural tube birth defects by as much as 70% with daily supplements of at least four milligrams of folic acid beginning before conception.

Frequencies of neural tube defects worldwide — before the discovery of the near-magical results of folate supplementation — have been found to be correlated strongly with frequencies of total twinning. The more total [liveborn] twins in any population, the more neural tube defects were found in the same population. They were not correlated with Weinberg-method estimates of fractions monozygotic in the same populations, which would be expected according to the orthodox common knowledge that excess malformations among twins just do always belong to monozygotic twins.

Folate supplements have been found, in some studies and not in others, to increase frequency of twin births. In other papers, it has been reported and argued that most of that increase is due to excess twins among the products of technologically assisted conceptions. Survival efficiency of induced and artificially assisted pregnancies is improved by folate supplementation to an even greater degree than that of natural conceptions.

Craniofacial Developmental Anomalies

Orofacial clefts are failures of closure of the structures of the mouth and the face. Like the development of the central nervous system from the neural tube, the face is formed from bilateral half-structures which must come from each side to fuse in the midline and then be reshaped into their final structures under guidance from cranial neural crest cells.

The mandible (lower jaw) comes from the midline fusion of the paired mandibular processes. The maxilla (upper jaw) mostly comes from the paired maxillary processes, but they will not properly fuse in the midline without the piece that forms the pleat in the middle of the upper lip (the philtrum). That piece comes down through the middle of the face as the frontonasal process. On its way to forming the philtrum and supplying the point of midline fusion for the maxillary processes, the frontonasal process splits the eye field (without which there may be only one eye) and forms the main part of the nose. The flesh around the nostrils comes from separate lateral nasal processes on each side.

The mandibular processes rarely fail to meet and fuse properly. A "weak" or receding chin may be a departure from normal function in the formation of the lower jaw from the mandibular processes, but failure of those processes, and thus the bony structures of the chin, to close in the midline appears to be quite rare in live births.

Fusion of the maxillary processes is much more likely to malfunction than that of the mandibular processes, most commonly by failure of the left maxillary process to fuse with the frontonasal process. This causes a unilateral (on one side only) left-sided cleft lip. This is the defect commonly but not very nicely called "hare-lip" (*not* "*hair-lip*"!) because of a vague resemblance to the face of a rabbit.

About 80% of unilateral cleft lips are found on the left side. A right unilateral cleft lip predicts greater risk of recurrence (higher probability of additional cases) in the family — which we customarily attribute to a higher multifactorial genetic loading, equal to that among close relatives of an individual with a bilateral cleft lip.

Unilateral cleft lips leave the philtrum attached to the half-lip on the opposite side from the cleft. In bilateral cleft lip, usually a shortened and distorted philtrum hangs between the two clefts.

Cleft lip may or may not also involve a cleft palate. Cleft lip and palate is a deeper version of cleft lip. The hard palate is formed from fusion of two palatal shelves that reach up over the growing embryonic tongue to meet in the midline, together with a section at the front of the mouth that comes with the frontonasal process, to form a Y-shaped joint. When lip and palate are cleft together, again much more often on the left side than the right, the corresponding branch of the Y is not properly fused. This, too, can happen on either side or both, with the same predilection for the left side, and the same evidence for increased multifactorial genetic loading when it occurs on the right side or on both sides.

Isolated cleft palate, without lip clefting, is a failure of midline closure of the secondary palate formed from the palatal shelves. As the palatal shelves are growing up and in toward the midline (anteromedially), the rest of the head is growing also. The tongue, over which the palatal shelves must reach, is growing larger, working against their progress by giving them farther to go. The rest of the head and face is also growing all the while, causing the roots of the palatal shelves to move apart. If growth of the palatal shelves cannot equal or exceed the progress of all the other processes working against it, the shelves may not meet in the midline in time to fuse before all the progress in other growing parts nearby pulls them out of reach of each other.

A medial cleft lip is, as its name implies, medial, central, in the middle of the upper lip. It is rare — less than half of 1% of all lip clefts, and concerns midline fusion of facial components in a totally different way — and it is as much a neural tube defect as a facial clefting anomaly. It differs from bilateral cleft lip, in that the bilateral cleft lip has a short philtrum, while the medially cleft lip has none. Medial cleft lip is one of the "milder" presentations of the holoprosencephaly sequence — the frontonasal process has stopped in its progress toward the mouth, short of creating the philtrum and fusing it with the maxillary processes.

In the most destructive version, alobar holoprosencephaly, the embryonic forebrain (the prosencephalon) never separates into the two lobes that should form the cerebral hemispheres — the left and right halves of the brain. There are severe skull and facial defects associated with this outcome. The semilobar and lobar versions are progressively less severe. In lobar holoprosencephaly, the brain may appear outwardly to be nearly normal.

Another lesson in here is that, in the head, closure of the neural tube depends upon, enforces and maintains, left-right distinctions in the half-structures, which is crucial to the normal continuation of brain development.

Embryogenesis of face and brain are highly interrelated. If the medial nasal process stops early in its progress toward the mouth, what there is for a nose may be a single opening in the middle of what would otherwise have been the forehead, above the one eye that results from the failure of the medial nasal process to split the eye-field. This is cyclopia — the "Cyclops" malformation. When the frontonasal process progresses a little farther and a little farther, the eyes may be separated, partially (occasionally, two eyes in a single palpebral fissure), then fully, with a one-hole "nose" just below. Other yet milder versions include a wider than normal gap between the central incisors (the two front teeth which are absent at least part of the year in most six-year-olds), or no gap at all — a single, more or less double-wide central incisor, arising from incomplete separation of the two maxillary central incisor tooth buds.

Many human anomalies in the holoprosencephaly series arise from various mutations in the human counterpart (SHH) of a gene that was named sonic hedgehog (shh) when it was first found in the fruit fly *Drosophila melanogaster*. The product of the SHH gene is known to participate in a number of functions at several levels of the establishment of these and other midline anatomical asymmetries. We will encounter it again soon in the development of the teeth.

Congenital Heart Defects

The congenital heart defects of interest here are malformations of the heart, including, among others:

- atrial septal defects, ("holes in the heart" between the smaller upper chambers of the heart — the atria — plural of atrium);
- ventricular septal defects ("holes" in the septum between the larger chambers of the heart, the ventricles);
- stenosis (narrowing) of the aorta (the major artery that carries oxygenated blood from the heart to the body) or of the pulmonary artery (through which the heart pumps de-oxygenated blood from the body through the lungs);
- transposition of the great vessels (the aorta and the pulmonary artery in switched places);
- underdevelopment of the left side of the heart (hypoplastic left heart), and Tetralogy of Fallot, which is a four-way combination of a large ventricular septal defect, stenosis of the pulmonary artery, right ventricular hypertrophy, and an overriding (dislocated) aorta.

The Fusion Malformations

Taken together, the neural tube defects, congenital heart defects, and orofacial clefts are the best known of the fusion malformations. Some people call these the "midline malformations," which is just fine; they involve midline structures and they show up in the midline — right where the fusion should have taken place. Taken together, they are the most common of the major malformations. They are the most common malformations in all human births, and they are more common in twin births than they are in singleton births. These malformations occur with excessive frequency not only in twins, but also among the siblings and the offspring of twins. These malformations are also all associated with other anomalies of embryogenic a/symmetry development including nonrighthandedness, in the victims of the malformations and in the parents, siblings and offspring of the victims.

The excesses of these malformations in the parents, siblings and offspring of twins do not differ significantly by the zygosity of the twins. In the case of every malformation in which there is a significant difference as a function of zygosity, such as the excess of twins among the parents of children with neural tube defects, the association is greater in the families of dizygotic twins.

Common knowledge says that the excess frequency of these common major malformations among twins is due overwhelmingly to the monozygotic twins. The papers attributing the excess malformations in twins to the monozygotics, or cited by others as authority for doing so, unanimously offer no better evidence than the results of sorting samples by the Weinberg Difference Method. Because the frequencies of these malformations are lower among boy-girl twin pairs than in same-sex pairs, the circular logic of the Weinberg method invariably assigns the excesses of these malformations in twins more or less exclusively to the monozygotic twins.

The individual malformations in this class of anomalies are sufficiently uncommon that it is prohibitively difficult to assemble statistically effective comparison samples, even if the costs could be covered — especially for the purpose of questioning something that has for so long been considered beyond question. The idea that these anomalies are due to side effects of the disruption of embryogenesis by monozygotic twinning is so firmly entrenched that any same-sexed twin pair in which *either member* has any of these malformations has been routinely declared to be monozygotic even in the face of sound genetic evidence to the contrary. Any evidence of dizygosity in such a twin pair has traditionally been ignored or dismissed as erroneous, in favor of the dogma which assigns the origin of dizygotic twins to double ovulation (where common knowledge says there are, of course, no oddities of embryogenesis or such consequences thereof as malformations).

All of these midline fusion asymmetry malformations are also associated with unusual brain function asymmetry. This is most readily represented as nonrighthandedness in the individuals surviving with the malformations, and among the parents and siblings of those individuals. Fusion malformations, further, are present in significant excess among the siblings and offspring of twins as well as the twins themselves. In these distributions, there is in general no zygosity difference. There is no statistically significant or otherwise interesting difference between monozygotic twin families and dizygotic twin families in frequency of the midline/fusion malformations among the sibs and offspring of twins.

It is possible and plausible, and reasonable, to propose that all victims of the fusion malformations are products of twin conceptions and of common subtle asymmetry-related errors in twin embryogenesis. Just as with nonrighthandedness, the strength and direction and mutuality of the associations, and the frequencies, are amenable to such consideration. The numbers would work. Again, to confirm or refute this notion requires (only) a way to identify accurately the sole survivors — the products of twin conceptions who were not born as twins. The only way I know by which that might be accomplished is multivariate statistical analysis of adult dental diameters — complicated and expensive (more about that in the next chapter, on twins and teeth).

The most fundamental components of the self-assembling organization of the embryo most importantly concern the establishment of the axes of development: where head and tail should go, which side is right and which is left, what's belly, what's back. Cells that might become eggs or sperm gather in a properly-assigned-and-labeled staging area and march together into tissues that they can turn into properly-layered ovaries and testicles. Cells that will build nerves, muscles, bones, blood vessels and hearts similarly gather in specified places and move along specified paths according to signals from other cells about where they belong and what to do when they get there.

All of this must begin in the egg cell — the largest and most complex of all human cells — because it cannot happen without a very well-laid-out and reliable plan. The sperm only has room for half a set of genes, a pair of centrioles as framework for the zygote's cell-division machinery, an outboard motor to effect the delivery, and a chemical drill to make a hole through which he can get inside when he gets there.

The egg cell has labels and triggers and messages and signals parceled out and sequestered in deposits which will be sorted out by early cell divisions into regions of its substance that will become specific areas where specific functions will be performed.

In ways that are only beginning to be understood, twin embryos become twin embryos by doing some of these things differently from single embryos. Those differences in embryogenesis have consequences that can be seen in the bodies of children and adults that develop from those embryos. The presently most bewildering thing about those differences, from which I hope to disperse the clouds of bewilderment farther along, is that dizygotic twins are at least as different from singletons as the monozygotics are, and in the same ways, and the ways in which the development of twins differs from that of singletons have to do most particularly with the processes by which embryogenesis establishes the structural asymmetries of the eventual human individual. And there is still more where that came from ...

Twins and Teeth and (the Realities of) Mirror-Imaging

> Adam and Eve had many advantages,
> but the principal one was that they escaped teething.
>
> *Mark Twain*

The teeth, like other paired structures in the vertebrate body plan, are approximately symmetrical from side to side. On casual inspection, teeth in the left halves of human jaws look very much like mirror images of teeth in the right half-jaws. It is easy enough to take that for granted as a standard component feature of the way things are, but the developmental implications of that fact far exceed the complexity of any statement of the fact itself. If you suppose that symmetry is the way things are meant to be, (as much of the early literature in developmental biology seems to indicate) then how do you suppose they are to get that way, and how do you propose that we should interpret deviations from that "ideal?"

Like everything else about the development of any complex organism, there is much less involved in saying how things are than there is to explaining how they get that way. Generally, however, the descriptive work has to come first. Maybe later we can frame hypotheses about how the system in question might come to be working the way it seems to be working.

I expect to set out in this chapter a somewhat simplified description of the embryonic processes for building a set of teeth, and then explain what I know about how twins do it differently, and what that seems to mean about how twinning must happen. There is no better way to learn or relearn anything than to try to teach that thing to someone else. For one thing, every time I go through this, I see something I missed before. One day I may find the very one thing that tells me I have been wrong about all of this all along, but so far, reviewing the material and catching up with whatever has changed since I last checked it has always confirmed and strengthened the idea for me.

Somehow, systematically interacting groups of cells in the embryonic jaws (when each left, right, upper, lower embryonic jaw is about half as long as one of these "i"s is tall) establish locations for each of the 28–32 different teeth. With the locations established, the cells of the embryonic primordia of the teeth begin to grow, in the right places, into the right shapes. The eventual product of that development will be a properly coordinated system of tooth structures. There will be a functional "*set* of teeth." There will be slicers (incisors) in the front, gradually morphing through canines for tearing, to grinders (molars) in the back. There will be appropriate matching from maxilla (upper jaw) to mandible (lower jaw) — so they can meet in a bite to do their jobs — and from left to right.

Development of the face, including the eyes, ears, nose, jaws, and the teeth, interacts with development of the brain throughout embryogenesis, fetal development, childhood and adolescence — throughout growth. Each of those structural subsystems is constructed from parts arising from multiple embryonic source tissues and involving at least most of the cell types and developmental mechanisms involved in any other aspect of development in the head and brain.

Only a handful of cell divisions back in embryogenic time, (the precursors of) all of these substructures were in a very few cells, quite close to the prochordal plate (around which the mouth will later form), harboring a very different future from that of most other cells in the embryo.

A body of work discussed at the British Association for the Advancement of Science Festival of Science in York in September 2007 showed that many complex developmental disorders can be recognized in or predicted from three-dimensional scans of children's faces — once the computer software has been given enough examples from which to "learn" to highlight the consistent differences. The algorithmic basis of this process is similar to the multivariate statistical methods I have used in the studies of teeth that I will discuss in this chapter. A large group of measurements provides the primary substrate for the analyses, but the whole system of relationships among those measurements carries a great deal of additional and more powerful information than the measurements themselves, no matter how numerous, especially for developmental considerations.

If we have N (say 10) randomly-variable measures, that may be a very good thing and a relatively great deal of information. An even better thing is that some robust and informative forms of statistical analysis can allow us to make productive use of $N(N+1)/2$ (= 55 when N is 10) additional structural variables, in the form of variances and covariance relationships among all pairings of those 10 variable measures.

Many unfortunate outcomes of brain development are reflected in anomalies of craniofacial or dental development. A prominent example is the holoprosencephaly sequence, as discussed in the previous chapter on twins and malformations. Probably, the outwardly mildest expression of the holoprosencephaly sequence is a dental variation: a single central incisor — the "two front teeth" are not separate, but grow from one duplex tooth bud to form one double-wide tooth.

A great many of the genes whose contributions to human development we are coming to understand are coded by DNA sequences that are homologous to the sequences of related genes that we first discovered in the fruit fly *Drosophila*. Although some genes in some organisms now control different developmental functions than they once did or still do in different organisms, more often than not we have found closely related gene products performing the same or a related function.

As early as the primitive streak stage of the bilaminar- and trilaminar-disk stages of embryogenesis, the sonic hedgehog gene product is normally suppressed on the right side of the primitive streak (by another signal molecule called activin) and is therefore active only on the left side.

The sonic hedgehog protein signal molecule produced over on the left side of the primitive streak, where its synthesis is not suppressed by activin, initiates a cascade of other such signal molecules. Not all organisms use the same systems and pathways of signal molecules for these functions, but sonic hedgehog, nodal, lefty-1, lefty-2 and Pitx2 gene products interacting seem to be fairly widely used components among us vertebrates.

Whichever set of switching molecules is used, without this system of inductions and repressions happening normally, the looping of the primordial heart is randomized instead of being reliably directed to form the mature heart to the left of the midline — and the rest of the normal a/symmetries of the internal organs are weakened, often to the point that they also may be randomized or reversed.

Where left and right come from, and how all the body's parts grow where they belong so they can do their proper jobs, are fundamentally important questions on which some very smart people have worked with great intelligence and diligence for a long time.

It will be a great day when one of the members can tell the rest of the group and the world how the cells on the right side of the primitive streak know that they should produce activin to inhibit the synthesis of sonic hedgehog on that right side, allowing the uninhibited presence of sonic hedgehog on the left side to set off a cascade of control signals that will result in "proper" looping of the primordial heart, to give the appearance of beginning the visceral asymmetry developmental cascade. [That sentence is too long by most people's standards, but ... if I break it up into three or four shorter sentences, it loses something.... .]

A recent advance I believe will prove to be a major one shows that serotonin is crucially active in the embryo as early as the two-cell stage, long before any nerve cells have differentiated, and appears to have important functions in directing proper installation of these developmental asymmetries. Serotonin is a close relative of the amino acid tryptophan, made in the gut, stored primarily in blood platelets, and known to most of us primarily as a neurotransmitter molecule.

Uptake of serotonin at serotonergic synapses (the connections between nerve cells that use serotonin as signal molecule to pass nerve action along from one cell to the next) is the target of the SSRI family of antidepressant and anti-anxiety drugs, the "selective serotonin reuptake inhibitors." Using various drugs, such as those SSRIs, to make specific changes in the ways that serotonin is handled is showing us a number of changes in embryogenesis in several different animals. Apparently, serotonin is carrying messages between embryogenic cells long before there are nerve cells among which to be transmitting communications. There will surely be more where that bit of news came from, probably soon, and I expect it to be exciting.

Behind (posterior to, toward the tail end from) the primitive streak, defects of sonic hedgehog are suspected in sirenomelia (from Greek, "mermaid limbs"), an almost invariably lethal malformation in which the legs grow into a single flipper-like extremity, and most of the urogenital and mid-pelvic structures are absent.

Overactivity of sonic hedgehog has been suggested as the cause of the double-face malformation known as diprosopus, in which separation of left and right halves near the anterior end of the body goes too far, such that the face-forming structures of the embryo may be duplicated to varying extent. Two-headed animals have been reported in many species, apparently most commonly in reptiles, but definitely including the human. There is at least one such human alive now in the USA. When we are discussing a snake or a frog, we speak of what we see as one animal with two heads. In this human case, it seems much more appropriate to think and speak not of one girl with two heads, but of twin girls sharing one body. There are two heads each with a brain and two faces each with a name, and it seems clear that there are two distinct personalities. There are two minds seeing the world with two sets of eyes and thinking and feeling differently about it.

Sonic hedgehog is important to a great many, perhaps most, of the final expressions of developmental a/symmetry, and we will see further indications of how that includes the head and face.

A "set of teeth" is composed of numerous parts integrated into a functional system — incisors in front for cutting off a bite, canines for tearing, premolars and molars for chewing and grinding at the rear of the jaw. These can be thought of as four different kinds/families/subsets of the teeth, reliably highly correlated throughout in sizes and shapes, in left and right half-sets. It is easy enough therefore to see the development of a "set of teeth" as a subsystem model for all of the building of the head and brain and face.

The four "front teeth," the maxillary (upper) incisors, arise from the leading end of the frontonasal process. The rest of the teeth arise from interactions most prominently featuring the maxillary process (forming the upper jaw, the maxilla) and the mandibular process (forming the lower jaw, the mandible). The maxillary and mandibular processes both arise from the first (most anterior) of the branchial arches, which form the gills in fish and look like gills in the early part of human embryogenesis.

Feathers, scales, nails, hair and teeth are all derivatives of embryonic epithelium (epithelia are tissues composed of layers of cells, typical of skin, mucous membrane and other boundary surfaces which separate the mass of the body from the environment, inside and out).

> Bear in mind ... topologically (topology is the branch of mathematics that deals with surfaces and shapes), you are a torus — a donut, a tube (your gut is the hole that goes all the way through and whatever is in the gut is actually outside of the solid part of your body). Oh, and by the way, if we do think of the gut contents as "inside," then the great majority of the (number of) cells "in" your body are not human at all, but microbial.

The teeth grow from interactions between dental epithelium and mesenchyme ("mezz-in-kime" has been called "embryonic connective tissue," but connective tissues occupy only a small fraction of the list of cell types that are derived from mesenchyme cells). The dental mesenchyme is derived in parts from neural crest cells and from cells of mesodermal origin. (The mesoderm is the middle layer of the three-layer disk (gastrula) stage of embryogenesis, largely dedicated to future muscle and bone.) In dental embryogenesis, mesenchyme cells have been shown to differentiate into several different types of cell, including at least: odontoblasts, fibroblasts, chondrocytes and osteoblasts.

Odontoblasts are cells of neural crest origin that produce the outer part of the dental pulp. Their function is the creation of dentin, the main bulk of the body of the tooth under the enamel. Fibroblasts form connective tissues (fibroblasts are most common among the cells that move in to close a wound). Chondrocytes and osteoblasts, respectively, produce cartilage and bone.

Skin along the future line of the jaw-bones thickens to form the dental lamina (layer). Ten "buds" of the first set of teeth (the deciduous dentition, "baby" teeth, or "milk" teeth) form at regularly-spaced intervals along each maxilla and each mandible. [These deciduous teeth are the ones destined to fall out, like the leaves fall off of a deciduous tree.] The dental lamina later extends beyond the last deciduous tooth bud and forms buds for the permanent molars.

At about three months of gestation, the dental lamina forms a full second set of tooth buds deeper in each developing jaw, the buds of the permanent teeth. They develop in the same way as the deciduous teeth but rather more slowly (after all, they have about six years before the time comes for the earliest of their number to reach the surface).

Epithelial tissues [tissues composed of layers of cells, like skin and mucous membrane] overlying the tooth buds grow inward from the enamel knot at the tip of each bud to form a cup-shaped enamel organ, which will function in the formation of enamel, the initiation of dentin formation, the establishment of the shape of the tooth's crown, and in the establishment of the structural relationship between the tooth and the gum.

The enamel organ includes the inner and outer layers of enamel epithelium, separated by a network of star-shaped cells that provide a matrix for connective tissue. The enamel knot, in the center of the inner enamel epithelium, is a signaling center for governing the development of the final shape of each tooth. It produces molecular signals from several of the major signaling-molecule families — fibroblast growth factors (FGF), bone morphogenetic (meaning structure-building) proteins (BMP), hedgehog (HH) and wingless (WNT) — to direct the growth of the surrounding epithelium and mesenchyme.

Sonic hedgehog in particular is strongly expressed in the enamel knot, where it is required for proper formation of the tooth's enamel. In particular, SHH promotes the differentiation of the enamel-producing ameloblast cells, and up-regulates the expression of amelogenin and ameloblastin, which are proteins responsible for giving shape to the enamel as it is formed and hardened by mineralization. (This is where the variations are that provide the data in which I found the twin differences we will discuss later.)

Each of the primary enamel knots will be removed by programmed cell death, and secondary enamel knots will appear that regulate the formation of the future cusps of the teeth (the "points" on the crowns). Under the enamel knots, connective tissue cells outline the structure for most of the body of the tooth. The cells of the enamel knot grow and begin to produce the enamel while some cells of the papilla build up the dentin.

The hard layers are deposited beginning at about 20 weeks of gestation, as bone of the jaws forms a cup surrounding each developing tooth bud. Production of enamel and dentin continues until the crown of the tooth is complete. The time required varies depending on the type of tooth, but, in general, when the crown is complete, the tooth is pushed through the gum surface by the continuing growth of the root.

The roots grow until they are completely surrounded by the jawbone growing up around them. Cement is produced by the tissues of the papilla when the root begins to grow. When the root is fully formed, the opening to the pulp cavity narrows until very little nutrient transfer can occur. Growth then ceases although the tissues still receive enough nourishment for maintenance.

The permanent teeth continue to develop slowly from a second layer of enamel knots behind the baby teeth. When the crown of each permanent tooth is fully formed, its root begins to grow. This increases pressure on the base of the temporary tooth. The periodontal (around the tooth) membrane and the cement and the root of the temporary tooth root are softened and degraded by enzymes and removed by macrophages (cells of the immune system whose primary function is to remove foreign, extraneous and damaged materials throughout the body).

When the cement and the membrane have been digested, the temporary tooth becomes loose and falls out to leave a clear path for the permanent tooth. When the permanent tooth has reached full size and is firmly embedded in its socket, the tooth ceases to grow because the opening of the pulp cavity closes just as it did at completion of growth in the temporary teeth.

Taking a Breath and Moving On

Yes, that is a lot of stuff there! This bit of the story seems to have more polysyllabic names in the cast than one of those great thick old Russian novels; not people here, but cells and molecules with long names, and systems of tiny cellular bits of anatomy. It does not seem to have more structural subsystems than the brain, but neither does it seem to have a great many less!

This is important for our present purposes primarily because the systems of processes involved in the building of a set of teeth overlap extensively with the lists of processes for building all the other substructures of the head, including the brain. A major element of the organization in all of them is bilateral a/symmetry — the strong but not exact or perfect similarities between members of left-right pairs of structures, from maxillary central incisors to mandibular second premolars, to ears and eyes and all the parts inside them, to parietal lobes and superorbital frontal lobes (the part of the brain in the forehead above the eyes, involved primarily in executive functions, decision making, behavior control) and hippocampi, and so on.

> "Hippocampi" means more than one hippocampus (we normally each have two) — a pair of "sea-horse-shaped" structures (I don't really see seahorses, but it is from that shape that the hippocampus got its name; I see more like commas) "down in the middle" of the brain, essential to the normal processing of memories, especially in the early, short-term part of the processes. [Memory still functions to some extent with only one, but with both gone, no new memories can be formed.]

Pretty much every piece of each of us is bilaterally symmetrically paired. We each only have one heart, for example, but it was made from left and right half-structures that still have importantly different structures and functions even in the adult heart.

The pineal, pituitary, spleen, thymus and thyroid occur to me as structures that seem to be single, unpaired, more or less in the midline, and not obviously composed of left and right parts. It does not, however, take much investigation to find that all of those structures are fundamentally asymmetric in one or more structural features or phases of their developmental processes.

Covariance (as) Structure; Multivariate Multifactorial Analyses

It is a fundamental but seldom mentioned assumption of the genetic twin study method that monozygotic, dizygotic and singleton groups have the same covariance structure, the same distribution of sources of individual variation. Without that equality, we should not expect to find the same origin and development, the same structure of causes for individual final outcomes for the trait in question, and we would have no reason to expect results from twin studies to be meaningful for the general population.

If we are to be able to believe the results from genetic twin studies, if those results are supposed to be meaningful for the rest of us who are not twins, then we are obliged to make the fundamental assumption that monozygotic, dizygotic and singleton groups have the same distribution of sources of individual variation.

What we are doing when we have drilled down into the mathematical innards of the modern genetic twin study or any other multifactorial genetic analysis is — never a squad of drummers around when you need them! — fractionating the phenotypic variance (variation in specific traits — "phenotypes," in individual appearance or structure or behavior). People are different, right? And we have long been doing genetic twin studies to discover how people come to be different with respect to some particular characteristic, "phenotype" or "trait." We want to learn how much of all that variation is genetic in origin (dependent upon DNA sequences that are segregating in families in Mendelian fashion) and how much of any particular kind of variation is produced by other influences.

To use the genetic twin study as a method to do that, we set out to determine whether monozygotic co-twins (who are supposed to have all the same versions of all their gene DNA sequences) are more like each other than dizygotic co-twins (who share on the average the same 50% fraction of gene DNA sequences as in non-twin sib-pairs and parent–offspring pairs).

If we want that approach to work properly and honestly, it is really not okay if monozygotic twins (as a group) and dizygotic twins (as a group) go about developing the trait in question by different processes, or if either of the twin groups develops that trait by a different process than is used by the single-born majority of the "general" population — the rest of us — whose biology we have set out determined to understand.

Unless monozygotic twins and dizygotic twins and singletons have the same variance/covariance structure, to indicate the same origin and development and the same structure of causes for individual final outcomes of the trait in question, then the logic of the genetic twin study is profoundly and probably irredeemably flawed. The basic notion [that finding monozygotic co-twins more alike than their dizygotic counterparts indicates genetic contributions to the trait in question] is intact, but any hope of precision in any estimates thus derived is without foundation.

We have already discussed some developmental differences between twins and singletons, with respect to motor hand preference (handedness) and the most common major malformations — the fusion, or midline, malformations.

The same malformations are in excess among the siblings of twins, just like the nonrighthandedness — the excess of these malformations among the siblings of twins does not differ between the sibs of monozygotic twins and the sibs of dizygotic twins. In newer work with more careful assignment of individual zygosity, the only zygosity differences in frequencies of these traits show greater associations with dizygotic twinning than with monozygotic.

Children with neural tube defects have an excess frequency of twins among their parents — significantly more so dizygotic than monozygotic. It has recently, finally been soundly acknowledged that twins have an excess frequency of schizophrenia, clearly statistically significant among the dizygotic twins, not quite statistically significant among the monozygotics. And so on. The point that must be understood here is that twins [of both zygosity groups, equally except when things are worse for dizygotics] show developmental differences from singletons, in traits that have to do with body-plan asymmetries that are established early in the embryo.

As per the Orthodoxy, as "everyone knows," monozygotic twins simply must have taken an anomalous path through embryogenesis, but dizygotic twins are supposed to have come from separate, independently ovulated egg cells, which should not cause them to have any involvement with any odd embryogenesis.

What Can Teeth Tell Us about Biology of Twinning?

I used sets of measurements of tooth sizes — called dental diameters. Two measurements are taken from each tooth, the maximum diameter along the jaw from "the middle to the outside," i.e., from the incisors to the molars, from the front teeth to the back (the mesiodistal diameter), and the maximum diameter across each tooth from cheek to tongue (the buccolingual diameter).

From the normal set of 28 teeth, this gives 56 measurements. This is exclusive of the "wisdom teeth" (technically, the third molars), because many people do not grow any, and many people who do have them need to have them pulled because when they do grow they often grow in badly and make a mess of things. Each measurement is the maximum diameter of the tooth in its particular direction, taken for this work with a digital electronic caliper. The measurements were taken at the Indiana University School of Dentistry in Indianapolis under the supervision of Dr Rosario Potter.

The multivariate statistical math I used does not work reliably with missing values (from loss, decay, or fillings affecting a measured surface). Therefore, I estimated missing values using iterated minimum variance least squares multiple regression. (What that means in plain talk is that the strong system of relationships among all the measurements allows me to find an estimate for any missing value which gives the best fit with all the non-missing values. In other words, the program finds numbers that "look right" among the other numbers of a set.) This was always done within the group to which the individual set of measurements belonged, for example, male monozygotics, female singletons, etc. To check on the performance and accuracy of the method, I tested the procedure on sets of measurements which had no missing values and found that it always gave very good approximations of known values that I deleted to simulate missing values.

I began the first cycle of each estimation process with either the partner of the missing measurement from the other side (the "antimere"), or the group average; replaced that with the first computed estimated value; and ran through the process again until the estimated value did not change any more than 2.5% between cycles. With no more than three missing values in a set, the estimates always converged within three or four iterations. If any set of measurements included more than three missing values, or the estimating process for the missing values did not converge that readily, I did not use that set of measurements for further analyses.

I used several multivariate statistical methods [mainly multivariate analysis of variance, linear discriminant function analysis, quadratic discriminant function analysis, canonical correlation analysis and principal components analysis]. All of those methods work, in slightly different but analogous ways, on the covariance matrix (the whole set of 56 dental diameters, plus the whole set of relationships among all pairs of those measurements, for a total of $N \times (N+1)/2 = 56 \times 57/2 = 1596$ variables), as a model representing the whole system of structural relationships among these parts.

I compared twins vs singletons, males vs females, monozygotic vs dizygotic twins, first- vs second-born twins, left-side teeth vs right-side teeth, and left-right differences, as well as the single measurements.

The simplest results have to do with sex. That is the only difference in all of the dental measures that I can "see with my own two eyes." Just by reading the individual lines of measurements, I can correctly guess the sex of the individual about 80% of the time. That part is easy — human males are usually bigger than females, and that includes the sizes of their teeth. The extensive strong correlations among the sizes and shapes of a set of teeth are such as to make the whole line of measurements tend to move up or down in size together.

The multivariate statistical programs were always able to classify the individual sets of measurements with very high accuracy as to their sex, (but *only*) so long as the measurements came from singletons or same-sex twin pairs. A linear discriminant function that correctly classified 42 of the 44 same-sex dizygotic twins misclassified 12 of the 20 boy-girl twins, and a quadratic discriminant function that correctly classified 123 of all 128 same-sex twins of both zygosities misclassified 16 of the 20 boy-girl twins.

The sets of dental diameter measurements from members of boy-girl twin pairs could not be classified according to sex. Boy-girl twins are intersex with respect to patterns of developmental relationship among their teeth. There is good reason to suppose that the same system of relationships will be found involved with other sets of measurements from faces and heads, because a number of studies of behavioral and functional variations have shown girls with twin brothers to be shifted in a masculine direction. In some measures but not in most of them, the boys with twin sisters are correspondingly "feminized."

> More to the point, and meshing more importantly with the handedness results I told you about earlier, the dental diameter investigations showed very clearly that twins are different from singletons in asymmetries of embryonic development, and that dizygotic twins are at least as different from singletons as the monozygotics are and in the same directions.

Among the males, the monozygotics and dizygotics are about equally different from the singletons. Among the females, the dizygotics are clearly more different from the singletons than the monozygotics are.

It is possible to identify the zygosity of *individual* twins (with no reference to within-pair similarities), with clear statistical significance. However, the multidimensional distance between the monozygotic and dizygotic group centroids (the multidimensional counterpart of univariate averages) is a great deal smaller than that between the singletons and either twin group, in both sexes.

The multivariate separation between the twin zygosity groups is almost entirely independent of (i.e., uncorrelated with, or in a different multidimensional direction from) the separation between twins and singletons. The developmental differences between the twin zygosity groups are much smaller and of a different kind than the differences between twins and singletons.

Just as the excess of nonrighthandedness among twins is a loss or reduction of the normal asymmetry of right-handedness, in the multidimensional structure of a set of teeth, twins of both zygosity groups equally are more symmetrical than singletons, with no clear differences between zygosity groups.

Average left-right differences show that left-side dental diameters are in general larger in both male and female singletons. Maxillary (upper jaw) dental diameters are more variable on the right in male singletons; mandibular (lower jaw) variability is larger on their left sides. All of these differences are much smaller (and not statistically significant) between monozygotic and dizygotic male twins than between twin and singleton males.

The overall pattern is one of substantially reduced asymmetry (of both "directional" and non-directional "fluctuating" types) among twin males relative to singleton males. There are no interesting differences between dizygotic and monozygotic male twins.

In both sexes, twins show clear and substantial reduction in non-directional (random, or "fluctuating") asymmetry. In directional asymmetry (significantly non-zero average left-right differences), male twins show a clear reduction. Just as with the twins' reduction of normal asymmetry with regard to handedness, both twin and singleton females are more symmetrical in their dental development than their male counterparts. Differences between twin and singleton females are smaller than the corresponding differences among males, but they are clearly in the same direction. No monozygotic vs dizygotic difference in either sex is comparable in magnitude to the corresponding twin–singleton difference.

Left-Right Correlations

Another different but related way to consider these relationships is by way of the correlations between the antimeric (left vs right) measurements (the tendency of side-to-side paired measurements to vary together and proportionately). All 28 left-right pairs of measurements are more highly correlated among twin than among singleton males (23 of the 28 in females), and this is equally true at the multivariate level ("canonical correlations" between sets of variables).

Comparing the monozygotic and dizygotic male groups in the same way shows only four of 28 significant differences in left-right correlations, all in upper-jaw measurements, all higher among dizygotics, and all showing greater difference between dizygotics and singletons than between monozygotics and singletons. Twenty-four of the 28 dizygotic-singleton differences are larger than the corresponding monozygotic-singleton differences. Again, dizygotic twins are at least as different from singletons as the monozygotics are, quite contrary to the common knowledge doctrine, quite contrary to the equality of singleton and dizygotic embryogenesis required by the double-ovulation hypothesis.

The *Real* Twin Mirror-Imaging

The left-side and right-side covariance matrices, representing the whole system of size relationships among tooth sizes on the two sides, are clearly and significantly different among singletons. The hypothesis that singleton left and right sides develop to the same pattern, as mirror images, must be rejected. The left and right halves of singleton sets of teeth are not mirror-symmetrical.

Among twins, in total or in either zygosity group separately, in both sexes and in all comparisons, the same tests yield probabilities in excess of 0.98 that greater differences could be expected due to chance sampling variation alone. In all twins, but not singletons, left and right sides of the adult dentition develop to the same pattern.

The system of structural relationships among the sizes and shapes of a set of teeth clearly differ between the sides of the mouth in singletons, and clearly do not do so in twins.

Here is the real "mirror-imaging" of twins. Similarities from the left side of one twin to the right side of the other, in dimples and such, are trivial and coincidental side-effects, little eddies in a great stream, of the matching from side to side within each individual member of a twin pair. It happens in the same ways and to the same extent in monozygotic and dizygotic twins. Right-side and left-side structures of the relationships among the tooth shapes and sizes are mirror-imaged in twins, in all twins, and definitely not so in singletons.

You and your dentist working together almost certainly cannot tell the difference just by looking at the teeth, but a fraction of a millimeter here going this way and a different fraction there going that way in the relationships among the sizes and shapes of all 28 teeth, and in the relationships among those measurements, add up to a clear and regular difference in patterns detectable by multivariate statistical analyses. Remember where those differences are made, back in the building of the enamel. Subtle left-right differences in tooth sizes are strongly consistent in singletons and are equally clearly not present in the teeth of twins.

Twins do not develop the left-side and right-side structural patterns of tooth shapes differently, as the singletons do. It is rather as if they use the left-side pattern inside-out for the right side, or vice versa. Again, and probably the most important feature of all this, whatever it is that they are doing differently from singletons, the dizygotics and the monozygotics do it differently from singletons in the same ways — the same distances in the same multidimensional directions.

Because of the high correlations between left-side and right-side measurements, and therefore between their respective covariance matrices, that comparison by discriminant function analysis is biased. (This is an important fundamental detail of statistical methodology, which may be even more important in multivariate analyses). (My statistics mentor told me that) S N Roy's union-intersection test is the proper unbiased test, so I performed that test as well, although it was not available in any of the statistical software packages I could find and I had to program it from scratch — as a result of which at that time I probably understood how that procedure worked better than I did any other of the multivariate analyses I was doing.

Results of the union-intersection test agree in all testable comparisons with the results of the discriminant function analyses. Left-side and right-side structural patterns are clearly and significantly different in singletons and not statistically different in twins.

Twins of both zygosities equally are substantially more symmetrical in craniofacial development than singletons. Normal asymmetries of development in singletons are substantially reduced in twins, and the differences are at least as great in dizygotic twins as in monozygotic twins.

In these processes of building mouths and faces and heads, dizygotic twins do not develop like singletons, as if their embryogeneses are independent and undisturbed. These dental comparisons give us results that are strongly inconsistent with the double-ovulation hypothesis for the origin of dizygotic twinning.

There are clear sex differences in craniofacial development, among singletons and twins in same-sex pairs. Both male and female members of girl-boy twin pairs are statistically so firmly intermediate between normal gender-specific structures that they cannot be distinguished by measurements which can sort the sexes in singletons or same-sex twins almost perfectly.

- The developmental histories of dizygotic co-twins are not independent.
- Boy-girl dizygotic twins are not developmentally representative of all dizygotic twins.
- Dizygotic twins are not the developmental equivalent of singletons.
- The development of dizygotic twins is not compatible with an origin in double ovulation.

Chapter Fifteen

The Oddest of Couples: Boy-Girl Twins

The first thing Society wants to know about each of its new members is
"Is it a boy or a girl?"
Then, they ask if everybody came through the delivery alright.
The truly worst thing that can happen on such a day
is for the real answer to the *second* question to be "we don't know yet."
It usually *seems* even worse if
that is the only right answer to the first question.

From the lecture on Sex Differentiation & Development
in my Human Genetics course

Male and female represent the two sides of the great radical dualism.
But in fact they are perpetually passing into one another.
Fluid hardens to solid, solid rushes to fluid.
There is no wholly masculine man, no purely feminine woman.

Margaret Fuller

When I hear "opposite-sex" twins, sometimes I have wondered what kind of sex that "opposite" sort might be — "unlike sex" gives me similar problems. So, lately, I normally alternate between "boy-girl" and "girl-boy." It says what I mean to say, and alternating avoids any unseemly primacy in the long run.

Boy-girl twins are at least as special as anyone has ever thought they might be. There are several things about their particular kind of biology that provide very plausible reasons for this, any one of which might be considered sufficient.

For one thing, which might in passing seem the most obvious: a girl-boy twin pregnancy is by far the most intimate possible kind of relationship in any pairing between a human male and a human female. In no other circumstance in any pair of normal human lives are male and female so close for months at a stretch. So close and yet ... they never touch skin to skin because they are separated by the gestational membranes.

The common knowledge has it that only monozygotic twins ever share a single amnion. However, we were for a very long time adamantly and absolutely wrong to believe the same about sharing a single chorion. Although nothing in the new understanding unfolding here forbids it, I had not been able to find any monoamnionic boy-girl twins in the literature until I recently found a pair of twins conjoined at the head and chest with "discordant genitalia." Inadequate testing failed to demonstrate the extent of their chimerism (Kim *et al.*, 2007). Woods (2009, Table 1.1, p. 6) cites a parish register from 1655, reporting an aborted conjoined boy-girl pair. Perhaps we should give it time ... the field in general has not been able yet to digest knowledge of the existence of monochorionic dizygotic twins.

This closeness does not, however, have much if any bearing on the relationship that is generally imagined.

Entirely by limping analogy with mixed-sex multiple pregnancies in other mammals (primarily cattle and mice and marmosets), there has developed an extensive lore in the literature about the influence of testosterone from a male twin fetus on the development of his female co-twin fetus. There are a number of papers out there about how females are masculinized or defeminized in the womb by twin brothers, with such results presumed to have been caused by her exposure to his testosterone while they were so close in the uterus.

A number of other papers refer to those findings as if they constitute evidence that common variations in sexual development, such as homosexuality, may be shaped or caused by anomalous variation in prenatal testosterone concentrations. Prenatal testosterone concentration is assumed to be the cause of any such developmental sex difference. It is all nonsense because there is no sound evidence supporting any such connection in human twin pairs to begin with.

In cattle, the female member of a male-female twin pregnancy is usually imperfect in her sexual development, often ambiguous in the appearance of her external genitalia, and usually infertile. She is called a freemartin. I have no idea why she is called a freemartin. According to all sources I have so far discovered, aside from some vague and tentative suggestions, the etymology of the word is in fact unknown. The freemartin condition has also been reported to occur in female members of male-female twin pairs in sheep and in some species of deer, where connections among the placental blood vessels similar to those in cattle are routine.

It seems that most of the thinking about the freemartin phenomenon clustered around the idea that the oddities were caused not by mixing of cells, but by the heifer's exposure to the testosterone in the blood from her fetal brother, circulating through her female fetal body. Most of the literature about bovine freemartins talks about how hormones from the fetal bull calf have circulated in the body of the fetal heifer calf and caused aberrations in her sexual development. In most such writings, there is no mention of any effects this arrangement might have on baby brother bullock's sexual development.

The calves from mixed-sex bovine twin pairs are usually blood chimeras. That is to say, they usually both have blood cells of each other's genotype, as well as their own, acquired by exchanges through shared placental blood vessels. Because of their having exchanged cells before their immune systems settled in to recognize "self" and kill off the white blood cell lines that would have attacked their own tissues, they have developed immunological tolerance of each other's cell-surface antigens. The antigens of each are recognized by the other as immunological self. Tissue transplants are therefore straightforward. There is no immune rejection, as would be expected in any normal non-twin brother-sister transplant. This latter finding was part of Sir Peter Medawar's 1960 Nobel Prize-winning work.

Often, it seems, they are both sexually imperfect, but apparently it is almost invariably worse for the heifer. The heifer often has abnormal external genitalia intermediate or ambiguous in appearance, and she is usually sterile even if and when she appears superficially to have assembled the reproductive components of her fetal self correctly. Some reports mention no effect on the bull calf, some say there are none, and others report that the bull calf will usually not grow into normal reproductive productivity, even if and when he does show any interest in making the effort.

The freemartin heifer (the designation "heifer" is not about her age — a "heifer" becomes a "cow" after she calves) is often as likely to mount another heifer or cow in heat as her twin brother is. Some dairy-farm freemartins have been kept on hand for just such telltale functions, instead of going directly to slaughter. There is not otherwise much benefit in keeping, raising, and feeding her. Cows can only be milked after they have calved, and she will not be doing that.

This is therefore certainly at least not *all* about his testosterone masculinizing or defeminizing her. According to the weight of the evidence I have been able to find, the bull calf is usually also changed, and we do not generally tend to give the heifer's estrogens credit for being able to make such differences as that. After all, male mammalian fetuses are always gestated inside the body, and fed and watered from the blood, of a normal, fully functional female — surrounded by the female bodies of their mothers and bathed in typical female concentrations of estrogens.

As a result of that kind of thinking about these observations, I entertain serious doubts about nearly every part of that whole system of ideas, especially insofar as they might pertain to human twins.

In all human twin pairs, it is usual to find one of the twins, as children, to be more outgoing and more engaged with other people. (Of course, unless they are exactly the same in this or any other characteristic, then one of them must be more so and one must be less. Since they are actually never exactly the same, then they are usually different — how deep is that?) The more outgoing twin is often called the "socially dominant" twin. They may take turns at it, with one leading for a while and then the other, or they may share the labor and do the leading and following, like other chores, differently in different situations.

In boy-girl pairs, however, with rare exception the girl is the socially dominant twin and the leader in most social situations, and she maintains that position fairly steadily. It further seems that, in most cases, she achieves that position by verbal means: "You should do this for me because I am your twin sister ... ."

It is easy enough to suppose that this happens the way it does because girls in general develop their verbal skills earlier and are for several years better on average at more or less all verbal functions than boys of the same age. When Adam Matheny reported this observation about social dominance in girl-boy pairs, he attributed the difference to the long-known female advantage in the timing of verbal development. There are also reports in the literature to the effect that girls with twin brothers are more inclined to risk-taking than singleton girls or both-girls twins — so that could also be part of the basis for the usual difference.

There are a number of other developmental variations in respect to which it has been reported that girls with twin brothers are "masculinized" or "defeminized." Very importantly, the traits in which such differences have been reported are not all behavioral. There are some subtle anatomical differences as well, such as the relative lengths of the second and fourth fingers — according to some reports that is particularly the case on the left hand. There is no real question of this being learned or otherwise negotiated.

Now, it is true that a normal male fetus makes more testosterone than a normal female fetus. That is to be expected because a normal boy fetus has testicles — the primary function of which is to produce testosterone, a function which they perform better than any other tissue. Making testosterone, which will enable and encourage masculine developmental functions in target tissues, primarily in the form of its metabolite dihydrotestosterone, is arguably the most important part of all the purposes his fetal testicles serve, what they are there for at that point in his life. It has been reported that it is possible to make a very good guess about the sex of a fetus simply by measuring the testosterone in the amniotic fluid.

It is also true — in experimental animals — that experimental variations in prenatal testosterone level can shift the distributions of several sex-differential behaviors: more testosterone, more masculine, less feminine. In those situations — in experimental animals — we know that testosterone is what is causing the differences — because it was injected into the body of the mother, and there were experimental controls, such as some of the animals getting injections of only the oily vehicle in which the testosterone was delivered to the others.

However, in normal human pregnancies, there are several reports in the literature to the effect that the testosterone circulating in the mother's blood serum does not vary significantly with the sex of her fetus. So, tell me if you can, please, how testosterone could be moving from the testes of a boy twin fetus, by way of his blood, into the body and blood of his twin sister fetus, without showing up in his mother's blood on the way? I can find no evidence that there could be any way for testosterone from a boy twin fetus to get from his body into his twin sister's body other than by way of maternal circulation carrying testosterone from his placenta into hers.

Perhaps it is reasonable to imagine that testosterone might just soak through the membranes from one twin to the other in the womb? I cannot claim any certain knowledge, but that just does not sound very likely to me, and the data do not support it. When boy-girl twins have been assayed for the concentration of testosterone in their respective amniotic fluid volumes, the values reported were typical: the girl twins had completely typical properly girlish values, and the boy twins showed proper boy numbers. No support could be found for the notion that the girl fetuses in boy-girl twin sets were soaking up boy-fetus concentrations of testosterone, like the rodent pups in the corresponding experiments, with their very different placental structure.

> There is no reason to believe that boy fetus testosterone is responsible
> for the masculinization or defeminization of his twin sister.
> It happens, some other way.

> We humans, as mothers and fetuses, simply do not do placental
> circulation, especially twin placental circulation, the way cattle
> and deer and sheep do it.

> Simplex veri sigillum, indeed … believe it!
> … but here we go again:
> Not all simple things are true, and not all true things are simple.

Perhaps I told you that already? Did I tell you the one about stumbling over the truth and getting up and going on as if nothing had happened? Do you remember anything I said about checking the assumptions? This is certainly not the only sizable literature out there that is based on assumptions that can be easily dismantled with simple tools made of observations that can be found in other papers not far away. There are many things about development that we do in ways very similar to the ways mice do them, for example, but there are also things that just are not the same.

Backing up just a little, there is something special about relationships between sex and handedness among twins, which fits in well here: Twins and their siblings do not show the male excess frequency of nonrighthandedness that is characteristic in virtually all other populations.

> Sex and brain function asymmetries interact —
> and they do not interact the same way in twins
> as they do in singletons.

There are, further, a number of indications that the members of boy-girl twin pairs are not entirely like other members of their respective genders among singletons and same-sex twins.

In one study, girls with twin brothers showed a more "masculine" proneness to aggressive behavior than their same-sex counterparts. Other results show that females with twin brothers scored more like males on a test of willingness to bend or break rules.

In most tests showing such differences, the male members of boy-girl pairs are not significantly different from singleton males. Males with twin sisters did, however, gather more-masculine-than-usual scores on the test of willingness to "challenge the system."

The ratio of the lengths of the second and fourth fingers (pointer or index fingers vs ring fingers) has been widely reported to be generally lower in heterosexual males, particularly on the left hand. Girls with twin brothers have lower (more "masculine") 2D:4D ratios than girls with twin sisters or singleton girls. Boys with twin sisters do not appear to share any such effect, not differing from boys with twin brothers or from singleton boys.

Most people have never heard of this particularly strange bit of information, but the involvements of human ears with sounds are not limited to hearing them. Ears also *produce* sounds — called spontaneous otoacoustic emissions (SOAE). This is not the same as tinnitus ("ringing" in the ears). Unlike tinnitus, the sounds of spontaneous otoacoustic emissions are generally not heard by the ears that make them. The ears of female humans usually make more of these sounds than do the ears of male humans. Girls with twin brothers have fewer spontaneous otoacoustic emissions than their same-sex or singleton counterparts, i.e., in this way (with respect to SOAE), among several other ways, they are more like males. Boys show no interesting difference as a function of twinship or of sex-pairing among twins.

All of these differences are considered in the literature in the context of girls with twin brothers having experienced in the uterus concentrations of testosterone well above normal for females. This is treated, throughout the literature I have found, as if those researchers are certain of the existence of such a difference, and of its causal connection with these variations. However, I can find no published evidence that direct tests of that assumption have ever done anything but fail. There is no reason to believe any such difference actually exists in human girl-boy twins. To all appearances, the assumption that it is there has been transferred intact and unexamined from the bovine freemartin situation. It is baseless folklore, like double ovulation.

My studies of structural relationships among the sizes and shapes of teeth In twins and singletons (detailed in the previous chapter) show probably the clearest evidence of these unusual features of boy-girl twins. How such differences are installed in the course of development will remain an excellent question for further study. There is more there to be learned.

The members of "opposite-sex" twin pairs have other developmental distinctions as well.

Both male and female members of boy-girl twin pairs have lower prenatal and infant mortality than their singleton and same-sex twin counterparts. Whatever makes boys generally more vulnerable throughout the fetal period is somehow substantially reduced among boys born with twin sisters, and it is not contagious. Girls with twin brothers do not have their mortality statistics shifted in a masculine direction — in fact, they do better than their singleton female counterparts or their same-sex twin counterparts.

Both members of girl-boy twin pairs are somehow better off, in several (developmental) ways, than their single-born or same-sex-twin counterparts. It should be remembered that this observation concerns only live-born boy-girl twins and may not be true for the others.

Sex Ratio at Birth, and the Origins of Its Variations

In general, the typical singleton boy fetus fares less well in all losses and failures during *recognized* pregnancy, and males are widely considered to be at a developmental disadvantage throughout the rest of life. In spite of an excess of male losses throughout the duration of recognized pregnancy, there are almost always more males than females among live births. Not until after middle age (and half a lifetime of higher mortality rates in males) do females gain the advantage in numbers.

In spite of several well-performed research efforts, no evidence has been found and reported that this skew in "sex ratio" begins with or before fertilization. Every separate study has reported finding no statistically significant departure from equal numbers of X-bearing and Y-bearing sperm cells.

However, when all of the results from all such studies are considered together, in meta-analysis, there is evidence for a small but statistically significant (unlikely to be due to chance variation in sampling and therefore, arguably, "real") excess of X-bearing sperm. We have no evidence on the basis of which to know whether this has anything to do with the standard technique for performing those experiments, namely "fertilizing" hamster oöcytes with the species-specific zonae pellucidae removed to allow the human sperm to enter.

What intervenes, what might change, between insemination, when numbers of X-bearing and Y-bearing sperm cells are equal in every test that has been reported in the literature, versus all of life beyond the recognition of pregnancy? A great deal happens, actually. All of embryogenesis happens between those landmarks, between insemination and pregnancy recognition.

During embryogenesis, males have a large advantage in functionality. Male embryos progress and change more quickly than female embryos. There is something about a paternally-imprinted X-chromosome[a] that very substantially slows development of embryonic females in every mammal properly examined to date.

Timing is very important in embryogenesis because much of what happens in embryogenesis depends on chemical signals from one cell, or from one tissue primordium, to another. Relative timing is likely to be even more important than absolute timing, especially for such cell-cell signaling within the developing embryo. Without properly timed signals, many cells will not "know" where and when to perform the next step of what they came on the scene to do. Several very competent and effective subsystems of the living cell's biochemical systems can and will, on recognizing the absence of a required signal or the presence of a signaling error, call a halt to all ongoing processes and trigger the death and dissolution of all that has been built up to that point.

---

[a]We should discuss genome imprinting and the rest of epigenetics at greater length at some point. For now, understand that the DNA in the chromosomes that are packaged into sperm cells has been chemically manipulated in and by the process of building sperm cells in certain ways that differ from the corresponding processes in building egg cells. There are at least a few hundred genes that function differently in the embryo's paternal half-genome from the way they function (with the very same DNA base-pair sequence) in the maternal half of the embryogenic genome. Some of those genes are located on, or somehow affect the embryogenic functions of, the X-chromosome. A zygote with a paternally imprinted X-chromosome is, of course, female (if the father's contribution to the zygote's sex chromosome pair is not an X, but a Y, then the product is not female). We do not know yet what it is, but something about an X-chromosome that has come into the zygote from a sperm cell very substantially slows embryogenesis for females — reduces the speed, thereby causing events and processes to take longer and to end with different results.

Equal numbers of male and female zygotes at the onset of embryogenesis change during embryogenesis to an excess of males. That excess of males is sufficiently large to afford excess losses of males in every untoward event during all of pregnancy beyond embryogenesis (throughout the fetal period) — and still have an excess of males at delivery. There are excesses of females among the earliest failures, in embryogenesis, establishing a male excess for all forms of loss throughout the fetal period, and still a male excess at delivery.

When experimental chimeric mouse embryos are created from mixtures of male and female cells, the result at delivery is almost always a functional male with a normal male appearance, consistent with the male cells always very substantially out-growing or overgrowing the female cells in the mixed-sex embryo.

Girl-Boy Pairs in Genetic Twin Studies — It Just Isn't Done

Throughout all the decades of genetic twin studies, opposite-sex pairs have generally been ignored. Remember, the focus of the genetic twins study is the comparison, between monozygotic and dizygotic twin pairs as groups, of differences between the members of each pair. Sometimes there may be sex differences in the trait to be studied or in one or more variables that might be possible causes or contributions to that trait. When that is the case, then within-pair sex difference has always been considered to be extraneous and intrusive to the task at hand. Because no one ever figured out a way to isolate and understand any differences sex might be contributing to within-pair differences, the boy-girl pairs have been consciously omitted from Galtonian genetic twin studies.

We have "always" been quite certain that any pair of twins that includes a normal boy and a normal girl is a dizygotic pair. The Weinberg protocol for estimating the fraction of monozygotic versus dizygotic pairs in any sample of twin pairs has relied on that certainty and on the assumption that sex-pairing fractions at delivery are the same binomially distributed fractions as were established by separate and independent fertilizations. Results of such estimates, as applied in various circumstances, have been the very pillars of the orthodox understanding of the biology of twins and twinning.

When the question of estimating zygosity fraction involves any abnormality, the Weinberg protocol implicitly includes the additional essential assumption that the boy-girl twins are in every way absolutely representative and characteristic of all of the dizygotic pairs, in all normal and abnormal aspects of their development. We will be discussing the Weinberg protocol in greater depth in the next chapter, including the reasons for the need for this assumption. I think it should be clear to you already from what we have covered in this chapter that this assumption is unfounded — and even clearly false. The members of girl-boy twin pairs differ substantially in many aspects of their development from their gender counterparts among singletons or same-sex twins.

Monochorionic Girl-Boy Twins

Boy-girl twins have recently provided an enormous advance in my understanding behind this whole story, in the form of monochorionic boy-girl twins.

Normal girl-boy twins are absolutely certainly dizygotic. (Monozygotic twins can be boy and girl, but only by way of the chromosome anomaly known as Turner syndrome. A Turner syndrome girl experiences unusual development in a variety of ways, as described earlier.)

According to the orthodox tradition, monochorionic twins are all, always, monozygotic, also with absolute certainty.

So, just the very suggestion of the existence of monochorionic boy-girl dizygotic twins is cause for uproar. Proof of that fact changes the world, changes the paths of the stars in the heavens, and shatters foundations of ancient creeds. That is simply not acceptable!

Monochorionic dizygotic twins are necessarily chimeric, and most of those reported to date have been reported as such. There are reports suggesting reasons to suppose that they are more likely to occur in ART (artificial reproductive technology) pregnancies, primarily IVF-ET (*in vitro* fertilization and embryo transfer) and ICSI (intra-cytoplasmic sperm injection).

Same-sex twin pairs that are diagnosed monozygotic on first impression because they are same-sex and monochorionic, but who have subsequently been more closely investigated because of discordance for a major anomaly and thus found to be dizygotic, are also much more common in ART pregnancies.

The anomalies found in these cases tend to cluster in a class of such problems caused by errors in genomic imprinting, leading to the supposition that the manipulations of the gametes in preparation for ART conceptions is disturbing either the imprinting processes of gametogenesis, or the execution of the imprinted variations of developmental programs early in embryogenesis.

The development of male and female embryos and fetuses together as twins is a unique set of circumstances, representing a process that differs in many ways from same-sex twinning of either zygosity. The use of boy-girl twins as a reference point to represent all dizygotic twins has been and remains logically and biologically ridiculous.

The Weinberg Tautology and The Lingering Perfume of Red Herring

In any collection of data, the figure most obviously correct,
beyond all need of checking, is the mistake.

"Finagle's Third Law"

Error does not become truth by reason of multiplied propagation
nor does truth become error because nobody sees it.

Mohandas K. Gandhi

A good place to apply scientific leverage is on an implicit assumption
that everyone makes and that is so implicit that
no one would even think to mention it to students entering the field.
Negating that assumption may lead
to new and interesting ways of thinking.

Rodney Brooks

Tradition is what you resort to
when you don't have the time or the money to do it right.

Kurt Herbert Adler

As I have mentioned here and there earlier in this writing, much of the lore of the biology of twinning is based on conclusions drawn by the use of the "Weinberg Difference Method" or the "Weinberg Rule."

That formula was a bequest to twin biology about a century ago from Wilhelm Weinberg, a German physician and a serious hobbyist in the mathematics of population genetics. His "Method," or "Rule," is an algorithm (a recipe or procedure for calculations) for the purpose of estimating what fractions of any given sample of twin pairs are monozygotic vs dizygotic.

In the time when Weinberg lived and worked, no one had yet invented the kind of genetic testing we might routinely use today to determine accurately whether any given pair of twins is monozygotic or dizygotic. It was generally understood that there were very important differences in cellular origins and developmental consequences between monozygotic and dizygotic twins, and that knowing the relative numbers of the two sorts of twins in various circumstances would be crucial to understanding their different biologies. Against that background, this contribution from Weinberg was seen as a great, brilliant light in the dark.

There was a general understanding then that there were two "kinds" of twin pairs. Pairs of one "kind" were understood to comprise otherwise ordinary (except for their twinship) pairs of siblings. The other kind appeared to be much more like two very good, if not quite exactly perfect, copies of the same person. It was imagined that the two kinds of pairs must have very different cellular origins. There seemed to be various situations among which the proportions of the two kinds in the mixture seemed to vary, and there was some curiosity as to the meaning of that variation. The question about the fractions of each kind seemed to some workers to be an important and worthwhile one to try to answer.

Examples of situations in which the question of zygosity fractions has been considered to be a matter of profound concern have included:

- the variation of the biology of twinning over continental subpopulations,
- variation in the frequency and biology of twinning after discontinuing the use of chemical contraceptives,
- variation in frequency and biology of multiple pregnancies associated with the use of artificial reproductive technologies,
- variation of the biology of twinning as a function of such variables as age of either parent, or age of gametes as timing of insemination varies relative to maturation of the egg cell.

Most of the answers we have gathered about such questions as these (on which to base the stories we wish to tell each other about how twinning happens) have been assembled from Weinberg Method results. These studies have not been redone since more exact approaches have become available, because the old answers have been repeated so many times as if they were facts that most people in the field seem to have come to believe that they *are* facts instead of unconfirmed hypotheses and that there is no reason to re-examine anything.

How the Weinberg Trick Works

The formula, the Weinberg algorithm, appears to be simplicity itself. All we have to do is to count the same-sex pairs and the boy-girl pairs, then subtract the number of girl-boy pairs from the number of same-sex pairs in the sample. The *difference*, the result of that subtraction, the produce of the "Weinberg *Difference* Method," has long been considered to be a very good estimate of the number of monozygotic pairs among the same-sex pairs in any given sample.

The Binomial Distribution

Sometimes, people and other sorts of things show up in groups, which may be all of one kind or of two (or more) different kinds, and the nature and distribution of those combinations can be of interest. Whenever, for example, there are two kinds of things in a particular universe, such as males and females or other such binary classes and sorts of people, or heads and tails on coins, and I want to know what to expect when dealing with some number of groupings of those things, the binomial distribution describes the probability distribution of the likely results.

Over a few decades of learning, thinking, teaching and writing about the ideas that surround and support understanding human biology and development, it has seemed to me that probability is the hardest part of it for most people to understand. Coin tosses provide a ready and classic example, and people are often surprised at how differently things happen from the way that they would first suppose.

If I flip two coins, for example, there are *three* possible outcome *combinations*: both heads, both tails, and one of each. Many people incorrectly suppose at first thought that those three outcomes should be equally likely. If I bother to record first vs second, or if I distinguish between the quarter and the dime in some other way, then it becomes clear that there are two ways (permutations) to generate the condition (combination) "one of each." "One of each" can happen in either of two ways: "heads first, then tails," and "tails first, then heads"; OR "heads on the quarter, tails on the dime" vs "tails on the quarter, heads on the dime."

Taking into account the order of tossing or any other distinction between the two coins, there are *four* possible outcome *permutations*. If the coin tosses are truly random and independent, the binomial distribution predicts that, over the long run, one-fourth of the double tosses will be both heads, one-fourth both tails, one-fourth HT and one-fourth TH, for a total of one-half of all combinations of two coming out to include "one of each." "Both heads" and "both tails" each can only happen one way, which leaves one-of-each being twice as likely as either "both heads" or "both tails."

This is the mathematical basis of the Weinberg Difference Method of estimating the number of the monozygotic pairs in any given sample of twin pairs. Simply replace heads and tails with X-bearing versus Y-bearing sperm cells, which might succeed in forming a zygote and thus initiate the building of male vs female embryos and perhaps on to liveborn babies, and the binomial distribution will show you that one-of-each, boy-girl pairs [at fertilization] are about half of all the pairs the sexes of the members of which were *independently*[a] determined at fertilization.

Among monozygotic twin pairs, both members of each pair are of the same sex because they derive from cells descended from a single zygote cell that is the product of a single fertilization. A single sperm cell provided either one X-chromosome to go with the X-chromosome from the mother to cause female development, or one Y-chromosome to cause any resulting offspring to develop as a male. Dizygotic twinning requires two sperm cell nuclei: either both with X-chromosomes, both with Ys, or with the two sperm cells carrying different sex chromosomes, one X, one Y.

Consider what it means when we assume the binomial independence of sexes of dizygotic cotwins: If *and only if* dizygotic co-twins arise independently by the independent fertilization of independently ovulated oöcytes, then they may be expected to be independent in sex, just like any pair of individuals derived from independent fertilization of independent oöcytes — such as non-twin siblings from separate pregnancies.

Now, we have all the pieces at hand: If half of all individual members of such pairs are male and half female because half the sperm cells bring an X-chromosome and half bring a Y, and if the events in question satisfy all of those

---

[a] "Independently" is a crucial concept here. Statistical independence requires that no one of a set of events under consideration has any influence on any other one of those events. Independence is fundamental to the valid use of the binomial distribution. When the events in question are functions of living organisms, this logic absolutely requires that same independence in the things that cells must do to generate those events, just as it does in a fair coin toss.

necessary assumptions of independence, then and only then ... may dizygotic twin pairs, at fertilization, be responsibly expected to assort binomially as one-quarter each two males or two females (half same-sex), and one-half mixed-sex, boy-girl pairs.

As above, if $m = 0.5$ and $f = 0.5$ and $(m + f) = 1$,

[i.e., where m is the fraction male, $\frac{1}{2}$, and f is the fraction female, $\frac{1}{2}$,

and those mutually exclusive conditions add up to all 1.0, 100%, of

reality ... in other words, everybody must be one (f) or the other (m)

and no one can be both ...]

then the distribution of randomly sorted *pairs* is:

$$0.25(mm) + 0.25(mf) + 0.25(fm) + 0.25(ff)$$
$$= 0.25(mm) + 0.5 \ (mf \text{ or } fm) + 0.25 \ (ff)$$
$$= 0.5 \text{ same-sex} + 0.5 \text{ not-same-sex}$$

Of course any two kinds of things need not exist in equal proportions. Short runs may be substantially skewed and real departures from those $\frac{1}{4}, \frac{1}{2}, \frac{1}{4}$ proportions may be expected if the proportions of the two kinds of states or events depart significantly from 50:50.

Plotting the expected fraction of boy-girl pairs ($2mf$) vs the fraction male (m) or the fraction female ($f = 1 - m$), from 0 to 1, easily makes it clear that the fraction of boy-girl pairs among dizygotics has very little sensitivity to deviations in sex ratio within a range likely to be observed. Papers have been published offering "correction" for departures from 50% of each sex, departures from $m = f = 0.5$, observed at birth. This is negligible in effect and off the point. Even a 60:40 or 40:60 sex ratio, each of which is a very significant departure from 50:50 however imposed, predicts 48% boy-girl pairs [$= 2 \times 0.4 \times 0.6 = 0.48$], a difference from 50% which could not reach sound statistical significance in any sample smaller than 2500 pairs.

Much more critical is the question of independence, whether each sperm is drawn from a 50:50 pool, randomly and independently with respect to any variable which might affect viability or development. If the paired events are not independent, the nice mathematical exercise of the binomial fractionation is irretrievably inappropriate. Bill James (WH in the literature), for one, has on several occasions published cogent reasons to doubt that the sexes of dizygotic cotwins are independent. Since there is no evidence at this beginning for or against that necessary independence, and no way to get any until we see some results, our need to believe in what we are doing requires the assumption of independence.

The mathematical basis for Weinberg zygosity-fraction estimates is this very same one and only binomial distribution. This is one of the simplest and most straightforward of all such major chunks of reasoning about statistical probability distributions. For these uses, that straightforwardness and simplicity is at least as misleading as it is familiar and comforting. Even this relatively easy path has rules upon which its appropriate uses depend absolutely, and ignorance of the rules is no excuse for doing it wrong.

The derivation of the formula is equally simple at first glance. Weinberg's algorithm assumes that the girl-boy dizygotic pairs are equal in number to the same-sexed dizygotic pairs [at birth, born as pairs of living twins]. Supposing that to be the case, then subtracting the number of boy-girl pairs from the total number of same-sex pairs should leave as the difference the number of same-sex pairs that are not dizygotic. The same-sex pairs who are not dizygotic are of course the monozygotic twin pairs.

The Weinberg Method cannot tell us which ones are which. When we have finished the algorithm, we have an estimate of how many there may be of each — what fraction of a sample of same-sex pairs belong to each zygosity group — but we have nothing contributing to individual identification.

The assumption that girl-boy pairs and same-sex dizygotic pairs are equal in number follows from the (absolutely necessary) prior assumption of random binomial sorting of pairs from a 50:50 pool of X-bearing sperm cells and Y-bearing sperm cells at — robustly independent — fertilization.

If half the sperm carry X-chromosomes, as we believe they do, and will cause the development of female embryos if they produce any embryo at all, then the probability of randomly drawing two X-bearing sperm cells to fertilize two maternal pronuclei, to initiate development of female twins, is $0.5 \times 0.5 = 0.25$, one-fourth. To pick two Y-bearing sperm out of two, to initiate the development of male twins, the probability is the same, 0.25. To produce a boy-girl pair, as we saw above, there are two ways: either X for the first, then Y for the second;

or Y for the first and then X for the second. Each of those two outcomes has 0.25 probability, for a total probability of twin pairs with one member of each sex = $0.25\,X_1Y_2 + 0.25\,Y_1X_2 = 0.5$.

It Is a Long, Rough, Narrow Trail from Conception to Birth

Since the (Weinberg) question at hand concerns the zygosity fraction in a sample of *liveborn* twin pairs [to be calculated on the basis of what we assume happened at insemination], it is also necessary [here comes the part that nobody seems to have noticed] to assume that the distribution of sexes among the liveborn twin pairs is the same at delivery that it was at the time when sperm and egg cells were meeting. We count same-sex and boy-girl twins at birth and we are absolutely required to assume — if we want to apply the Weinberg algorithm — that those counts accurately reflect what happened at fertilization.

Weinberg considerations seem never to have included adequate reflection on the biological distance between the beginning and the end of pregnancy, between which occasions a very large fraction of all conceptions will fail.

For that particular fundamental and necessary assumption to be true, if the sex-pairing in liveborn twins must have the same distribution that it had back at their fertilization, then we further must assume that all of prenatal development, including all those losses, will have been the same for males and females, and the same for members of same-sex pairs as for members of girl-boy pairs. If either of those assumptions is false, then we must suppose that the male vs female fractions, or the same-sex vs opposite-sex fractions, could have changed and probably did change during gestation. If either of those assumptions is false, the whole procedure has been a waste of time. But that's not all!

We must now, also, further assume that the members of girl-boy pairs are perfect developmental representatives of all of the dizygotic twins, if we are to be able to believe that any difference (in number, condition, whatever ...) between boy-girl pairs and same-sex pairs is due only and entirely to the members of monozygotic pairs. That was from the beginning and still is the whole reason for all of this, after all. We never had any reason to hope to get more out of it than that.

If, for example, girl-boy pairs or the members thereof were more, or less, prone to failure during embryogenesis or fetal development than same-sex pairs and their members, then these required fundamental assumptions would be false and we should expect that the fractions of boy-girl vs same-sex pairs may have changed over the course of prenatal development, and the fractions at birth are not those that were established by random independent fertilizations.

If there should be any significant difference between males and females in their vulnerability to any form of prenatal loss, those required fundamental assumptions would be proven false and we should expect the fraction of females vs males, and the fractions of same-sex vs girl-boy pairs, to change in the course of prenatal development.

In case of the failure of those assumptions, we cannot believe the results of Weinberg method estimations, and we really have not learned anything about the relationships we set out to understand.

As a matter of simple, repeatedly demonstrated fact, failures during the fetal period of prenatal development include more males than females, and include more members of same-sex twin pairs than members of girl-boy pairs.

This, and other, sound evidence has long been available to show that each and every one of those assumptions mentioned above as being fundamental and necessary for the logic of the Weinberg Difference Method is in fact forever untenable.

And yet, papers have been published even in the last year or two in which the Weinberg Method has been used to estimate zygosity fractions of samples of twins, and the results have been put forward as if they have important meaning that can be used as basis for further explorations.

The logic of the Weinberg Difference Method is inherently circular, and every attempt at practical application of the protocol is deeply flawed in its logic. This is most particularly true with respect to questions about the distributions of anomalies as a function of zygosity. In most of its applications, the procedure unavoidably has its conclusion as part of a premise. The problem is right back there where we had to assume that girl-boy pairs are perfectly representative, with respect to every aspect of development, of all the dizygotic pairs. We are compelled to assume that assumption, if we are to allow ourselves to believe that every difference between boy-girl twins and members of same-sex pairs is due to the members of monozygotic pairs ... .

If, by any chance or necessity, by any mechanism, for any reason, girl-boy pairs are not perfectly representative of all dizygotic twins (such that every part of development happens for all twin girls and for all twin boys the very same way that it happens for boys and girls who are twins together in the same uterus), then we cannot expect to learn anything from the Weinberg Difference Method exercise.

Knowing (estimating, really, and acting as if we know, from which simulated knowledge to draw further conclusions) the zygosity fractions in any one given sample of twins is of no particular value anyway. Even if we could be certain of the zygosity fraction in a given sample by any means other than genotyping all of them, we would know nothing useful. All of the real, practical questions are about differences between or among samples/groups/populations of individual twins. For example, are the zygosity fractions different in samples of twins from African mothers vs European vs East Asian mothers? Roughly a century of published work says that is the way things are ... If that is true, then what does it mean?

Are the zygosity fractions different among twin pairs of which either member has a malformation, or a chromosome anomaly, or is afflicted with any of a number of other traits of those general sorts? Roughly a century of published work says that is the way things are ... . If that is true, then what does it mean?

Do the zygosity fractions change with changes in mothers' ages? When the frequency of twinning increases among older mothers, are all "kinds" of twins equally becoming more frequent? What do older mothers do more of, behaviorally or physiologically, to cause their twinning frequency to increase? Is there more "splitting," or more "double ovulation"? Or both? Or neither: Could it simply be that a twin pair, once conceived, may be more likely to survive to term in a more experienced womb? Or in an older womb just because it is older, even if it has never been lived in?

Weinberg results have been used to give the appearance of confirming the common knowledge understanding about how monozygotic twinning is a developmentally dangerous thing to do because of the odd "splitting" embryogenesis that must be involved. The other face of that same outcome, at that same cellular level, is the understanding that dizygotic twinning is simple and easy and safe and developmentally straightforward. In that case the totally unfounded prejudice confirmed by cranking through the Weinberg tautology is that dizygotic twinning is simply a double, side-by-side execution of the ordinary process of singleton pregnancy, two parallel instances of singleton embryogenesis and gestation. There should be none of that "splitting" mess and none of its untoward consequences.

It was decided and decreed in those days (and has been held close ever since for most of those who concern themselves with twins) that for lefthandedness and asymmetry malformations to be excessive in frequency only among monozygotic twins is perfectly compatible with the idea that "splitting" causes the departure from normal brain function asymmetries that leads to lefthandedness. And the same must be true with all the other anomalies excessive among twin births, of which dizygotic twins will not partake because they come from two independent egg cells and do not have any splitting to do.

Assuming as we must that the members of girl-boy twin pairs are perfectly representative developmentally of all dizygotic pairs, then of course any difference observed between girl-boy pairs and same-sex pairs must be due to the monozygotics.

If [and *only* if!] the same-sex dizygotic twins truly do develop exactly like the girl-boy pairs, then every difference between same-sex pairs and boy-girl pairs must be caused by the monozygotic twins among the same-sex pairs.

The common knowledge lore of twin biology has always been wrapped tightly around this question of different "kinds" of twin pairs and branches of that question, even for a good while before Weinberg provided his way of estimating an answer for that question. Suppose the Weinberg Difference Method estimate were credible and accurate. In fact, results from application of the Weinberg Method have been reported to yield a quite acceptable approximation to the genetically determined zygosity fractions among the twins of the East Flanders Prospective Twin Study in Belgium (almost all being of white European ancestry and developmentally normal).

Even if Weinberg estimates of zygosity fractions were routinely accurate and exact all the way out to any arbitrarily large number of decimal places, the answers thus derived would be absolutely non-informative with respect to anything we might actually stand to learn.

With no perceived need for evidence, we have been confident from the beginning, without the benefit of any calculation, that the opposite-sex pairs are dizygotic. The combination of sexes in each pair is the closest approximation there is, anywhere in sight, to an actual bit of evidence. Opposite-sex pairs of normal twins are dizygotic. The rest — any reckoning or conclusion beyond that — is founded on one or more gratuitous assumptions. We learn nothing we can believe and nothing that would do us any good even if we could believe it.

The Weinberg algorithm cannot tell us whether any given same-sex pair is monozygotic or dizygotic, such that we could actually use that knowledge to learn anything sound or direct about the distribution of abnormalities, for example.

Conclusions from the Weinberg algorithm have provided, however, the whole basis for each of several cardinal components of the Orthodox Common Knowledge version of twin biology.

Weinberg estimates of zygosity fractions provide the whole basis for the constant insistence that the excesses of malformations, nonrighthandedness, and every other untoward outcome associated with twinning, are all due more or less exclusively to the monozygotic twin pairs.

Weinberg estimates provide the entire basis for the repeated assertion that variation in twinning frequency among racial groups ("major continental subpopulations" in a more contemporary and politically correct phrasing) is due entirely to variation in dizygotic twinning frequency. For decades, that populational variation in frequency has been offered as proof that dizygotic twinning must be genetic. The same results allowed workers in the field to convince themselves that the frequency (and — *mind the gap!* — therefore the whole biology) of monozygotic twinning is constant over the world's populations, as is appropriate for the kind of non-heritable, developmental accident that they have steadfastly assumed that monozygotic twinning must be.

Many of the fundamental elements of common knowledge twin biology have no basis other than conclusions drawn from the use of the Weinberg tautology. This pillar of the orthodoxy of twin biology arose in good faith a century ago as a means of estimating zygosity fractions in samples of twin pairs, in the then-prevailing absence of any practical way to sort twin pairs by zygosity accurately.

Even now, with highly accurate genotyping readily available (still not trivial in cost, but not too bad in the general context of biomedical research reagents and diagnostic supplies), papers are still published with arguments based on the erroneous principles of the Weinberg calculation.

Flaws in the logical structure of Weinberg's idea as applied to these purposes have been widely reported, all of which are serious, but almost trivial relative to that irredeemable circularity that remains almost universally unrecognized. The most fundamental flaw of the Weinberg Difference Method is the circularity of its logic, the tautology itself. What some consider to be answers to questions at issue in applications of the Weinberg procedure are among the initial assumptions necessary for any of those uses.

Wilhelm Weinberg died over 70 years ago. It does not matter now whether or how he himself questioned or even realized the assumptions required for the use of his zygosity-fraction estimation method. The responsibility and the problem now lies with each of today's investigators, in every ongoing use of the Weinberg approach, or of any conclusion that depends on it, which is done without proper attention to its untenable fundamental assumptions. Perhaps the worst of it is: that includes all of the major elements of what we have been passing along for a century or so as "the biology of human twinning."

When any exercise in deductive logic draws a conclusion that repeats its own premises, its conclusions are logically false regardless of any bits of truth which happen to have wandered in to take a place among those conclusions. If no other validation is available and the circular logic of Weinberg estimation is the only source of understanding, then in fact there is no understanding.

Those "givens" are essential to the utility of the Weinberg zygosity-estimating protocol. They are "If and only if" conditions. The Weinberg Rule zygosity-fraction estimates cannot be correct unless each and all of these following things are true:

- two normal, independent ovulations;
- two normal, independent fertilizations;
- two normal, independent embryogeneses;
- two normal, independent processes of fetal development;
- complete genetic and epigenetic independence for all dizygotic cotwins, and, *only by virtue of all of the above*;

- complete developmental equivalence for same-sex and boy-girl dizygotic pairs and singletons, yielding no change of sex fractions or sex-pairing fractions throughout gestation.

According to orthodox Weinberg calculations, all developmental anomalies occurring among twins in excess of normal background population frequencies are thereby forever and irrevocably reserved for monozygotic twins only.

The appearance of this notion as a conclusion of Weinberg Method analyses carries no logical weight at all because it is a premise of the argument, and the result can never be otherwise.

These premises are assumed, whether or not stated, throughout the whole scope of twin biology orthodoxy, as if they were known, proven facts supported by repeated scientific observations. They are nothing of the kind. Double ovulation is hypothesis, born full grown, christened "common knowledge," and never tested. This is not a matter of this one man's opinion — it is a matter of definition, of the essence of the thing.

Unsupported repetition can elevate hypothesis to myth, but not to theory, let alone to enshrinement as common knowledge as if fact. Only by making predictions which are borne out by critical testing can any hypothesis, no matter how simple or how popular, earn elevation to theory.

Being the simplest of alternatives, if alternatives are appropriately considered, often enhances the credibility of an idea. This double-ovulation hypothesis has implications and makes predictions which are not really all that simple, most of which have never been tested, and several of which have been tested and failed.

For a practical example: there were 616 twin pairs among more than 53,000 pregnancies that were followed from first prenatal visit through the first year of life in the National Institute of Neurological and Communicative Disease and Stroke (NINCDS) Collaborative Perinatal Project. About 80% of those pairs were blood-typed for zygosity diagnosis.

The original investigators were stymied by the minority of pairs who could not be diagnosed for zygosity because at least one member (was dead and therefore) could not be blood-typed. They set aside the chorionicity and blood-typing data they had collected at great expense of taxpayers' grant money and investigator effort, and they retreated to Weinberg estimates, which ground out the expected result attributing the excess fetal and neonatal mortality among the twins to the monozygotics.

I found it reasonable, and overdue, to compare the Weinberg-estimated sorting of monozygotic and dizygotic twins, alive and dead, with the corresponding results from numbers identified by chorionicity and blood-typing. Instead of throwing up our hands and falling back on the Weinberg method, let us put to good use the real data that are available to us because of clinical efforts that were paid for with taxpayers' money.

In that sample, the Weinberg estimates predicted more monozygotic twins than had been identified by blood-typing. This was not because more monozygotic twins were dead and unavailable for blood-typing, as those authors would suggest (using their Weinberg results and their preconceptions about the excess vulnerability of monozygotic twins). A strong majority of the Weinberg-estimated monozygotic twins were in fact identified by genotyping and/or chorionicity data, and were in fact alive (as they had to be for blood-typing, if they were dichorionic).

I ran those Collaborative Perinatal Project data through three different alternative methods for estimating the zygosity fractions, without using the Weinberg approach and without requiring anything logically or biologically impossible. Those three different approaches gave mutually-consistent results showing that same-sex-dizygotic twins suffered fetal and neonatal mortality much greater than that among boy-girl pairs, and at least as great as the fetal and neonatal mortality among the monozygotic twins.

At the center of our concerns are the estimated-but-unidentified monozygotic pairs [the difference between the Weinberg-estimated number and the number identified by blood-typing or presumed identified by monochorionicity]. The results made it clear that those estimated-but-unidentified monozygotic babies had to account for three deaths each — *they had to die three times apiece to match the Weinberg estimate* of the zygosity distribution of the observed fetal and neonatal mortality. To me, that just did not seem fair. Humans appear to be the only organisms that will kill or die for an idea; no fetus or infant should be required to experience more than one death in the service of any hypothesis.

This bizarre distortion is too much to ask of sampling error, unless a sizable fraction of the "monozygotics" identified as such by blood type were actually dizygotic. That prospect is unlikely unless genetic diversity is much reduced among families who have had dizygotic twins. In other words, if parents of twins are much more likely than parent-pairs randomly chosen from the population to match for particular versions of the set of genes used for genotyping, then that would raise the probability of matching between pairs of their children. This prospect would be a dramatic surprise if it should ever be documented, with implications impinging upon a great many studies using twins to estimate heritabilities. One would think that such a departure from the assumptions underlying those methods should have been noticed, but in fact the genotypes of parents have very rarely been collected or included in genetic twin studies.

You may recall a similar result, with similar questions arising, with respect to trying to fit Weinberg results with blood-typing in the Nigerian samples mentioned in Chap. 7. Twenty-five percent of the boy-girl pairs in those samples had blood genotype results matching for all markers, consistent with their being monozygotic. We will look at that again soon. ...

Here the logic of the Weinberg process is most obviously circular because of its very worst necessary fundamental assumption. Application of the Weinberg algorithm to sort developmental anomalies (up to and including fetal and infant death) by zygosity most especially requires the assumption that boy-girl twins are exactly developmentally representative of all dizygotics, with particular respect to the anomalies in question, so that differences in anomaly frequencies between girl-boy twins and same-sex pairs must be due to the monozygotics among the same-sex pairs.

Only if the anomaly in question is exactly equally likely and exactly developmentally equivalent in girl-boy pairs and in same-sex-dizygotic twins can Weinberg estimates be imagined to yield valid zygosity fractions for twins with anomalies of development. Nevertheless, the expectation that monozygotic twins provide the lion's share of all anomalies, and that dizygotic twins are simultaneous but otherwise developmentally normal singleton womb-mates, has become so firmly entrenched in "common knowledge" that sound data indicating the contrary are routinely and blatantly ignored. The Weinberg way of thinking is clearly false with respect to fetal and neonatal death of twins. Both members of boy-girl pairs are in much less danger of fetal or neonatal death than the members of same-sex pairs of either sex, of either zygosity.

Because of all of these things, it seems clear to me that

- It is simply not true that girl-boy twins are developmentally representative of all dizygotic twins.
- It is simply not true that all of development occurs for same-sex dizygotic twins in the same ways it does for boy-girl twins.
- It is simply not true that the development of dizygotic twins is identical to the development of singletons.
- It is simply not true that monozygotic twins are responsible for the overwhelming majority of the anomalies that are excessive in frequency among twins.

The 616 twins pairs in the Collaborative Perinatal Project sample are too few to extend this investigation with statistical confidence to all of the developmental anomalies which occur more often among twins than singletons. Frequencies of specific individual malformations are small enough that most of them do not appear even once in this sample. It is, however, common and reasonable to consider all developmental anomalies up to and including infant death to be part of a continuum. It seems unlikely that proper analyses with samples of sufficient size should give different answers for non-lethal anomalies than it gives us here for fetal and infant mortality.

With few exceptions, the entire literature of twins and twinning reads as if anyone with sense enough to find his supper dish should know that the double-ovulation origin of dizygotic twins and the corollaries dependent upon that assumption are facts beyond dispute.

I have read the literatures related to twins and twinning and their uses for genetics research for about 40 years. Every source I have ever seen either directly states the double-ovulation assumption as a given, as a fact, or includes a reference to some earlier writing as authority for the pronouncement of double ovulation as a given. The included reference may include an earlier reference, which in turn may include another, and so on. Regardless of the length of the chain of references, the other end of the chain is always a statement, without physical evidence, that dizygotic twins just do arise from multiple ovulation. Nearly everyone has been so comfortable with this notion that no one ever bothered to collect any evidence or to pay attention to evidence which showed up when no one was looking for it. It has been a matter of faith. Relinquish that comfort.

There is no sound reason to believe it, and there are a number of observations offering good reason to know it is false.

Populational Variation in Frequency of Twinning

It has long been reported and repeatedly verified that the frequency of twin births varies over human "racial" groups. Twins are significantly more frequent in families of sub-Saharan African descent and significantly less frequent in families of East Asian descent than they are among families of white European descent. The facts have a structure which is rather less simple than this story sounds. The meaning assigned to this variation by the traditions of orthodox twin biology does not merit its long standing among the pillars of the orthodoxy.

By way of Weinberg estimates of zygosity fractions from large public-record samples of twin births in various racial subpopulations, it long ago became "common knowledge" that the frequency of monozygotic twinning at birth is approximately constant over human racial subpopulations and that racial variation in twinning frequency must therefore be due entirely to variation in live-born dizygotic twinning rates.

Dependent upon the assumptions that dizygotic twins just do arise from independent double ovulation, and that all of development proceeds just the same for boy-girl pairs as for same-sex-dizygotic pairs, this particular conclusion from Weinberg estimates is the basis for the general understanding that there is a racially-variable (therefore inherited) tendency for double ovulation and consequently for the birth of dizygotic twins, while monozygotic twinning is an accidental developmental anomaly because it happens at approximately constant frequency in all human subpopulations. These are fundamental tenets of the orthodox biology of human twinning.

Very few populational samples of statistically useful size and structure have had zygosity fractions estimated by any analysis more credible than Weinberg estimation. The largest such sample is the collection of Belgian twins in the East Flanders Prospective Twin Study, where twinning frequency approximates that of the rest of Europe. In that sample, the Weinberg estimate yields a zygosity-fraction approximation that is not statistically significantly different from the results of genotyping. This is the largest sample ever more-or-less fully genotyped (monochorionic pairs and boy-girl pairs were not genotyped; "everyone knows" their zygosities), and still much too small for statistically valid consideration of all the anomalies more frequent in twins.

Among twins in Nigeria, where liveborn twinning frequencies include the highest of any presumably random populations ever measured, genotyping results and Weinberg estimates have been reported, compared with the results from Belgium, and have been repeatedly used to argue for the soundness of the Weinberg Method and of the rest of the orthodoxy. But that is a flimflam. In fact, the results in those two populations disagree substantially in several clear and important ways.

The frequency of twins per live birth and the fraction of girl-boy pairs among liveborn twin pairs indeed appear to be greater in Nigeria than among Europeans, but those measures vary considerably among ethnic subgroups in Nigeria, and show no evidence inside of Nigeria for the Europe-vs-Africa correlation that is the common-knowledge basis of that pillar of the orthodoxy.

Nearly one-fourth of Nigerian boy-girl pairs shared all of the markers used for genotyping in that study, just as they should have if they were monozygotic pairs. This is a very reasonable "experimental control" [just to make sure that the set of genetic markers used does in fact have a high probability of soundly diagnosing zygosity] that was not included in the Belgian study.

The authors of the Nigerian study then quite reasonably supposed that a similar fraction of same-sex dizygotic pairs must have been misdiagnosed as monozygotic. The authors "corrected" these problematical results with a statistical manipulation which was not used with the Belgian sample. They used Weinberg expectations to "correct" the genotyping results. That statistical manipulation does make some sense. It also makes a substantive difference in the result: The logic of that adjustment was, quite intentionally, fully compatible with the philosophy of the Weinberg tautology. The adjusted answer is no more plausible, no more scientific, no more justified by findings of fact than if none of the genotyping results that were available to them had ever been gathered.

One must wonder at the inefficiency of the chosen set of markers for this particular sample, and/or the reduced allelic diversity of this population, compared to the Belgian sample — if in fact there is no such oddity among the Belgian twins.

Because the Belgian and Nigerian answers were arrived at so differently, they cannot both be correct. This substantially diminishes the argument for the universal soundness of the Weinberg Method. The conclusion that the frequency of monozygotic twins per live birth is constant among all the major subpopulations, and that variation in the frequency of dizygotic twinning accounts for all inter-populational differences, is unfounded, and untenable as a basis for any scientific consideration.

Although the Nigerian genotyping results, thanks to the boy-girl-pairs control sample, can be "adjusted" to an approximation of Weinberg results that makes it credible to the motivated believer, the conclusion that dizygotic twinning frequency differs greatly between European and African births, but that the monozygotic twinning frequency does not, clearly is falsely made — however deeply cemented into the pillars of the common-knowledge orthodoxy.

There is another difference here that deserves to be considered. The observed and recorded differences among populations in chorionicity fractions among the monozygotics urge the consideration that there is more to the biology of twinning than its frequency, however estimated or counted. The Nigerian monozygotic twins are about half monochorionic. Monozygotic twins of European ancestry are about two-thirds monochorionic. According to our best understanding of the implications of chorionicity, African monozygotic twinning events happen earlier in embryogenesis than the European variety. The biology of monozygotic twinning is not the same in Europe and Africa, even if one clings without evidentiary support to the belief that its frequency is the same. Interestingly, several other reproductive events and processes happen earlier for mothers of African ancestry: earlier menarche, earlier first pregnancy, earlier delivery in her average pregnancy (mean, median and mode — not a skew toward majority-defined prematurity), and earlier last pregnancy.

Japanese and Chinese mothers have the lowest observed total frequency of twin births, the lowest fraction of boy-girl pairs, and roughly the same Weinberg-estimated frequency of monozygotic twins as European and African samples. East Asian twins identified by genotyping as monozygotic are, however, about 80% monochorionic.

Neither the frequency nor any other aspect of the biology of monozygotic twinning is in fact constant over human subpopulations.

There is no reason to believe that dizygotic twinning is responsible for all variation in total twinning frequency among groups or situations.

There is no reason to imagine that the binomial distribution of sex-pairing assumed to occur at conception should be or could be maintained through embryogenesis and gestation all the way to live birth.

The orthodox folk science of human twin biology is, at best, not founded in reproducible/refutable observation. Because Weinberg-Method results are so fundamental to all of the most basic (mis)understandings of the biology of twinning, it "all falls down" and there is nothing left to believe when that prop is removed.

The difference matters a great deal to the potential that twinning may have to add to our understanding of human developmental genetics in general.

Excess Malformations in Twins

The frequency of malformations and other developmental anomalies is elevated among twin births. This is another repeated observation. Again, the orthodox interpretation leaves the facts well behind. Because of results from Weinberg estimates among samples of affected twin pairs, the orthodoxy holds that the excesses of these anomalies among twins belong overwhelmingly, if not perhaps entirely, to the monozygotics. This belief is so firmly a part of the orthodox twin biology credo that we have seen monozygosity assumed without question for any same-sex twin pair either member of which has any of the malformations known to be more common in twins. Twice I have stood and denounced that assumption being made on the podium in international meetings. At least twice in papers sent to me for peer review, I have managed to keep instances of that false presumption out of print (at least in the next issues of the journals for which I did the reviews). In those papers, observed differences in genotype were even mentioned and dismissed as lab errors not worthy of retesting because they disagreed with the orthodox presumption and must therefore be certainly wrong.

In spite of high multifactorial heritability values for these developmental anomalies from singleton pairs-of-relatives estimates, and in spite of the prevailing assumption that they concentrate overwhelmingly in monozygotic pairs, the malformations that are excessively frequent among twins are seldom concordant in the affected twin pairs.

Discordance in "monozygotic" twin pairs [they must be monozygotic because they are same-sex and there is a malformation, and everybody knows malformations in twins are peculiar to monozygotics], in spite of high heritabilities computed from singleton pairs-of-relatives analyses, has been widely considered a great puzzle, and is often recited as evidence for "environmental" contributions to the causes of the anomaly in question, which "environmental" contributions in general remain unidentified after generations of intelligent effort.

Although the orthodoxy holds that monozygotic twinning is itself an accidental developmental anomaly which occurs at approximately constant frequency over all the world's populations, and the malformations associated with twinning are — according to Weinberg-shaped traditions — overwhelmingly due to the monozygotics, those malformations which are associated with twinning are not themselves constant over those same subpopulations. Neural tube defects, for example, vary substantially over subpopulations, and are highly correlated with total twinning frequency [variation in which is attributed by the orthodoxy to variation in dizygotic twinning frequency only], but they are not significantly correlated with the Weinberg-estimated fraction "monozygotic."

Monozygotic twinning *is* more or less obviously and directly an anomaly of embryogenic a/symmetry development. Two body symmetries unfold from a group of cells that should normally be expected to build only one body.

Nonrighthandedness is a departure from the normal asymmetry of brain function development. According to the observed distributions of brain function asymmetry, dizygotic twinning *is* — exactly equally with monozygotic twinning — a symmetry anomaly, and one which travels with equal ease through either parent. This is also impossible to reconcile with the assumption that dizygotic twins universally arise from independent double ovulations and independent embryogenesis just like that of singletons except for simultaneity. The embryogenesis of dizygotic twins is at least as different from that of singletons as is that of monozygotics. Naturally conceived dizygotic twins do not arise from perfectly ordinary embryos of which there just happen to be two at a time.

The fact that mothers and fathers share quite equally the relationship with handedness in their twins of both zygosities is difficult to reconcile with the double-ovulation hypothesis, given that only mothers ovulate. In a sample of some 400 Mormon families with repeated boy-girl twinning in their genealogies, it was found that almost exactly half of those related twin pairs were related only through males.

A Scottish family with twins in every generation for over 200 years provides further evidence of paternal transmission of dizygotic twinning tendency. The answer to the question of how the father can "cause" dizygotic twinning will not be simple, and we may be certain that it will not be readily believed when we find it, but there is no room left for believing that dizygotic twinning can come only from the mother, as the double-ovulation story would have it.

The malformations which are excessively frequent in twins are the same ones most common in all live births: the midline, or fusion, malformations, including particularly the neural tube defects [NTD], cleft lip +/− cleft palate [CL/P], and congenital heart defects [CHD]. These are disorders of embryogenic asymmetries of development, anomalies of structures built from embryonic left and right halves which must meet in the embryonic midline, fuse and remodel to form the final structures. In NTDs, the edges of the neural plate do not properly roll up and close to form the neural tube; brain and/or spinal cord is left open and malformed. In orofacial clefts, parts of the face that should meet in or near the midline and fuse do not complete the process. Most CHDs are failures of fusion and remodeling of various parts of the cardiac tubes.

These malformations share associations that might make compelling contributions to understanding normal human developmental biology through understanding the departures from those paths which are excessively frequent in twinning. These a/symmetry malformations are excessive not only in twins (where their traditional attribution primarily to monozygotics has arisen entirely from repetition of conclusions from Weinberg estimates), but also among the sibs and offspring of twins. Among the sibs and offspring, there are in general no significant differences in these associations as a function of twin zygosity, but the excess of twins among the parents of children with NTDs is significantly concentrated in dizygotics.

Just as with twinning itself, these malformations have excesses of NRH among people with these anomalies and their first-degree relatives. Children born with cleft lip +/− cleft palate, and their parents, are NRH more often than the general population, and unilateral CL/P far more commonly affects the left side. Right unilateral CL/P is equivalent to bilateral CL/P in increased genetic risk to siblings or offspring.

Genetic and developmental relationships between minority brain function asymmetry and fusion malformations, between minority brain function asymmetry and twinning, and between twinning and fusion malformations, are not distinguishable by zygosity. All assertions to the contrary are without sound supporting evidence, arising entirely from Weinberg assumptions. Anomalies of embryogenic a/symmetry development are not at all peculiar to monozygotic twinning processes. The only hope of getting it altogether right is to genotype critically every pair of twins born from now on, giving priority to those in which either member is born with any anomaly. Even that will be kept from perfection by chimerism, by virtue of which dizygotic twins — by sharing each other's cells — might show the same genotype in some samples.

The twin-singleton differences observed in these studies have more to do with asymmetries established in embryogenesis than with anything else — exactly the focus of the problems long supposed to be characteristic more or less exclusively of monozygotics.

The Weinberg Difference Method for estimating zygosity fractions in a sample of twins is wrong and misleading.

The phrase "red herring" as applied to a misleading fallacy is said to derive from the lore of fox hunting, where a dried smoked herring (which is a dark coppery red) dragged across the trail of the fox could be depended upon to cause the dogs to lose the scent. The Weinberg fallacy as applied to the biology of twinning has pulled us off the scent and misled us outrageously for a century. Every result we have incorporated from it has been wrong, constantly wrong in ways that have kept us from following the trail of our observations toward the truth.

Chapter Seventeen

Twinning and Spontaneous Chimerism

> Because people understand by finding in their memories
> the closest possible match to what they are hearing
> and use that match as the basis of comprehension,
> any new idea will be treated as a variant
> of something the listener has already thought of or heard.
> Really new ideas are incomprehensible.
> The good news is that, for some people,
> failure to comprehend is the beginning of understanding.
>
> *Roger Schank*

Including, and perhaps especially, inside complex living organisms:

> "The universe is not only queerer than we suppose;
> it is queerer than we can suppose."

For years I carried that quote around in my head without proper attribution. I loved it and was inspired by it. It spurred me to new efforts in my science and I knew it must be right. When I first found it, however, I did not make myself remember who had said it first. I felt certain that it must have come from a cosmologist or a theoretical physicist — Einstein, Feynman, Gell-Mann, maybe Sagan — at any rate, someone used to thinking in terms of universes. But, no ... it seems to have come from the maverick polymath population geneticist and wearer of several other very roomy hats, John Burdon Sanderson (Jack) Haldane. None of this was happening in his day, but, had he known, I feel certain that he would have had his brain frozen in hopes of lighting him up again when it came time to see it. He and Feynman could probably be called kindred spirits, in the "rogues" column.

This is the same Haldane who is said to have been asked a question by some gathered theologians about what he had learned about the Creator from his studies in biology. He is said to have answered the question to the effect that the only thing he knew for certain was that the Creator must have an inordinate fondness for beetles (of which there are at least a few hundred thousand different species, several times the membership of any other Order of multicellular creatures in the whole of phylogeny).

There are a few different versions of this quoted thought:

> ... the "world"/"Universe" is
> not only "queerer" than/"stranger" than
> we "imagine"/we "suppose" ...

I have recently come to appreciate that at least a few people think that it was after all an astrophysicist — Arthur Stanley Eddington — who gave us this thought. There are some who think it was Haldane who said "... the world is ... queerer than ..." and that Eddington said that "the Universe is ... stranger." And there are some who seem certain that it was all Haldane. These ways of stating the same sentiment are in my opinion too much alike to be independent. If both of these men said something of this sort, then I must believe one of them was promulgating the other without reference to the source. Among the sources I have located, Haldane has the advantage in numbers of citations, but I have found nothing with reliable dates by which I might assign priority. Haldane had the kind of creative imagination that fits the story, and I have never known enough about Eddington to make that supposition. However, I must further suppose Haldane would also have been quite capable of stealing an idea that good, so I really just don't know. But I do love the line ... .

And this is a good spot for applying such a knack for synthesis to the prospects of explaining these facts. ...

You should recall my telling you earlier that I began this work believing that we could learn from developmental differences between monozygotic and dizygotic twins as groups. We should be able to learn about the developmental consequences of the odd embryogenesis of monozygotic twins. There is a strong control comparison readily at hand in the embryogenesis of dizygotic twins (which has always been presumed to be perfectly ordinary and just like that of singleton embryos, against which notion stand all of the observations assembled in previous chapters). Such comparisons should greatly improve our understanding of how "normal" singleton development occurs.

The results of my investigations of those questions over the past 30 odd years (it is closer to the truth without the hyphen) have clearly shown that the embryogenesis of dizygotic twins is not "ordinary." The embryogenesis of dizygotic twins is not the same as singleton development. In fact it much more closely resembles the development of monozygotic twins than it does the development of singletons. Not a bit of all of this system of observations I have spread out for you here makes sense if we insist on believing that dizygotic twins arise from double ovulation.

"Double ovulation" is generally taken to mean that all of what happens at an ordinary ovulation happens twice at about the same time: two separate and independent ovulations, coming from two separate and independent ovulatory follicles, in the form of two separate and independent egg cells, with two separate and independent structural plans for embryogenesis carried in the two egg cells, proceeding through two separate and independent processes of syngamy and two separate and independent processes of embryogenesis. They are womb-mates; living in the same womb at the same time, but they have not been introduced.

There is nothing in that idea which should cause, or even allow, dizygotic twin development to differ at all from that of singletons, with the possible exception of physical demands that challenge uterine resources late in pregnancy. (That prospect cannot explain the differences we see, arising from the cellular beginnings of the gestation.) There is nothing in there to suggest that dizygotic twins could proceed through conception and gestation and begin life developing as if they were independent singleton conceptions, all the while somehow growing to differ from singletons more or less exactly as much as monozygotic twins do, in more or less exactly the same ways (with the exception of a few circumstances in which they actually differ from singletons more than the monozygotics do — in the same ways but more so).

I have laid out for you here several different repeated and mutually consistent observations which are not compatible with the predictions of the double-ovulation hypothesis for the cellular origins of dizygotic twins.

Dizygotic twins arise from two zygotes. That is the definition. I have no quarrel at all with that, and have given you no cause to have any quarrel with that. That definitional factoid — dizygotic twins, by definition, arise from two zygotes — *neither says nor implies anything about how the zygotes are formed*. We cannot be certain about the cell membranes at the very earliest stages, but we are certain that there must come to be two nuclear genomes — two different diploid sets of chromosomes must be assembled. There need to be two different maternal half-genomes to go with two different paternal pronuclei from sperm, to compose two zygote nuclei with different sibling genotypes.

In every way we have been able to examine the question, dizygotic twins do not go through embryogenesis the same way singletons do. The twinning process leaves developmentally visible residuals of subtle anomalies of the asymmetries of embryogenesis. Those subtle structural anomalies are the same in dizygotic twins as in monozygotic twins. The results clearly and consistently defy essential predictions of the hypothesis of independent double ovulation. Dizygotic twins have the same kind of strange embryogenesis that monozygotic twins have.

This is not news. For a long time already, from results that I have explained in earlier chapters and in earlier decades, it has been clear for all to see that dizygotic twins do not do their embryonic development like singletons. They develop more like monozygotic twins than like singletons. Double ovulation is a fallen hypothesis, every essential prediction of which that has ever been reasonably tested has failed.

And the world of people interested in the biology of twins and twinning gathered around and sat there with a puzzled look, and shrugged — "Well, if not double ovulation, then what ...? Of course it's double ovulation. Everyone knows. Boklage doesn't really seem to be deluded or stupid, but he must be ... ."

Absent an image at least as simple to replace the one that is ruined, they cannot absorb the loss; they pretend that none of it ever happened. They gather to comfort each other, and some of them write letters and papers telling the old fables anew.

The Monochorionic Dizygotic Epiphany

Into that long dark night of my scientific soul, the very existence of monochorionic dizygotic twins brought a new light by which we can see the path to the only credible and worthy answer. The very fact that monochorionic dizygotic twins exist, a fact that has been well and repeatedly documented over several decades, is compelling news, and the biology of their nature and structure is enlightening. Monochorionic dizygotic twins turn the world of twin biology upside down and inside out, and they totally ruin the old dogmas — which, as I have told you earlier, no one had ever given us any reason to believe anyway.

I should probably tell you why this so excites me: As I have explained earlier, monochorionicity can no longer responsibly be declared proof of monozygosity. That simply is not the way things are. It seems most of the world does not yet believe or even understand that. We might continue to suppose, in a slightly relaxed circle, that monochorionic placentation is more common among twins believed to be monozygotic than it is among the dizygotics. Realize that a majority of the twins believed to be monozygotic have not been genotyped and are believed to be monozygotic (only) because they are monochorionic, and you can feel the breeze from the circular logic going around again. About half of all "monozygotic" twin pairs of African ancestry are monochorionic [don't forget that 25% of the girl-boy pairs in that sample came up "monozygotic" in blood typing results]. About two-thirds of "monozygotic" twin pairs of white European ancestry are monochorionic, as are over 80% of the "monozygotic" twin pairs of East Asian ancestry.

You should remember: the cover story has been that monozygotic twinning is a developmental accident, like any of the (other) malformations that are associated with twinning and blindly assigned to the monozygotic twins.

The standard story, the party line, the common knowledge, heedlessly reprinted in a prominent journal within just the past few months, is that monozygotic twinning happens the same way at the same frequency all over the world, as a developmental anomaly, while total frequency of twinning varies among populations only as a function of the frequency of dizygotic twin birth across populations, which is said in turn to vary entirely as a function of the frequency of double ovulation, a tendency to which can be inherited (only) in the maternal line. That has been the story, for decades, and there remain entirely too many people who are so certain of the truth of that story that they have not paid attention to the various reasons that have arisen over and over again for doubting it.

The last time I checked, it was still much too easy to find people who are supposed to know more or less all there is to know about the biology of twinning who will tell you that monochorionic twinning is absolutely exclusive to monozygotic twinning, that no dizygotic twins are ever monochorionic and that no monochorionic twins are ever dizygotic.

I rarely go as long as a month without a call, a letter or an e-mail from a twin or parent-of-twins who is confused about an incongruity between observed similarities or differences on the one hand and, on the other, their expectations from an uninformed zygosity diagnosis they had been given previously. It is easy to find mothers who were told by their obstetricians that their twins are "fraternal" entirely because they were delivered in separate chorionic sacs, or that they are "identical" entirely because the placental tissue came out in a single mass (neither of which observations indicates any such thing).

According to the newest numbers I have seen, separate, dichorionic placentas fuse into single placental masses a little over half the time, regardless of whether the twins in question are monozygotic or dizygotic. So, clearly, the number of separate placental masses carries zero information about zygosity.

The literature includes not a great many — but far too many to ignore — reports of monochorionic dizygotic twins. More than half of those reports are recent, but some of them have been in the literature for decades. Many of the reported cases have involved some sort of pathology. The presence of such pathologies has allowed most of the workers in the field an excuse to dismiss these as freakishly rare anomalies, more in the nature of bizarre exceptions that prove the rule than as exceptions that demonstrate the failure of the hypothesis underlying the false rule. Is it not still the case that a single clear contradictory example confidently negates any absolute statement?

> One cannot say of something that it is and that it is not
> in the same respect and at the same time.
>
> *Aristotle*

However ... the pathologies involved have no causal relationship, in either direction, with the number or structure of the fetal membrane layers. The pathologies involved have served only to draw closer investigation, that would not have occurred otherwise, in the course of which closer study the dizygosity of the monochorionic pair was discovered. Only by investigating some such odd outcome have we discovered twins sharing a single outer gestational membrane who are not monozygotic like "everyone knows" they are supposed to be.

We have seen this "anomaly" primarily among abnormal twin pairs because the abnormal twin pairs are the only ones we have examined closely enough to see that the reality and the dogma have nothing in common! We have no idea how many more such monochorionic dizygotic pairs we have missed because they were unremarkably normal and gave us no reason to investigate them closely enough to recognize their departure from what we have always been told to expect.

> Hold that thought: The fact is that we have no useful idea how frequent monochorionic dizygotic twins may be. We have only two bits of evidence to indicate even that *most* monochorionic twins might be monozygotic. One sample from Scotland found 75% and a sample in Taiwan found not quite 70% of monochorionic twin pairs to be monozygotic. Except for these direct genotyping studies, the monochorionic dizygotic twins we have known about have been found entirely by surprise, usually because of something wrong that made us look much more closely than usual. We have no idea how many more there might be who have been delivered with no anomalies that would have made those managing the delivery examine them closely enough to discover their dizygosity.

In general, the pathologies at issue here have gotten that closer investigation because the twins — confidently assumed at first sight to be monozygotic because of monochorionicity and matching sex — are discordant for an anomaly known to have substantial genetic contributions behind its occurrence.

Various departures from normal sexual development ("ambiguous genitalia") have been among those visible pathologies, often leading to the diagnosis of hermaphroditism, wherein one or both of the twins is found by closer examination to carry both male and female gonadal tissues. This often takes the form of an ovotestis, with both ovarian and testicular tissue in the same organ.

This fits right in with the situations of all of the fusion malformations that are excessive among twins. Most workers in the field have been telling one another for decades that the fusion malformations are peculiar to (same-sex and therefore, according to Weinberg) monozygotic twins. Consistent with that is the insistence that monozygotic twins with these primarily genetic anomalies are quite often discordant because of the interference their embryogenesis must have suffered from "splitting" and "mirror-imaging." Therefore, those common discordant malformation pathologies are considered not to count as evidence against monozygosity. And therefore, such twin pairs are very seldom investigated closely enough to show that monochorionic same-sex twins are not all monozygotic, or that a pair of twins (only) one of whom has a fusion malformation is not necessarily monozygotic.

Many of the monochorionic boy-girl pairs recently reported have been products of technologically-assisted conceptions. Those ART procedures have elsewhere been reported to involve substantial degradation of epigenetic controls on development. Those epigenetic controls are normally responsible for the excess of males present throughout the fetal period of gestation and up to live birth, and for the variation among race-group subpopulations in the frequencies of girl-boy twin pairs.

Epigenetic disorders, disorders of parental genomic imprinting such as Beckwith–Wiedemann syndrome, have been reported in several papers to be excessively frequent in monozygotic twins and almost always discordant. (Beckwith–Wiedemann is one of the major developmental disorders discordance for which has triggered closer investigation, which has led to discovery of monochorionic dizygotic twins.)

Question: By the way … just a thought: If the members of a pair of twins are obviously discordant with respect to some developmental anomaly known to be substantially genetic in origin … then how did they get to be so confidently called monozygotic in the first place, when the papers reporting these strange twin pairs usually do not mention any sort of genotyping for zygosity diagnosis — and some that do, dismiss results indicating dizygosity as lab error? Answer: Almost invariably, as far as I have been able to discover, they got that designation because of being of the same sex and monochorionic — that's how. "Everybody knows" we already know the zygosity under those circumstances, so "everybody knows" it is a waste of technician time and lab supplies to genotype monochorionic twins like that … .

It seems that the easiest way through this on behalf of the old way of thinking is that monochorionic boy-girl twin pairs are considered to be a sort of pathology caused by the perturbations of the reproductive system that are inherent in technologically-assisted conceptions. Because of our traditional understandings about embryogenesis, male and female have no more business being together inside the same chorion than they have among reproductive structures or tissues in a single body.

The apparent bias against mixed sex in monochorionic twins and other chimeras is entirely compatible with what we have recently come to understand about the male vs female difference in growth rates during embryogenesis. A paternally imprinted X-chromosome (present only in females; males are males because their fathers gave them Y-chromosomes, not Xs) slows the development of female embryos and makes them more vulnerable to failure in embryogenesis. From this situation, we get the normal excess of males at birth, and mouse mixed-sex experimental chimeras almost always show up at delivery as functional males.

At least very nearly all of the monochorionic dizygotic twin pairs reported to date, that have been appropriately analyzed, have been found to be chimeric, with one or both of the twins having in their tissues some cells derived from the other twin's embryo. Monochorionic dizygotic twinning and chimerism are fully congruent outcomes, generally considered without evidence to be extremely rare and pathological, perhaps sharing the same pathologies, which are perhaps related to disturbed epigenetic controls.

Chimerism Itself Has No Phenotype

Chimerism is common, but it is only rarely discovered. Virtually every writing to date that mentions human spontaneous chimerism speaks of the extreme rarity of the condition. All refer at that point to stories about there being in the world literatures only a few dozen cases that have been identified by anomalous mixed-field agglutination results in blood-typing laboratories, or by investigation of discordant developmental anomalies — most of which to date have involved sexual development; almost all of which are characterized by high heritability values.

When deliberately investigated directly, chimerism is in fact found to be quite common. In one direct study, conducted on blood samples from dizygotic twins and triplets born alive as such, with exquisitely sensitive fluorescent antibodies for five different blood cell surface antigen markers (which made it possible to see one labeled cell in 10,000), 32 of the 415 twin pairs (8%) and 12 of the 57 triplet sets (21%) in the sample included members who had some cells in their blood that were of the other twin's genotype. Some had very small fractions of cells of the second cell line; some had very substantial fractions. Some were reciprocal, each twin having some cells of the other twin's genotype; some appeared to have shared cells in only one direction.

> Why the twin-triplet difference? [The difference has enough statistical significance to allow us to believe it is real.] I would suggest that this is probably because each dizygotic triplet set (a monozygotic pair A1A2 plus the odd B) includes two (A1-B and A2-B) and each trizygotic triplet set includes three (A-B, B-C and A-C) dizygotic *pairs* of genotypes. In a triplet set, there are more opportunities available for sharing cells between more pairs derived from overlapping embryonic cell masses.

Even at those frequencies (which are truly astonishing compared to everything that had ever been reported before and the still-current belief), this observation identified only a fraction of the chimerism present. Chimerism is substantially more frequent than indicated by that result. I can say that with complete confidence because:

(i) these twins and triplets born alive together represent only a small fraction of all the products of twin conceptions;

(ii) with only five antigens detectable, the probability that any given pair of siblings will have matching versions of all five gene products is not negligible;

and most importantly because

(iii) only blood was examined. Chimerism, when it is present, is not at all certain to appear in the blood.

Another study specifically and directly tested for chimerism in autopsy samples from normal women. *Over one-third* of the women sampled showed foreign (male) cells (with Y-chromosome DNA sequences; tissue-specific cells — not just blood cells caught passing through) in heart, lung, kidney, or spleen. In no case was chimerism found in every organ from any sampled individual.

Women who had delivered one or more sons had no greater frequency of chimerism — indicating that the chimerism found was not due to fetomaternal transfer (unless it was from male conceptuses whose tenure was brief enough to pass without being detected, which is plausible).

Because the tissues were tested only for Y-chromosome DNA sequences, however sensitively, any of the samples which carried a female chimeric second cell line went undiscovered. There is no logical alternative but to suppose that the instances of chimerism found and reported in those samples were only the male half of all that were there.

> These results together tell us that there is a large difference between the presence of chimerism and the discovery of chimerism. Chimerism is necessarily far more common than the discovery of chimerism has been.

The First Reported Human Spontaneous Embryonic Chimera

The oldest report I have found in the literature of the discovery of a spontaneous human chimera was "Mrs. McK." She appears to have started the whole spontaneous human embryonic chimera story with her donation of blood at a clinic in northern England in the spring of 1953. The very existence of human spontaneous chimerism was discovered within my lifetime! The fact that it has become no better understood than it is arises from several layers of difficulty.

The blood-typing laboratory in Sheffield found that Mrs. McK seemed to have two blood types. Her blood behaved in their tests as if it included cells of both type A and type O:

(i) There was no agglutination in type-A serum (which should cause any type-B cells to fall to the bottom of the tube). This is consistent with type A or type O. There are no type-B cells.

(ii) In type-O serum (which should agglutinate type A and/or type B, but not O), some cells agglutinated, some did not. Therefore, because type B is already eliminated by the failure of her red cells to agglutinate in type A serum, there must be some type-A cells in there (which agglutinated), and some type-O cells (which did not agglutinate).

(iii) In type-B serum (which should agglutinate type-A cells, but not cells typing O or B), some cells agglutinated, some did not.

This result is called a "mixed-field agglutination," consistent with the presence of some type-O and some type-A cells.

They repeated the tests to rule out accidental mixing of either blood samples or the agglutinating antiserum reagents they used. Then they sent a letter with a sample of the mysterious blood to Robert Russell Race. He was then the hero of the science of blood cell surface antigens. Race was Director of the MRC Blood Group Unit in London (the Medical Research Council is the UK counterpart of the USA's National Institutes of Health). There it was confirmed that the blood circulating in Mrs. McK's body contained cells of two types: two-thirds type-O and one-third type-A cells.

It was known before that time that dizygotic cattle twins routinely exchange blood cells through connections between their placental blood vessels. When such cattle twins are male and female, the heifer calf of the pair is usually a sterile "freemartin." They questioned Mrs. McK about being a twin: indeed, she had a twin brother who had died.

Thus, the "obvious" next questions: Is Mrs. McK "normal" in structure and function? Yes. Does she exhibit a normal femininity in appearance and behavior? Yes. Has she any children, i.e., is she fertile, is she a reproductively functional female? Yes.

In the writings about her case there is no mention of any evidence to suggest that her chimeric second cell line was male. Cytogenetics was not at all then what it is today, to have routinely examined the chromosomes and identified the sex. In fact, this was all happening a few years before we were even able to visualize chromosomes well enough to be certain of the normal number of chromosomes in human cells. This happened in the same year that Watson, Crick and Wilkins deduced the double helical structure of DNA. Of course, the molecular genetic tests for DNA sequences characteristic of X- and Y-chromosomes also did not yet exist in 1953.

It is reasonable that we always have to work from what we know, and that those cattle twins were the only creatures known about at that time that might be circulating mixtures of blood cells. The manner in which the cattle twins came to be circulating a mixture of blood cells was all that anyone knew about how such a situation might come to pass.

We are told that "dispermia" was considered and dismissed because Mrs. McK was normal and symmetrical in appearance, with both eyes the same color, hair the same color on both sides of her head, and so on.

Clearly, the investigators were expecting that anyone who had two different cell lines originating in prenatal development should be an obvious mixture of parts from what "should" have been two different bodies, most probably in large blocks in multiple tissues, something like the classical portrayals of the mythical beast of prominent parts. By the way, "dispermia" is another name for what later came to be called the "dispermic" sort of chimerism, arising from double fertilization of a single egg cell (which must have "retained its second polar body" to provide the half genome for the second maternal pronucleus), to distinguish it from "blood" chimerism or "twin" chimerism, imagined to have arisen later in pregnancy from twins sharing blood cells through placental anastomoses.

This clearly exemplifies the situation that the whole idea of human chimerism "has an image problem." From these beginnings, the idea has been plagued by such terminology problems as these. People who are talking about the same thing, each without knowing much about what they are talking about, often do not know that they are talking about the same thing. As far as it concerns human chimerism, most of those problems come from problems of visualizing the subject of the discussion.

The idea that the structure of any chimeric organism by definition includes parts from more than one biological source is quite correct in the abstract. The body of a chimeric human does indeed contain cells and tissues of more than one genetic type. However, it seems that many, maybe most, people cannot handle the abstraction without an image, and the image most commonly abstracted from the verbal definition of the mythical Chimaera monster is counterproductive at best.

In the case of human chimerism, the truth is stranger than the fiction entirely because the fiction has mythically been given a face (actually, three faces) in classic art, and the human truth does not come close to matching the picture thus provided. From there, the intimately limiting relationship between what we expect to see and what we might actually see takes over and leads us far astray if we are insufficiently careful.

The use of the heifer-bullock twins as a transitional idea was of obvious utility in moving toward the understanding of twins sharing cells of different genotypes to generate chimerism. However, the temptation to incorporate the form and function of the bovine multiple placenta as the mechanism by which human twins might exchange cells is still strongly misleading today. Human multiple pregnancies never have placentas like those that cattle twins always have.

Too Many Blood Types

Over the subsequent 55 years, a few dozen additional cases of human spontaneous chimerism have been discovered in blood banks by mixed-field agglutinations like the outcome that resulted in identifying Mrs. McK as chimeric. All of those mixed-field agglutination results pointing to two-blood-type cases have necessarily involved very substantial admixtures of the second cell line. Below a threshold which seems to be in the neighborhood of 20% representation of the less numerous cell line, mixtures of blood types are not at all likely to be detected by the methods routinely applied in blood banks. Even with deliberate, close attention, it seems that *any admixture of less than 20% is nearly certain to be missed.*

Since the most often entertained notion of a mechanism for generating chimerism involved placental anastomosis between twin fetuses (as is the case with the mixed-blood cattle twins), there was no reason for those investigators to imagine that they were routinely missing cases with large departures from equal 50% contributions.

Serological typing is not a highly precise process. It does not in general need to be. There is, however, at least one case in the literature where a chimeric second cell line, present as a minority population of only about 5% of the cells in that unit of blood, went undetected in the typing lab and then caused a life-threatening immunological mishap when transfused into a surgical patient. The rarity of the cell antigen type that caused the problem in that particular case could be taken to suggest that more benign such differences might be both common and commonly undetected.

Some People Cannot Do Sex (Development) Right

Most of the cases of human spontaneous chimerism that have been discovered by any route other than mixed-field agglutination in a blood-typing lab have been found by close investigation of people whose sexual development was abnormal. In general, that anomaly arose from the presence of both male and female reproductive structures (by definition, "true hermaphrodites") grown from male and female chimeric cell lines.

Certainly not all chimeric individuals are hermaphrodites, nor have all hermaphroditic individuals been shown to be chimeric (nor have they in general been investigated with chimerism in mind). Note well: almost all such fruitless inquiries have been limited to blood samples, and I will continue to remind you that chimerism when present need not be present in blood.

This is a salient example of the problem of expectation vs recognition:

(i) they extrapolated — then and for decades to come — from the bovine mixed-sex twins to human twins,

(ii) they postulated placental anastomosis (as it happens routinely between the cattle twins, with placentas very different from those of humans) to be causing the mixing of (only) blood cells, so

(iii) they tested only blood, by the only (serological) means they had available, and as a result

(iv) they generally found nothing because they only looked for what they expected where they expected it to be, and that was and is almost always either not what was happening or not where it was happening.

Chimeric individuals with both cell lines being of the same sex would have no expectation of being hermaphroditic. On the other hand, most of the true hermaphrodites reported in the literature have not been investigated for the possibility of chimerism, so no reasonable or responsible statement about their frequency is possible at this stage of the development of our understanding.

Tissue Transplant Chimerism

Any successful transplant of a solid tissue generates an artificial chimerism. A blood transfusion is also a tissue transplant, but unlike a transplanted organ it can serve its purposes without establishing a permanent presence. Some blood transfusions will include stem cells capable of establishing persistent or permanent colonies. Fetal cells routinely enter the maternal circulation, where they have been found occasionally to have established colonies still active over 50 years after the pregnancy from which they originated.

Population Frequency of Spontaneous Chimerism

I have no excuse but prejudice for my reluctance to offer *two-thirds* as a sound and plausible estimate of the frequency of chimerism in the population at large. That *is* the number for which the autopsy survey for chimerism described above provides sound direct evidence [over one-third of women tested had XY cells; the other half of all second cell lines which might have been expected to have XX genotypes could not have been found].

There is no evidence against that two-thirds number, and no plausible alternative interpretation of those results comes readily to mind. In the same survey, women who had had transfusions or who had (delivered) sons did not have a higher frequency of detected chimerism, which would have been expected if any large fraction of the detected chimerism was from transfusions or from fetomaternal transfers.

Although I consider it a significant stretch, I would like very much to see that survey repeated, confined to autopsy samples from *virgin females*, against the prospect that many of those chimeric women might have been colonized by cells from male conceptuses they never knew they had, or from some sort of invasive stem cells in semen. Given the high fraction (more than two-thirds) of human conceptions that fail even before recognition, about a third of all conceptions would be males that would never be known to have existed.

The prospect that over a third of all women might be chimeric sole survivors of unlike-sex twin embryogenesis is a possibility that must be considered and could be a plausible explanation of the reported observations, but I find it difficult to go that far in one jump. Because I have as much trouble believing it as I have after having thought about it for years, I would expect that you will also have at least that much trouble. But the observations in question appear straightforward and sound, and in fact I believe I have exhausted right here before your very eyes all of the plausible alternative explanations.

I remind you here that in none of those cases was the observed chimerism found in every organ examined from any given woman. This is entirely consistent with experience from experimental chimeras, where the distribution of cells of distinct genotypes appears to be random, showing no pattern and no priority among tissues that have yet been recognized and reported as such.

Chimerism is still generally considered freakishly rare and pathological, primarily because of the pervasive but unacknowledged huge difference between the existence of chimerism and the discovery of chimerism. By far most human chimeras have been and will remain undiscovered because they have two normal cell lines of the same sex [or of mixed sex with cells of one sex present in an invisibly small fraction] and they will never give the rest of us any reason to examine them closely enough to discover that they are chimeric.

I have found no good reason to want to know the exact frequency of chimerism in the human population, only partly because I am certain that knowing that frequency exactly is impossible. If one supposes that such knowledge could only come from a count of all individuals (or at least a representative sample statistically sufficient in number) and scoring them accurately for chimerism, it is not going to happen. We have good reason to believe that most chimeras will never be discovered by any simple observation.

Even if we were to slice every suspect very thinly and examine every slice, it would still be possible to miss many instances of chimerism.

Given the estimates available from the two direct surveys mentioned here, and the limited reasonable expectations from the methods used in those studies, there is no room for supposing that the frequency of chimerism in the general human population can be any less than 10%.

Chimerism from Fusion of Embryos or Fetuses? In a Word … No

Historically, as outlined above, it has been imagined [no evidence, ever] that human spontaneous chimeras must arise from the fusion of placental circulations between dizygotic twins during gestation. Of course, they are outrageously rare, so the logic goes, because fusion of placental circulations between dichorionic twins (incubating in separate sets of fetal membranes, with initially separate placentas) is quite rare.

It is in fact the case that connections between circulatory systems of dichorionic placentas very nearly never happen. Reports of placental anastomoses (connections between placental blood vessels) between dichorionic placentas are at least a thousand-fold less frequent than even the lowest reported frequency of discovered chimerism mentioned above. Placental anastomosis very nearly never happens between dizygotic twins *or* monozygotic dichorionic twins. Even when separate placentas have fused into a single mass, which they do a little over half the time whether the twins in question are monozygotic or dizygotic, it is extremely rare to find circulatory connections between the two placentas of any dichorionic pregnancy.

In his studies of a great many twin placentas, Professor Benirschke never found human dichorionic twins with placental anastomoses. Accepting the general belief that they were necessary to explain chimerism, he surmised that they must be so very rare that he had simply missed seeing any. [Just for perspective ... if an event has a probability of one in a thousand, the probability of seeing at least one in the course of examining a thousand possibilities is almost two-thirds.]

In my earlier writing about chimeras and chimerism, I have written about their origin in terms of the fusion of embryos, to distinguish my vision of the process from the fusion of placental circulation systems. That was a mistake, to the extent that anyone could take that to mean that embryos from separate and independent ovulations might be apart and then come together to fuse. I should have anticipated it. Apparently, that particular misperception is quite widespread, and now that is at least in part my fault! There are very few, if any, beliefs in this world that are held so strongly as the belief in double ovulation as the origin of dizygotic twins. No experimental chimera reported in the literature has ever been made between pre-hatching embryos without first taking off the zona pellucida. Only by that means can cells of the separate embryos touch, and then only before the differentiation of the trophoblast.

All experimental chimeras made from embryos that have developed beyond differentiation of the inner cell mass from the trophoblast have been achieved only by fusion of cells from the inner cell mass removed from one blastula and injected into the other. With electrofusion, it has been possible to fuse trophoblasts of two separate blastocysts, but only inner cell mass cells (that came within reach to touch) have been found to contribute to development of the chimeric embryo after such manipulation.

The fact of monochorionic dizygotic twins requires that the chimeric dizygotic twinning event can and sometimes does happen after the onset of trophoblast (future chorion) differentiation. As far as we know, that always happens before hatching. Dichorionic dizygotic twin pairs undergo their twinning events and become twins earlier than their monochorionic counterparts, and therefore certainly still inside the zona pellucida of a single egg cell.

We normally have two copies of each of our DNA molecules, except for differences between DNA sequences on the X- and Y-chromosomes in males. With those exceptions, we necessarily have either two different alleles/versions of each gene DNA sequence or two copies of the same allele. A proper set of genes in a normal individual of a diploid species can include only one or two versions of any given gene DNA sequence. If there is an extra set of genes from a second cell line, there may be four times as much signal representing only one version of that DNA sequence ("allele") in all four copies, *or* there also might be roughly equal amounts of four different signals representing four different versions of that DNA sequence. The presence of three or four different alleles, or copies of the same one, instead of the normal one or two, if it shows up at multiple loci, involving gene DNA sequences on several different chromosomes, says there is more than one normal set of genes in the sample at issue.

This sort of discovery also occurs only when cells of the blood tissue include cells of the second cell line in a substantial fraction. Any less than perhaps 20% of a second kind of cells will almost certainly not be noticed in serological typing. Detection of chimerism in a single sample by DNA typing has so far not been much better. To extend the example and further define the scope of the problem: In the only major high-throughput genome-scan laboratory with which I have been directly involved, where hundreds of individual DNA samples can be analyzed with respect to hundreds of different DNA markers simultaneously:

- extra signals less than 30% of the corresponding main signals are dismissed as background noise, and
- extra signals at 30% or more of the strength of the main signals, at multiple markers, causes the whole sample to be discarded as contaminated ("obviously" mixed samples), without even recording the data so that the investigator can consider the possibilities.

In short, there is no level of chimeric mixture
that has not been seen and routinely ignored or dismissed
as being either erroneous or imaginary.

We do not look for chimerism in the situations where we might expect to find it because we do not in fact expect to find it. Therefore we will not find it until we stumble over evidence we cannot ignore. Thus, the mystery is maintained and perpetuates itself.

Many thousands of artificial chimeras have been produced for experimental purposes in genetics and developmental biology — between strains of mice, between mice and rats, between different species of birds, between different ungulate species, between sexes of the same species. Nicole Le Douarin, for example, has conducted a very productive career in developmental biology largely by following the behaviors of cells from quail tissue transplanted into (thereafter chimeric) chick embryos. The quail cells can be followed microscopically to illuminate pathways of cell migration and differentiation. Andrzej Tarkowski, Beatrice Mintz and Richard Gardner are also among the earliest and most prolific of those who have done so much with chimeric mice.

Where the Spots Are

From all of the work of the many people who have used experimental chimeras, we know that embryonic chimerism in vertebrates is patchy and random. As far as we know now, a chimeric second cell line can be anywhere within a body, forming part or all of any tissue or organ. We do not know enough yet to know whether or not any one tissue in the body of a chimeric individual is more likely to be mixed than any other. However, it already seems clear that, more often than not, the second cell line is not present in blood. When the blood is not chimeric, then of course chimerism cannot be discovered in a blood sample. Even when the blood is chimeric, discovery in blood samples is not routine, let alone easy.

Some chimeras have been discovered in the course of investigating patients with genetic problems, when samples are found to contain some cells with the genetic anomaly characteristic of the problem in question and also some normal cells. For the most part, this happens when the problem in question is a chromosomal anomaly, and the mixed-cell situation is discovered by cytogenetic microscopic analyses of chromosome structure and number. Such an individual is normally called a "mosaic" rather than a chimera, and is imagined to have become as s/he is by way of a chromosomal error in cell division early in embryogenesis, resulting in continued development with some normal cells and some abnormal. Full genotyping, which could show that the differences between the two cell lines involve genes on many chromosomes, is never done unless the two cell lines obviously differ in more than the one extra chromosome.

If the two cell lines in a chimeric individual are of different sexes, sometimes sexual development may be abnormal, causing investigation in which two kinds of cells are discovered. Experimental mixed-sex mouse chimeras almost always arrive at birth as functional males. It may (or may not) be true that mixed-sex human chimeras are less common, or less commonly discovered as such, than same-sex chimeras (we know too little about either to say anything intelligent about differences). It would probably be reasonable to expect such an outcome because male cells grow much faster in embryos than female cells as paternal imprinting of one of the female's two X-chromosomes slows embryogenesis in females of every mammal properly tested to date.

It may be that the majority of human chimeras are spontaneous and embryogenic, not caused by pregnancy or by any form of tissue transplant. Except for mothers colonized by fetal cells, spontaneous human chimeras are dizygotic twins who have shared cells in embryogenesis. One or both of the twins have in their bodies some cells that "belong to" the other twin. The mixture may be anywhere from only a handful of the co-twin's "foreign" cells in the whole body, to a 50:50 mixture. The "foreign" cells may be anywhere. Finding or failing to find the chimeric cell line in any given tissue sample tells us nothing about its presence in, or absence from, any other tissue sample from the same body.

If we should ever find that particular tissues are more or less likely to be mixed in chimeric embryogenesis, we might have to revise that statement. We will not be able to do that until we have critically examined multiple tissues in a statistically useful number of cadavers of chimeric individuals.

Chimerism may, further, be present in a particular tissue at one time and not be detectable in that same tissue at another time. Sperm cells and blood cells would be particularly susceptible to variation over time because both are constantly produced from stem cell lines.

There is every reason to believe that most human chimeras have two normal cell lines of the same sex, and therefore that most chimeric individuals are entirely normal in appearance and behavior and will never be discovered because they will never give reason for the kind of investigation necessary to discover their chimerism. Even when there is clear cause for investigation — such as an obvious chromosome anomaly — chimerism can easily be missed.

Our interests here must focus on the "how" of chimerism: how pairs of dizygotic twins manage to include in their bodies some cells that "belong to" the other twin. This offers us much to learn about how "normal" embryogenesis works.

Chimerism is Not Itself an Abnormality

Although widely associated with various abnormalities because it has rarely been discovered except in the course of investigating an anomaly, chimerism itself is not a developmental abnormality and does not imply the presence of any anomaly. When the two cell lines are both normal and of the same sex [which will be the case in the great majority of chimeric individuals], the dizygosity of the chimeric monochorionic twin pair or the chimerism of any given individual will be discovered only by accident. That will usually involve DNA genotyping for some totally unrelated reason. A few illustrative cases have been widely reported in the past few years.

A then-52-year-old teacher in Massachusetts needed a kidney transplant. Her husband and three sons were the first donor prospects tested. Her husband turned out to be a suitable donor, and that is, of course, a happy ending — not a very likely result, but not especially newsworthy. What ended up in the newspapers, the news magazines, the science news and the web logs was that, according to the DNA profile results, two of her sons clearly were not her sons!

Any number of kin, friends, and physicians can provide credible witness that she carried and delivered and raised those men. According to the DNA profiles, her husband is their father, but some other woman is their mother. The results allow for the possibility that the other woman who is their "real" mother might be a close relative of the woman who raised them, but there are no close relatives to match. There certainly is no one who could have passed through any of the events or processes of her motherhood in her stead.

Searching through frozen samples from some surgeries she had in the past, the investigating crew eventually found a second cell line in some of her tissues that is genetically compatible with her being the mother of those two men. The "other woman," who is the genetic mother of those two boys of hers, might be said to have been her unborn dizygotic twin sister — living today only as her chimeric second cell line in parts of some of her tissues, at least for a while once upon a time in at least one of her ovaries.

This "other woman" language is at best an unfortunate way to think or speak of this. Most of the strangely threatening feelings that the thought of chimerism gives some people comes from thinking like that. There was in fact never another person to whom the cells of a chimeric individual's second cell line once belonged. You might want to go back over the questions about defining personhood in Chaps. 2 and 10 and see where it fits in. It is simply not reasonable to suppose that the zygote is a person.

Our heroine has had some medical issues, and this kidney problem certainly has to be considered a serious one, but her problems have not been especially rare or bizarre, and we have no reason to imagine that they are causally related to her chimerism in any way. She is an entirely "normal" person (which never did mean "perfect"). It took DNA genotyping of transplant-surgery quality, involving her whole family, to uncover the oddity. Except for the accident of her kidney failure, causing her whole family to be genotyped in search of a kidney donor, there would never have been any reason to suspect her chimerism or investigate her body in ways capable of demonstrating it. Even given that level of investigation, if one of her sons is compatible with being the son of her primary cell line, the other two might have come from the same cell line instead of the second one, and no one would ever have known. Her chimerism would never have been discovered except for the intersection of two unlikely trajectories: her kidney failure and that accidental sorting of egg cells.

If you want to check my telling of the story, send your Web search engine after "Keegan chimera." Peer-reviewed scientific findings from her story are referenced in my 2006 Human Reproduction paper, listed in the bibliography here.

Not long after that, another woman, on the other side of the country, 26 years old, unemployed, with two children and pregnant with a third, had separated from the father of all three. Lydia Fairchild needed some public assistance for her new start. As part of the application process for the public assistance, the county required DNA from all members of the family, to make sure all responsible parties would be doing their respective fair shares. The focus, of course, was the father. We hardly ever make any mistakes or have any serious doubts about who the mother is. Sure enough, he is the father of the two children, and he would have been unpleasantly surprised if that had come out any other way — but the DNA profiles say that she is not in fact the mother of her two children!

Testimony of the doctor who delivered her children and watched them grow in her care was given no weight because "DNA doesn't lie." She was getting into some serious difficulty, defending herself against questions, and charges, of welfare fraud for trying to get taxpayers' money for children who do not belong to her. Worst of all, she was threatened with having her children taken away from her at any time, at the convenience and whim of officials of the county and its courts. Additional, worse charges might have been contemplated. How, after all, did she come into possession of two children belonging to someone else, children of her man and some other woman?! Have they kidnapped these children? Has perhaps something very bad been done to their *real* mother in the process of misappropriating that other woman's children?!

One of the attorneys involved in her case heard about the case in Massachusetts and wondered if perhaps some similar such thing could be happening in her case as well. The judge sent an agent of the court to attend the delivery of the third child and to see to it that samples for DNA were collected from mother and child on the spot. Her third child — the one they just watched her deliver — is a full sibling to the other two. This child has the same father, the same mother, and DNA-wise, genetically speaking, that mother is still not the woman from whose belly this child was just officially witnessed by an officer of the court to have emerged.

This child and both of the others are all, by the way, genetically quite compatible with being the full-sibling grandchildren of their mother's mother.

This woman did not have frozen samples stashed away from old surgeries, like the woman in Boston had, so new samples for testing were gathered from various additional reasonably accessible parts of her body. Some of those samples did contain a second cell line which was genetically compatible with being her sister and the mother of all her children.

Through no visible anomaly of her own person, having given no reason for anyone to investigate her body in such ways as to reveal her situation, she was revealed by DNA genotyping of her whole family, done for a reason totally unrelated to her own health and fitness phenotype, to be a chimera, carrying some cells from a second cell line in the embryo from which she grew.

If the third child, sampled at delivery, had shown a good match to the mother, as it happened for the other son of the woman in Massachusetts, we might wonder how far around the bend that case might have gotten. If it turned out that way, the genetic origins of her other two children could still be unexplained and criminal charges might still be within reach of being considered.

In one particularly interesting case that came across my desk, there were neither child support nor medical issues. Father and mother were both 20 years old in Seattle in 1954, as he was stopping over for some military training on his way to Korea. For the last six of his eight weeks there, they were "inseparable." No answer ever came to her from the letter she sent after he shipped out, in which she told him of her pregnancy. She quietly went on to raise her son and did not tell him the story of the circumstances of his conception until he was 40 years old. From that point, beginning with a polysyllabic Eastern European sort of name his mother was no longer certain how to spell, it took the son 10 years to find, out there in the world, the story of a man who might be his father.

He wrote that man a letter, explaining the situation and wondering … The man immediately recognized the story and knew without question that he himself was the very one and only man who had played the male romantic lead. This could not have been made up. He was there. This story cannot be fiction. These men are father and son.

As it happens, the father's wife of many years, and his daughter, cannot accept this stranger as a long-lost member of their family just like that. They believe that DNA might make it better. Having nothing to hide, and no concept of anything to fear, the men agree. According to the testing laboratory, the DNA does not match; they cannot be father and son.

The son lay awake at night staring at the results. He came to believe there were too many matching alleles between the profiles to believe they really represented two randomly chosen unrelated people. Then he brought his question to me.

He is correct in his conclusion. The profiles are not compatible with having come from random, unrelated individuals. They are entirely compatible with an uncle-nephew relationship, but there is no uncle. The putative father never had a brother who could have been there to take his place with that girl that night in Seattle in 1954. The father's only living relative is a sister, who turns out, on further DNA testing, to be a perfect candidate for the genetic position of paternal aunt. At several of the loci where the son has no allele to match his father (the ones that cause the lab report to say "exclusion") he does have an allele that matches the putative father's sister.

Other tissues were sampled from the father. All samples show the same DNA profile — none of them fits with the son. It is now, after all, more than 50 years later, and we know that blood and sperm cells, for example, are generated by differentiation from stem cell lines in waves. We also have long been aware that the relative proportions of the two cell lines in people who have two cell lines change over time, sometimes steadily in a consistent direction (as an abnormal line dies out), sometimes back and forth.

Chimerism can be proved by finding the second cell line in another sample. On the other hand, no amount of failing to find it constitutes evidence against its presence in some other tissue or even in the same tissue in another sample at some other time.

The configuration of DNA markers in the father, his sister and his putative son is several billion times more likely if they are in fact related in those ways than if the son is an unrelated individual randomly chosen from the population. If the possibility of chimerism is acknowledged at all, rather than being dismissed out of hand, then paternity is not in fact excluded by these DNA results, and the non-DNA evidence is overwhelming — that story cannot be fiction. They are family.

Spontaneous human chimerism is real, and not rare. It is common, but rarely discovered.

From here on, I hope we can dispense with arguments about how very rare chimerism is. All such assertions in the past have been entirely baseless. The only direct evidence available comes from the two direct samplings I have mentioned above, of the twins and the autopsy specimens, both of which make it quite clear that chimerism is not at all rare when assessed directly.

Spontaneous Embryonic Chimeras Are Twins

Spontaneous chimeras are usually twins of the "twins, but not twins," sort — all products of twin conceptions, but for the most part not products of twin live births. Except for mothers colonized by cells from their embryos or fetuses, the path to spontaneous human chimerism begins with a dizygotic twinning event.

Dizygotic twinning events that lead to chimerism sometimes happen inside a morula of which the outer layer of cells have already committed to differentiation as the future trophoblast. This — and no other credible path — leads to monochorionic dizygotic twinning. At that point in development, when twinning can happen inside a single chorion precursor, the blastocyst is still inside a single zona pellucida.

Not one of the many thousands of experimental chimeras that have been made has been made from two embryos in intact separate zonae pellucidae. Either the zonae were removed so that cells of the two embryos could make contact in order to fuse, or the work was done with cells from embryos that had already hatched from their respective zonae. Since hatched embryos have already differentiated their trophoblast precursors to the chorion, chimeras made from those embryos must be made by putting cells of the inner cell mass from one blastula into the other blastula in contact with the other inner cell mass.

There is no physically possible way in which these things can happen unless dizygotic twinning often happens in a contiguous mass of cells that have divided from the substance of a single secondary oöcyte (the only way they can be twins inside a single zona).

Clearly, it does happen that way; therefore, equally clearly, it can.

Dizygotic twinning can happen in a single confluent mass of cells, progeny of two contiguous zygotes formed inside a single zona pellucida, product of a single ovulation. We know it can because it does. If it happens that way sometimes, then why not every time? Nothing but the myth of double ovulation says otherwise, and the direct evidence from the teeth discussed above found no twins with the developmental pattern of singletons.

Chapter Eighteen

Finding the Right Answer is Much Easier
After Finding the Right Question

Our great intellectual revolutions are never simple
infusions of knowledge into a previous void;
they are always exercises in destruction and replacement.

SJ Gould, "Columbus Cracks an Egg"

Because virtually everyone was against us ...
I knew we were on the right track.

Marshall T. Rose

When things "all hang together" you have either gotten the joke, solved the
puzzle, argued in a circle, focused your chain of logic so narrowly that
you will be blindsided — or discovered a hidden pattern in nature.
Science, in large part, consists of imagining coherent solutions
and then making sure that you weren't fooled
by a false coherence, as in astrology.

William H. Calvin

When I began this work, I had not spent much time thinking critically about
the biology of twinning and I knew of no good reason to doubt what I had
been told. So, it was for a while my understanding and expectation that the
development of dizygotic twins would be — should be, must be — the same as
that of two simultaneous, side-by-side, but separate and independent singletons,
except that the dizygotics could serve as a control sample against possible effects of
twin gestation and delivery and childhood as twins. The dizygotic twins stood in
for singletons and held the honorary logical position of experimental controls for
any differences that might be caused by direct physical effects of twin gestation,
of beginning their lives as twins.

Quite soon after that beginning, the results I have described for you in earlier chapters began to make it clear to me that dizygotic twins do the building of each embryo differently from the building of singleton embryos. The differences are not about crowding in the womb, not about any reaction to any sensibility of there being two instead of one. It is about embryogenesis, about things that happen in the first few cell divisions — when there can be no crowding unless crowding can be a matter of molecular interactions among small numbers of cells. I gradually developed many and profound doubts about the independent double-ovulation hypothesis for the origin of dizygotic twins.

Those doubts grew with every result I gathered, eventually leading to my firm conviction that the hypothesis of independent double ovulation as the origin of spontaneous dizygotic twinning is completely and forever untenable.

It has long been clear that the monozygotic twinning process, with all that "splitting" and such going on, is an anomaly of embryogenic symmetry determination, so that its association with all those other such anomalies is no surprise. On the other hand, every new result I got kept telling me that dizygotic twinning is equally anomalous with respect to all of those same asymmetries. How can two-egg twinning be a symmetry anomaly?! In fact, it cannot. But dizygotic twinning carries with it all the same symmetry anomalies as monozygotic twinning! That leads, in turn, of course, to questions about how dizygotic twins might actually be doing embryogenesis instead of doing it the two-egg way everybody but me seems to believe it is done … .

At some point along that path, I did a thorough literature study of the historical development of the story about double ovulation. I told you about that earlier. The answer is that there is no evidence. I have on several occasions challenged readers of my papers with this observation, and no one has yet brought to my attention any such evidence I might have missed.

No one has ever reported any observation of any pair of naturally conceived human dizygotic twins ever actually developing from two separately ovulated egg cells. Nor did anyone report having noticed the absence of such evidence; no one reported having any doubts or concerns about that state of scientific affairs.

On the other side of the question, I have laid out for you in previous chapters several independent and fully consistent lines of evidence to the effect that dizygotic twins are at least as different developmentally from singletons as monozygotic twins are. Dizygotic twins are developmentally different from singletons in the same embryogenic ways that monozygotic twins are different from singletons. That makes it obvious that we truly understand no more about the cellular and molecular details of the embryogenesis of either "kind" of twinning than we do about the other.

The developmental differences I have explained here between twins and their first degree relatives vs singletons and their relatives — in handedness and other biometric and behavioral asymmetries, in the fusion malformations, in the structural developmental integration of complex multivariate craniofacial variations — all have to do with fundamental symmetries and asymmetries that are normally established at molecular and cellular levels early in embryogenesis.

The excess frequency of nonrighthandedness in twins is a reduction from a normal level of developmental left-right asymmetries of brain function. The excess frequency of malformations in twins is concentrated in the fusion malformations, also known elsewhere as midline malformations. These are distortions of the asymmetric development of midline structures formed by fusion and remodeling in the midline between approximately mirror-image-symmetrical bilateral half structures. The brain and nervous system, the heart, craniofacial configurations of the mouth and face, and placements of unpaired midline visceral organs are the best-characterized examples. All of those malformations are in excess among twins (of both zygosity groups equally except where the differences are greater in dizygotic twins), and in first-degree relatives of twins (of both zygosity groups equally, except when the difference is greater in families of dizygotic twins). All of these malformations represent functional anomalies of normal left-right developmental asymmetries.

Twins differ from singletons in the multidimensional developmental structure of their sets of teeth, and dizygotic twins differ from the singletons in the same ways that the monozygotic twins differ from singletons. Monozygotic and dizygotic twin groups overlap extensively when they are plotted according to their scores in these multivariate comparisons, but there is negligible overlap between twins of either sort and singletons on the same plot.

As might be expected from other parts of what we know about single-born sole survivors of twin conceptions, a few of the "singletons" in those samples had scores on those multivariate classification functions as if they might really be twins. On the other hand, the data available and all of our analyses thereof to date have never shown me any twins (defined in all of those studies by live birth as twins) who have developed in patterns like singletons.

It is not the case that some dizygotic twins do indeed develop just like simultaneous singletons and that others, some fraction of the whole group, are developmentally equivalent to the monozygotics. According to my results, all twins sort together in their developmental differences from singletons: Whatever the dizygotic twins are doing differently from single embryogenesis, they are *all* doing it. They are all doing the same thing and doing their structural development [the system of relationships among the sizes and shapes of their parts] the same way that monozygotic twins do.

> At the cellular and molecular levels where the foundational asymmetries of the vertebrate body are set, the cellular and molecular bases and mechanisms of dizygotic twinning events are the same as those of monozygotic twinning events.

Our understanding of the mechanisms for defining and establishing differences between left-side and right-side structures in embryogenesis is hugely advanced over what it was only a few years ago, but it remains a long way from a complete understanding, and from some viewpoints it can arguably be considered negligible. This is the sort of situation for which the old, old idiom about "just scratching the surface" is entirely appropriate.

> "[Development] … is a scientific territory which biologists of our day
> will see as Moses saw the Promised Land;
> only from afar and without power there to enter.
> *my translation, from Alfred Giard, l'Oeuf et les Debuts d'Evolution, 1875*
> [*The Egg and the Beginnings of Development*]

Our toolkit has burgeoned over the last few decades, along with our feel for the ground over which we travel, and the effectiveness of our explorations in developmental biology has grown dramatically. We have made a great deal of progress, and it seems clear that we will continue to discover that most of the developmental concepts we are chasing are not events, but horizons, which can be seen as bright lines or sharp boundaries only from very substantial distances.

Every time we get to where we thought we were supposed to go when we started, we see that any resolution or clear definition for the place we set out to reach will need to be sought elsewhere. To see as clearly as we do now that the most fundamental of those mechanisms for establishing developmental asymmetries are different? disturbed? bypassed? weak? in the embryogenesis of all naturally conceived human twins must surely tell us many things that are very important ... more and more so as we will come to know what each bit of it means ... .

Several excellent research careers have been spent on trying to understand, in cellular and molecular terms, how the egg cell carries the vertebrate body plan from one generation to the next. With particular respect to human development, because of its crucial importance to brain and behavioral development, there is no part of all that which is more compellingly fascinating than questions about the steps in embryogenesis that establish the various modular differentiations between left and right. "Modular"? The whole system of left-right differences consists of multiple elements that apparently can — and must — be established separately at various times in embryogenesis.

We have long supposed that monozygotic twinning must be a symmetry-development anomaly, that — of course! — the embryogenesis of monozygotic twins must be strange, to build two body symmetries from a group of cells that would normally serve to build only one. But all along, the origin of dizygotic twins from double ovulation seemed so obvious that no one within the established belief system ever recognized any reason to imagine or to consider any alternative imaginable. It has just been presumed to be obvious and true. The presumption has nothing in its favor beyond its apparent (but ultimately false) simplicity. It has been cemented in place entirely by decades of unquestioning repetition.

> The consensus reality is the myth,
> but it remains as invisible to the majority as water is to a fish.
>
> *Joan Borysenko*

Simplicity is still the symbol, seal and shield of truth, and that remains a noble thought, a worthy principle and a valuable guide for logical progress. I can recommend it to you as an axiom and rule-of-thumb enthusiastically, but not entirely without reservation. My reservation is this: As I told you earlier, probably more than once —

> not all true things are simple, and not all simple things are true.

Simplicity alone is not enough.

Simplicity is not even more important than everything else.

It will always, in the conduct of science, be more important that a hypothesis should agree with all the available observational data than that it should be simple. And why, tell me, should we expect any solid understanding of the most complex of all phenomena we know anything about to be simple? Make me a list of everything within all the known realms of human knowledge or curiosity that is more emergently complex than building a new human. It should not take you long.

If Not From Multiple Independent Ovulation, then How?

The egg cell makes its appearance on the scene at ovulation by emerging from a ripened follicle that pops open on the surface of the ovary. A zygote might be produced from the fusion of a sperm cell with that recently ovulated egg cell. "If there are to be two zygotes," so the story goes, "then clearly doubling the recipe calls for two eggs ... and two sperm cells."

A binovular follicle containing two egg cells might be an alternative, and such structures do exist, but the frequency at which they have been found to occur is much less than the frequency of dizygotic twin deliveries, let alone twin conceptions. Besides that, there is another enormous little problem, in that no one has ever reported seeing either of the egg cells in a binovular follicle yield even one normal embryo, let alone a live baby, and most assuredly not one each from both of them.

So it seems that what we really need to do, in order to move forward from this level of understanding, is to puzzle out how dizygotic twin embryos can safely, reliably and frequently — in fact, routinely — be produced from the substance of a single secondary oöcyte, the kind of egg cell released from a normal single ovulation.

In all the time since I first began to understand that the facts of twin development do not support the double-ovulation hypothesis, there has arisen only one plausible suggestion for an answer to that question, only one plausible alternative origin for dizygotic twinning. I published my version of that idea over 30 years ago, with Robert Elston doing the lion's share of explaining the statistical implications and expectations of the model we proposed.

We could plainly see the process or mechanism that would be needed to explain the accumulated observations. It had to be some version of what had been called "dispermic monovular" (two sperm, one egg) twinning. The same process has sometimes been called "second polar body twinning." That is not an appropriate name because, in such a process, there really never *is* a second polar body. The genetic substance of what would normally become a proper second polar body must be put to a different use.

The normal second polar body is essentially a lightly wrapped discarded spare nucleus — one half-set of chromosomal DNA, one copy of each chromosome, wrapped in just enough membrane to pack it away outside the cell. It appears to have no cytoplasm to speak of. Most importantly, it is therefore completely lacking the system of cytoplasmic structures and organelles that is essential for a cell to conduct properly all the functions of living and growing and — most relevantly — duplicating the genetic material in its nucleus.

That "spare nucleus" which would normally become the second polar body must be kept inside the egg cell if it is to provide the genetic material for the second maternal pronucleus to form a second zygote, if it is to have the cytoplasmic accessories necessary for it to function as a nucleus.

One version of the process has been seen and photographed in experimental mammals in circumstances of delayed fertilization and thus overripe oöcytes — exactly the situation in which Harlap's work shows a tripling of human dizygotic twinning frequency. In those cases, the second meiotic cell division [which normally proceeds to completion upon receiving the signal indicating sperm penetration and in the course of that division pinches off and pushes out the minimal "daughter cell" that is the second polar body] happens in a different way.

The spindle for the second meiotic division in the oöcyte is normally paused in metaphase with the chromosome pairs lined up in the middle of the spindle. It is poised with one of the spindle poles directly against the egg cell membrane.

The second polar body would normally exit through that point in the membrane, through the little end of the most extremely asymmetric cell division possible in all of human biology.

After sufficient delay (if he's late, she'll start without him), that spindle moves to the center of the cell and executes instead a symmetrical division of the whole substance of the secondary oöcyte. It looks almost exactly like mitosis in an ordinary cell division, but there is only a half-set of chromosomes moving in to each daughter cell.

The daughter cells of that symmetrical division are two fertilizable haploid "egg" cells — daughter cells of a divided secondary oöcyte egg cell — for which reason I have long called them "tertiary oöcytes."[a]

The recipe for building two embryos, fetuses, then bodies for twins from the substance of a single egg cell requires two paternal pronuclei from two sperm cells plus two maternal pronuclei from a single secondary oöcyte. One of those two maternal pronuclei must be the normal product of the egg cell's second meiotic division (the "normal" maternal pronucleus) and the other must incorporate the half-set of chromosomal DNA that would have gone away in the second polar body in what has been understood as the "normal" version of these events.

The Recombination Problem

Most of the efforts prior to ours had suggested that any such "one-egg" dizygotic twins, if they exist, should be more alike than "ordinary" "two-egg" dizygotic twins. It seems all agree that such twins would have two different paternal contributions from two separate sperm cells. Several people who thought and wrote about the question before we did believed that the maternal half-genome contributions to single-egg dizygotic twins should be identical, coming as it seems they do from sister chromatid[b] direct-duplicate DNA molecules.

---

[a] I have learned that the term "tertiary oöcyte" has more recently been used elsewhere, in various fish, caecilians, and grasshoppers for example, apparently always with meanings somewhat different from my use for unfertilized daughter cells of a divided secondary oöcyte.

[b] The DNA of the second polar body and the maternal pronucleus were duplicates, as "exact" as replication could make them, when they finished the last DNA replication they went through.

In other words, the idea was that such twins should match for all of the DNA sequences they got from their mother. This would cause them to be more closely related — to have more genes in common — than any set of two siblings from two separate egg cells. That incurably mistaken idea has not gone away. Once again ... not all simple things are true. Do not forget that recombination has happened since that last DNA replication, and the sister chromatids of the first meiotic division have swapped parts! I still frequently get questions from twins and their parents about the possibility that the twins might be of the "third kind" — not "identical," but more alike than they would expect "fraternal" twins to be. (The great majority of those cases are cases of misinformation — they are monozygotic twins who were mislabeled dizygotic because they were dichorionic, or dizygotic twins mislabeled as monozygotic because their placentas had fused into a single mass, or because they were monochorionic.)

Within the past year, there finally appeared a paper proclaiming the discovery of one pair of twins with different paternal half-genomes and apparently identical copies of a single maternal contribution. It would seem, therefore, that this can happen. However, no one has ever (before or since) reported seeing a cell do what it would have to do to make that happen. The minimum essential explanation might take either of two forms, each requiring structures and functions never reported and not known to be possible.

One version would require a duplication of the DNA of the maternal pronucleus and segregation of the products thereof, after completion of the second meiotic division in the egg cell and before fusion with the paternal pronuclei. As we currently understand what normally happens in that cell, the necessary protein machinery of cell division is not available in that circumstance. The "necessary machinery" would particularly include the centrioles. The sperm must normally bring in a pair of centrioles as required to create a microtubule organizing center. The microtubule organizing center is required in turn to generate the spindle. The spindle must be built and operated correctly to execute a normal chromosomal segregation by drawing the chromosome duplicates apart into daughter nuclei.

The other alternative would require successful egg cell meiosis totally without recombination, so that the final chromosomal half-sets at the end of the second meiotic division might be identical replicates of one of the half-sets of chromosomes generated by the first meiotic division. This latter idea seems to have the advantage of requiring the absence of a process that normally must happen, as opposed to requiring the presence of a complex mechanism either created from nothing or transported into the cell from nowhere.

We have found in this apparently quite singular event no reason to suppose that any human conception ever happened that way before this one, or that it will ever happen that way again. Sounds like one of those "never say never" situations, doesn't it? Since apparently it *has* happened, then apparently it *can* happen. If it can happen once, no doubt it can happen again, so I will keep an open mind and keep looking for a mechanism to explain it, and I will certainly let you know if I find it.

The nature of that particular pair was discovered during investigation to learn the cause of ambiguous external genitalia in one of the twins. She had an uncertain external genital configuration more feminine than masculine. Her appearance was close enough to normal that they apparently had no great difficulty answering the "boy or girl?" question at birth, but different enough that somebody began to wonder only a short while later. (On closer investigation, she was found to be a true hermaphrodite, having both XX and XY cell lines as well as having both testicular and ovarian tissues.) Her twin brother is to all external appearances a normal boy. He has a mixture of XX and XY cells in the tissues accessible to reasonable sampling very similar to her mixture. This is a clear example of the random patchy distribution of chimeric cell lines that I told you about earlier, from the observation of experimental chimeras and the autopsy specimens. Apparently in the embryogenesis of this chimeric pair of twins, although their blood cells have almost exactly the same mixture of XX and XY cells, his gonadal ridges were colonized overwhelmingly by XY primordial germ cells, while hers must have gotten some of each.

The conception of this one very strange and special pair of twins, the formation of the pair of zygotes from which they grew, had to occur in a manner unlike any cellular event we understand. That clearly does not make it impossible.

Perhaps it is not so bad to *say* "never" about what any living thing might (be able to) do, but it is not at all a good idea to believe yourself when you say that. Don't believe everything you think.

There might well be many such events occurring every day without discovery among the great majority of chimeric individuals who have two normal cell lines of the same sex and never give us reason (such as the ambiguous genitalia in the one child of this pair) to conduct the kinds of close investigation that are capable of making such a discovery.

At any rate, no other twin pair or chimera reported to date has been found to have this maternal-identical sort of relationship between the cell lines. It would be notable, quite striking in fact, to find two cell lines matching for one allele at every locus (from their "identical" maternal half-genomes) and matching for the other allele as well at more or less exactly half of all the tested loci.

The normal pattern of sib-sib matches includes: neither allele matching at about one-fourth of the tested loci, both matching at about one-fourth, and (only) one matching at the remaining half. This would easily be seen to be different.

Professor Elston and I explained in that paper our realization that the idea of dispermic monovular twinning as it had previously been presented was wrong in neglecting the possibility of recombination in the building of the egg cell and ignoring the effects that should be expected from recombination.

Recombination in building egg and sperm cells is not in fact a "possibility." In the normal process of building the egg and sperm cells, recombination is a <u>necessary</u> component of the process. For a normal outcome of the egg- or sperm-building exercise, it seems that recombination must occur at least once in each chromosome arm, and will usually occur more than once in the longer chromosome arms.

There is a strong association between deficiencies of recombination and anomalies of sorting chromosomes in the process of making eggs or sperm cells. Extra chromosomes in trisomic cell lines [causes of Down syndrome and several other more disruptive aneuploidy conditions] have often gotten to be the way they are by failing at recombination, and human males who cannot do meiotic chromosome synapsis and recombination normally cannot make normal sperm cells.

There are few, if any, absolutes in all of biology. This is not one of those few. There are known to be a few organisms in which recombination seems not to be an essential part of normal spermatogenesis — but only for spermatogenesis. Our concern here is not spermatogenesis, but oögenesis, and I have found no report of any such exception in any organism with respect to the making of egg cells. There is every reason to believe that recombination is universally necessary for functional oögenesis.

Recombination profoundly affects our present considerations because the DNA strands that were exact duplicate sister chromatids, after the last DNA replication before going into making the egg cell, have necessarily undergone recombination since that replication. At least one of each pair of those DNA molecules, and in most cases more likely both of them, have crossed over at least once with either or both of the copies of the same chromosome that came from the other grandparent. A very reliable result is that they are no longer exact duplicates. Each of those DNA strands that were once sister chromatids is now carrying a mixture of DNA sequences from both of the mother's parents. The same thing happens in building sperm cells. Every DNA strand that may go into a sperm is about half from each of the parents of the man making the sperm cell.

We were further "certain" when we were writing that paper that meiotic recombination simply does not occur in or near the centromeres. That idea had been generally accepted as truth — over the whole field of the genetics of complex organisms, right up to today's textbooks and society meetings — because photomicrographs of what we believed then, and still believe, to be recombination in progress did not show chromosomes crossing over anywhere near their centromeres. The microscopically observed structures in question are called chiasmata — meaning "crossings" in Greek. Nobody ever reported seeing chiasmata at or very near the centromeres.

There are now available, however, a few serious reasons to suppose that that observation was not a valid foundation for all the further work we did:

(i) The highly repetitive alpha-satellite DNA concentrated in human centromeres has a hierarchical repeat structure the evolution of which cannot be explained without frequent unequal recombination within its lengths, as the source of the oddly high variability among individual copies of the DNA of each chromosome's centromere. Those frequent rearrangements in the alpha-satellite centromeric DNA might be mediated by mechanism/s different from those of meiotic recombination as we more generally understand it. It might happen without chiasma formation. We have no real understanding of whether it should result in exchanges between DNA sequences on opposite sides of the centromeres.

(ii) Most of what we know and believe about the mechanics of recombination in the human we have learned from studies of meiosis in spermatogenesis, in males. It is much easier, I feel sure you will understand, to obtain samples of future sperm cells at the stage in the process of gametogenesis where recombination occurs. Future egg cells begin and complete that same stage of their development in the female fetus, late in the first trimester and early into the second. I could wish, for research reasons, that it were not quite so, but I will hope that we will not grow accustomed to taking cells from fetal ovaries, so that those cells will continue to be as hard to come by as they have been.

From genetic evidence (re-assortments of parental genes in the offspring) we have long known that recombination events are considerably more frequent in the human female (in the making of the egg cells) than they are in spermatogenesis.

We also have long known that the probability of crossing-over events is not uniformly distributed along the length of each chromosome, and that the distribution of crossover probability along the chromosomes differs between the sexes.

In recent years (long after that paper of mine and Elston's), several different human chromosomes have been shown to experience little or no suppression of recombination in the regions near the centromeres in human females. Where the question has been appropriately tested, the female excess over male frequencies of recombination is greatest near the centromeres. I had misled myself by assuming that observations of chromosome behavior in spermatogenesis should be representative of the same processes in building egg cells (not all simple things are true). The fact that everybody I knew believed it as well is the closest I can come to an excuse [Boklage's 17th Law: all excuses are created equal].

With this understanding now available, there remains no logical basis for the predictions we made about differences between ordinary siblings and tertiary oöcyte twins, to the effect that tertiary oöcyte twins might be expected to match for markers near the centromeres more frequently than would non-twin sibs or "two-egg" twins. It was through those predicted differences that I once expected to demonstrate the frequent occurrence of tertiary oöcyte twinning. On the basis of those assumptions [obligate recombination in both arms of each of the chromosomes, with no recombination occurring in or very near the centromeres], I was confident that our understanding of the mechanism of tertiary oöcyte twinning could be demonstrated straightforwardly, if not perhaps simply or easily.

I felt certain that tertiary oöcyte twinning should give the appearance that dizygotic twinning is genetically linked to the centromeres in the maternal line because their maternal half-genomes would come from sister chromatids which would have experienced no recombination near their identical centromeres. This should give the appearance of genetic linkage between the trait "being dizygotic twins" and markers near all of the centromeres.

In what we call an affected-sib-pair linkage analysis — with dizygotic co-twins counted as affected sibs and all non-twin sib pairs as unaffected — affected sib pairs should (according to that logic) share markers near the centromeres substantially more often than should the non-twin sib pairs. Because (we believed then that) all of the chromosomes in the maternal half-genomes of tertiary oöcyte twins went through recombination attached to the same centromeres, and assuming that recombination is suppressed in and near the centromeres, then the "tertiary oöcyte" version of dispermic monovular twins should match much more often for their particular variations in DNA sequences near the centromeres.

Away from the centromeres, they could be confidently expected to match with the same probabilities as two-egg twins or non-twin sibs. That was the hypothesis. The plan was to demonstrate that, and thereby demonstrate directly that dizygotic twins must come from single egg cells.

After more than half a million dollars worth of experimental work (funded by the National Institute of Child Health and Human Development and the National Human Genome Research Institute, and done for me at the Center for Inherited Disease Research at Johns Hopkins University), and after a few more years of thinking about the results thereof and the differences between the observed results and the expected results, it has become clear to me that no one will ever see what I then expected to see. There is nothing in the results to support the idea that dizygotic co-twins share DNA sequences near their centromeres more often than the members of non-twin sib pairs do. It is now clear that there was never realistically any such expectation.

That will probably be taken by staunch advocates of the old orthodoxy of double ovulation to mean that I have been wrong all along, and that all dizygotic twin pairs do in fact arise from two independent ovulations. In fact and indeed I was wrong, but not that way, not about that. I was wrong in believing that we could demonstrate it in the way I tried to demonstrate it. The experiment was based on a flawed understanding of the mechanisms at issue. There was a hole in the logic, through which the truth could escape — and appears to have done so. These results do not in fact demand a conclusion in favor of double ovulation, and there remains a substantial burden of sound evidence against the double-ovulation hypothesis.

I remain satisfied with all the evidence against two separate and independent processes of ovulation plus syngamy plus embryogenesis. The adult traces of variations in asymmetry determinations during embryogenesis are all the same in monozygotic and dizygotic twins, and in the first-degree relatives of both. The embryogenic events by which a conceptus becomes a twin conceptus leave the same traces in adult twins of both zygosities. I have been able to see no path through that except to suppose that those cellular events are the same for monozygotic and dizygotic twinning events.

We have to back up a bit now and come at this from another angle. I just wrote, a few pages back: "The zygote is produced from the fusion of an egg cell and a sperm cell. If there are to be two zygotes, so the story goes, then there is a clear need for two sperm cells and two egg cells." This is how we are accustomed to bringing our understanding of singleton embryogenesis into our effort to understand twin embryogenesis — by simply doubling the recipe for all components of the process.

That is exactly what we must expect from double independent ovulation; two singleton embryogeneses at once. That is not, however, anywhere near what we actually see, as reported in all those previous chapters. Dizygotic twin development is different from singleton development at least as much, and in the same ways, as monozygotic twin development differs from singleton development. Where the embryogenesis of monozygotic twins is odd, the embryogenesis of dizygotic twins is at least as odd in all of the very same ways.

Here we can see how we misled ourselves: We do not in fact need two egg cells and two sperm cells to produce two zygote cells. What we really need is two maternal pronuclei and two paternal pronuclei to enter two processes of syngamy to produce two zygote nuclei. That does not in fact create a demand for four whole gamete cells, in particular for two whole secondary oöcyte egg cells from two independent ovulations. Each single secondary oöcyte, the "egg cell" which is normally the target for sperm penetration and subsequent fertilization, comes with the full genetic equivalent of the necessary two maternal pronuclei.

The normal secondary oöcyte erupts from the ovulatory follicle with the second cell division of meiosis in progress. This cell crossed the threshold between its time as the primary oöcyte and becoming the secondary oöcyte when it finished the first meiotic division and began the second. The first meiotic division ended with the pinching off of the first polar body, a nearly naked spare nucleus with just enough membrane to cover it, carrying away a full diploid complement of extraneous and dangerous DNA attached to a half-set of (23) unduplicated centromeres, and leaving behind the same — two copies of the DNA of each chromosome, still attached to the unduplicated centromere of each chromosome.

To finish preparing the maternal pronucleus, the maternal half-genome contribution to the zygote nucleus, there is a need to finish the duplication and segregation of the centromeres and separate the two copies of each chromosome in the second cell division of meiosis. The second meiotic division commences immediately upon the completion of the first, and normally at ovulation has reached the metaphase stage, with the chromosomes lined up between the poles of the spindle waiting to be sorted by spindle fiber traction into the daughter nuclei.

In the "normal" version of this process, penetration by the sperm generates the signal to complete this second meiotic division, with the spindle moving one half-set of single chromosome DNA strands out into the second polar body, leaving the same inside to form the maternal pronucleus. The daughter nuclei are normally expected thereby to become: the maternal pronucleus on the inside of the cell, and the second polar body outside the cell, right next to the first polar body.

When fertilization is delayed in experimental mammals, and preparation of the egg cell for fertilization overshoots its developmental target state of readiness for embryogenesis, the spindle in the secondary oöcyte has been seen to move to the center of the cell and to complete the second meiotic division without the usual signal from the sperm's penetration of the egg cortex. When it happens this way, the division is not the extremely asymmetrical division that normally takes the second polar body out of the egg cell during the fertilization process. The resulting symmetrical division leaves two roughly equal-sized daughter cells, both of which can be fertilized. I first called those fertilizable haploid daughter cells "tertiary oöcytes" in public and in print somewhere around 30 years ago.

This process has been photographed in progress in experimental mammals, but not yet to my knowledge in humans. However, there are very few reports in the literature of working with naturally-ovulated human oöcytes in any way, for any purpose. Susan Harlap's experiment-in-nature shows that delayed fertilization results in multiplying the frequency of human twinning, just as it does in several experimental mammals. The excess human twinning resulting from over-ripe human egg cells is clearly dizygotic (another huge little observation that cannot be explained by double ovulation).

Naturally ovulated human egg cells are very difficult and expensive to obtain, quite precious for all sorts of reasons, and are unlikely ever to be used in experiments where they would deliberately be allowed to over-ripen in a laboratory vessel. Artificially ovulated oöcytes, having already had their timing disrupted by the chemical induction of ovulation, are certain not to be fully informative, but when allowed to over-ripen they do show intracellular behaviors and changes very similar to those undergone by egg cells from those species which have been photographed doing the tertiary oöcyte divisions.

So, plausible cellular alternatives to independent double ovulations as source of dizygotic twinning would have them arising from daughter cells of single secondary oöcytes divided symmetrically before sperm entry ("tertiary oöcyte twins"), or from those same two half-genomes in an as-yet-undivided secondary oöcyte. Some have found it easy enough to think of this as a "rescue" pathway for over-ripe or otherwise compromised oöcytes.

In any plausible mechanism, there must be two paternal pronuclei. These most often come from two sperm cells, but diploid sperm are apparently not yet conclusively ruled out. The two paternal pronuclei must achieve syngamy with two maternal pronuclei arising from the second meiotic division of the secondary oöcyte nucleus, one of which "should have been" discarded in the second polar body.

Two maternal pronuclei may occur inside one egg cell membrane with an unfinished second meiotic division, or as two cells (tertiary oöcytes) with separate membranes after a symmetrical second meiotic division. All variations have the final common expectation of two syngamies producing two zygotes inside a single zona pellucida. This would generate a structure not easily distinguished microscopically from the product of the first mitotic division of the zygote. The two cells should, however, be distinguishable by molecular analyses because, here, these first two cells are of different genotypes. This structure would be the same size as the product of the zygote's first mitotic division, having as it does the mass and volume of a single secondary oöcyte, and it would still be inside the single zona pellucida. This was exactly the appearance of the twinned embryos observed by Bomsel–Helmreich and Papiernik–Berkhauer in rabbits after delayed fertilization. They declared these double embryos to be monozygotic twins, and no one argued with them because "everyone knows" that only monozygotic twins could occupy a single zona pellucida.

The repeated observation of monochorionic dizygotic twins makes it clear that this process is possible. The apparent origins and distribution of chimerism require that it be frequent. The distribution of developmental differences between all twins and singletons, as discussed over several previous chapters, makes it clear that any two-egg alternative would be the rarer one.

Triploidy

Here we need to consider triploidy. A triploid cell, cell line, conceptus, embryo has three copies of each of its chromosomes instead of the normal two sets in diploidy. Over three-quarters of all human conceptions fail between fertilization and term. Of those that develop far enough for their presence to be recognized and then fail in the first few weeks after recognition, about half have a chromosome problem. A sizable fraction of those chromosome problems are triploidies (only a small fraction of triploid conceptuses survive their first trimester). The extra half-set of chromosomes sometimes comes from the mother (*digynic* triploidy) and sometimes from the father (*diandric* triploidy).

Those from the father are probably always from dispermy (two sperm cells having penetrated a single egg cell) followed by syngamy with a single maternal half-set. I found one liveborn diandric triploid fetus reported to have arisen from a single sperm with DNA for two pronuclei — called "*diplospermy*" for a diploid sperm cell with two sets of chromosomal DNA molecules. I did not, however, find the argument compelling (that the two paternal half-genomes arrived in one sperm cell and not two). The digynic triploids require syngamy between a single paternal pronucleus and two maternal complements — for which no mechanism other than "retention of the second polar body" has been proposed.

One may find a sizable range of estimates as to the relative frequency of diandry and digyny among triploids, but the reports indicating that they do not differ greatly in frequency seem to have the best supporting data. The frequency of triploids reported (which does not include any that have failed before recognition) shows an ample supply of doubled pronuclear contributions from both maternal and paternal sources.

> The frequency of triploidy, both digynic and diandric, is clearly great enough that neither dispermy nor dipronuclear egg cells can be considered likely to limit the possibility of dispermic monovular twinning.

Fathering Dizygotic Twins

Further, if we wish still to assume that only mothers could influence any probability of twinning, by double ovulation, we would have to suppose that the well-documented paternal effects on probability of dizygotic twinning would be exerted through single-egg dizygotic twinning. (Things really can get messy when one tries to have butter on both sides of one's toast.)

Other major pieces of this puzzle include: suspension of the second meiotic division after ovulation pending sperm penetration, the dependency of syngamy and early embryogenic cell division on the centrosomal material and centriole/s provided by the sperm, the need for the oöcyte to conduct a major rearrangement of the sperm chromatin to transform it into a functional paternal pronucleus, and other changes in the oöcyte after ovulation. This system of interactive processes required to complete fertilization provides a plausible focus for questions of paternal influence and monovular dizygotic twinning.

For embryogenesis beginning with two zygotes in a single zona, a single chimeric offspring would seem at least as likely as the formation of separate twin bodies. [Experimental chimeras made by fusing parts of two or more embryos never become twins.] If separation is achieved (requiring the same cellular behaviors as monozygotic twinning), so that two concurrent embryogeneses may proceed in parallel beyond that intersection, the likelihood that the two embryos would carry souvenir cells of each other's genotype seems high. Although admixtures may be quite small, and may be unidirectional instead of reciprocal, completely non-chimeric dizygotic twinning is unlikely.

The existence of monochorionic dizygotic twins provides an unavoidable lesson: twin zygotes, same sex or different, do at least sometimes form a single mixed embryo from which they may emerge as viable twins, often carrying samples of cells from each other mixed in as they build their separate bodies. In such chimeric embryos, spontaneous internal definition of two body symmetries occurs, perhaps (but only perhaps) most commonly before (dichorionic) but at least sometimes after (monochorionic) cellular commitment to the differentiation of the trophoblast.

We have no evidence as to the relative frequencies of those two possibilities. To date, the number of discovered monochorionic dizygotic twins is small, but not really small compared to the probability of finding them without looking.

Any prospect a chimeric embryo will have for development to live birth as two separate individuals requires the very same cellular event as monozygotic twinning, namely, to create two systems of body symmetry axes inside a single mass of cells, so that they can begin and continue to grow out as two bodies. The twin bodies that may be built upon those cellular/molecular armatures are dizygotic; they must develop from two zygotes, which will require two paternal and two maternal pronuclei. They are clearly not, however, independent — they have been at least temporarily within the same single embryonic inner cell mass.

According to the evidence accumulated here, this occurs with much greater frequency than previously imagined, with many more cases remaining undiscovered for want of asking the necessary questions. Unless exchange of pluripotent stem cells between twin fetuses through the maternal circulation can be shown to be routine, such an outcome seems highly improbable for twin embryos from independent oöcytes. Chimerism, would, however, be quite ordinary for twin embryos that begin development within a single zona pellucida.

I believe that you should be certain by now, as I am, that double (separate and independent) ovulation is not a plausible origin for naturally conceived human dizygotic twins. It should also be clear that the dizygotic twinning event, whatever we may imagine it to be, frequently results in spontaneous embryonic chimerism. I can do no better at the moment than to propose dispermic monovular dipronuclear twinning as a plausible connection.

Chapter Nineteen

Normal Embryogenesis of Left and Right, and Twins and Chimeras

... wherein will be revealed the hypothesis that must now replace
the untenable belief in double independent ovulation
as the source of naturally conceived human dizygotic twins

A kind of synthesis, but with some elements
that perhaps you wouldn't have expected in advance.
I always like that when that happens,
when something comes that is more than the sum of the parts.

Evan Parker

It is easier to perceive error than to find truth,
for the former lies on the surface and is easily seen,
while the latter lies in the depth, where few are willing to search for it.

Johann Wolfgang von Goethe

The farther backward you can look,
the farther forward you can see.

Winston Churchill

Now that we understand that natural human dizygotic twinning has nothing to do with double ovulation, the time has come to offer a new hypothesis to replace the one I hope and believe I have thoroughly ruined in these preceding chapters. This hypothesis I now propose, unlike the old double-ovulation story that I now consider thoroughly discredited, does in fact fit all of the observations of which I am aware, at full face value. There are no discounts; no unconscious overreaching assumptions; no ifs, ands, buts or maybes; no untestable presuming.

> When you have eliminated the impossible,
> whatever remains, however improbable,
> must be the truth.
>
> *Sir Arthur Conan Doyle, via Sherlock Holmes*

According to the results and my interpretations thereof detailed in these chapters, monozygotic and dizygotic twins must go through their embryogenic a/symmetry development protocols in the same way/s, equally different from singletons in the same ways, having to do with the same pervasive variations in the elaboration of left-right asymmetries in embryogenesis.

We have always supposed that the monozygotic twinning event (somehow dividing up one mass of cells to lay out two body symmetries and form two embryos) must be the cause of these anomalies of developmental a/symmetries in monozygotic twins. Twin research as a whole has never considered such an explanation for anything about dizygotic twins because of the stubborn insistence that the dizygotic twinning process arises from two independent oöcytes (with independent syngamies and independent embryogeneses) and therefore involves no anomalies of embryogenesis.

The previous chapters here allow us now to understand that natural, spontaneous dizygotic twinning must also, at least very nearly all of the time, involve the partitioning of a single mass of cells to form two embryos, yielding the same distribution of subtly profound variations on the themes of asymmetric human embryogenesis.

> Human spontaneous dizygotic twins routinely arise from a single mass
> of contiguous cells, inside a single zona pellucida, from the substance of
> a single secondary oöcyte, by way of establishing two body symmetry
> plans, two sets of head-tail, back-belly and left-right definitions, within
> that mass of cells.

This hypothesis not only accommodates the observed existence and frequency of spontaneous embryonic chimerism, but includes and embraces it. We probably must even consider it anomalous if we could ever prove that any naturally conceived dizygotic twin pair ever had no history of chimerism at all.

To divide a cluster of cells of two different genotypes into two embryos perfectly, with no mixing of the cells at all, ending with only one genotype of cells in each of the resulting two embryos, would require a much more exacting, much more highly-ordered process and is therefore very much less likely than any result in which some degree of mixing of the two genotypes occurs. This hypothesis readily accommodates the occurrence of monochorionic dizygotic twinning, with a currently quite unknown frequency because so few monochorionic twin pairs, and so far — almost exclusively — obviously anomalous such pairs, have ever been appropriately checked for dizygosity.

The Timing of Twinning Events

Monochorionic twinning events must occur after the outer layer of the morula differentiates to become the trophoblast, and inside that single trophoblast, which the twins will share throughout the remainder of their gestation.

Dichorionic twins, regardless of zygosity, must differentiate their separate trophoblast layers before or during their cellular action of forming two body plans, so that each may have his own.

This hypothesis is compatible with all of those observations we have discussed in these chapters, all of which observations are incompatible with the prospect of separate and independent double ovulation.

Dichorionic twinning requires that the twinning event — the cellular onset of the fact that there will come to be two embryos — must redefine the cells of a single "morula" cluster as two groups of cells before the differentiation of the outer layer of the morula, the trophoblast precursor to the chorion, so that each of the two sets of cells will have its own subset of cells ready to differentiate into trophoblast.

The timing of that dichorionic kind of event is generally understood to be earlier than the monochorionic event and therefore even more certainly still inside one same single zona pellucida. This must happen within the first few cell divisions, before the trophoblast cells commit to their separation from the inner cell mass to change the morula into the blastocyst.

Cellular mechanisms contributing to the earliest known control process governing left-right asymmetry differentiation are clearly detectable in the four-cell stage of every vertebrate embryo properly studied, after only two or three cell divisions of embryogenesis.

> Three divisions — four cells? Yes. The early cell divisions ("cleavages") in human embryogenesis are not synchronous like those of the starfish embryogenesis we studied in high school — 2, 4, 8, 16, etc. There must generally be three divisions of the zygote and its first few descendants to reach four cells. Through the first several cell divisions, in general, no two cells are dividing at any given instant, and each of these first several divisions is asymmetric. In each of these divisions, one daughter cell is larger than the other. The second mitosis divides the larger of the first two blastomeres, and then there are three cells. The third mitosis divides the largest remaining cell, which was the smaller of the first two, and then there are four. One of those four — the "right ventral blastomere" — is already specialized for specific involvement in the early serotonin signaling pathways. Already when there are only three or four cells, one of them specifies and displays asymmetric directionality for the whole set. Already "symmetry is broken." The cells are quite capable of carrying that marking through all subsequent divisions. The belief that the embryo is quite symmetrical until the occurrence of some event that first defines left and right is not supported by factual observation.

Consequences of Defining Two Body Symmetries in One Morula or Inner Cell Mass

Both monozygotic and dizygotic twinning processes require the recruitment of cells from within a single contiguous mass of cells to establish two separate systems of body symmetry axes. There must be two separate asymmetric plans for where all the pieces and parts must go. It is reasonable to suppose that this happening could disturb the early stages of installing left-right asymmetries in the building of the embryo, and that the prospects for disturbing those processes will operate equally in both monozygotic and dizygotic twinning.

Nothing within current understanding offers any improvement over the old idea that these embryogenic asymmetry anomalies that are so common among twins occur because "cells that had begun to become parts of the right side of Willy must now rearrange everything toward becoming the left side of Wally." This still serves handily as a way to imagine that the fully proper specification of all the body's natural asymmetries may be disturbed by the division of a single contiguous mass of cells into two body plans.

To the extent that the earliest molecular events of defining the left-right aspects of the body plan have already begun before individual cells are recruited and assigned to either one of the two new plans, then some aspects of their asymmetry-defining functions may be disrupted. We know that asymmetric differentiations have begun by no later than the second or third cell division, with the allocation of the control of serotonin transport, and that perturbations of that setup routinely reach all the way forward to the randomization of visceral organ asymmetries after heart looping.

To the extent that any differentiation has begun in the direction of establishing the asymmetric patterns of later development at the time that any decisions have been made in the direction of establishing two asymmetric patterns instead of one, the molecular mechanisms for establishing the proper normal asymmetric patterns of differentiation may be expected to be compromised.

The "determinants," the developmental mechanochemical changes that will ultimately result in specifying the necessary left-right differences, have already begun to be expressed in the first very few cell divisions, and it is entirely plausible that some of the cells that had begun to differentiate toward various right-side specificities may now end up being recruited to the left side of one of the two body plans. This may be reciprocal, but some pieces of the evidence to date suggest that the right side may be usually defined first.

The earliest-acting presently known molecule in the cascade of gene regulators that specifies the asymmetries of embryogenesis is *serotonin*. Serotonin is a very familiar molecule, a derivative of the amino acid tryptophan long known primarily as a *neurotransmitter*, which carries signals across synapses between neurons of certain types. It is the neurotransmitter on which the "selective serotonin reuptake inhibitor" (SSRI) anxiolytic-antidepressant medicines work. Serotonin has been found recently to have cell-to-cell signaling functions in embryogenesis long before any nerve cells have differentiated. In frog and chick embryos, one cell of the first four is already detectably specialized to participate in the serotonin signaling pathway. After only two or three cell divisions, there is already a clearly defined asymmetry to presage the development of the whole body's asymmetries. Some of these programmed asymmetries may be added, subtracted, or modulated at each successive cell division throughout development. These changes constitute one or more layers of the combinatorial code of gene expression that constitutes the functional differentiation of all the various cell types of the body.

In bird and amphibian embryos, disruption of the function of specific serotonin transporter molecules (with selective serotonin reuptake inhibitors) in that one cell of the first four disrupts the proper establishment of left-right embryogenic asymmetries "downstream" farther along in development. This early embryogenic asymmetry-determining function of serotonin has not yet been demonstrated in any mammalian embryo, but it is clear that mammalian embryos must establish the same system of asymmetries of cell, tissue and organ structure and function.

All mammals tested produce those same serotonin transporter protein molecules and show association between anomalies of those genes and their products, and anomalies of craniofacial and brain function asymmetries.

There have been a number of reports about pregnancy outcomes for women who have used SSRIs during pregnancy. None of the studies has reported any statistically significant increase in major birth defects, but there have been reports of increased frequency of cases with multiple minor anomalies, and with circulatory anomalies leading to pulmonary hypertension (above-normal blood pressure in the main artery from the heart to the lungs). Proper structure and function of the pulmonary artery is asymmetric; it must be continuous with the right ventricle of the heart. Several of those studies, but not all of them, have reported greater frequency of miscarriages. The sample numbers in those studies are generally small enough that the absence of statistically significant differences cannot be argued to constitute evidence that the expected effects are absent. None of the studies have addressed early losses of pregnancy by way of very early spontaneous abortions or single missed periods. Under such protocols as those used to date, only anomalies subtle enough to have allowed survival at least into the fetal period will be seen at all. The background frequency of the most likely anomalies is low enough that only much larger samples than these have any likelihood of showing significant increases. [This is the same situation as that of early studies of artificial reproductive technologies — for quite some time there were papers that showed "no significant excess" of birth difficulties in IVF babies. Over the same time span, there were no papers that showed a reduction of the frequency of those problems — which should have been happening in half the samples if in fact the rates were the same!]

Further, none of the studies I have found so far has addressed behavioral teratogenesis, which is a likely focus for the subtler anomalies that might be originating in the determination of developmental asymmetries. Those studies will require long-term follow-up of large numbers of the children born of such pregnancies. This would seem to be an obvious priority with respect to studies of the reproductive effects of psychoactive drugs taken in the course of pregnancy.

Whatever else may be said about the whole collection of information to date, there is no sound evidence that mammalian embryogenesis does this part differently from the embryos in which serotonin's involvements in early embryogenic asymmetry development have been demonstrated.

This needs to be watched closely, and more questions need to be asked, particularly about the histories of efforts to conceive by women taking SSRIs, where any effects of serious disturbances in the first few cell divisions of embryogenesis might be concentrated.

The Plan: Fixing Axes, Fixing Asymmetries

Much of the research on the development of the a/symmetry of vertebrate bodies has been framed as "How does the symmetry of embryogenesis get broken?" How do the parts that belong on the left, for example, know which side the left side is meant to be on, so that they may know where to go to build themselves, or go to be used to build things with. When do they first know? Where do the cells involved get the information they need? The asymmetries of the anteroposterior (head-tail) and dorsoventral (back-belly) axes are obvious and easy compared to the subtlety of the left-right differentiations.

"Heart looping," the establishment of the asymmetric major curvature of the primordial heart, has been considered to be a very early step in that direction, even seen perhaps as the beginning of the establishment of the asymmetries of the internal organs. It *is* one of the earliest visible left-right asymmetries and is a useful example of what we need to understand.

The cellular and molecular determinants of the asymmetric directionality of heart looping have been traced, step by step, back to asymmetric expression of one or more of the signal proteins activin, PCL2, BMP4, and TGFβ. [We can do all the rest of this just as well, without loss of any real specificity, if I use *"activin"* to represent this whole class of candidates without all the qualification.] The specific expression of activin in cells on the right side of the primitive streak in the embryo inhibits the expression, there on the right side, of the sonic hedgehog SHH gene product. But sonic hedgehog *is* expressed on the *left* side, where it triggers a cascade of other cellular signals and cellular responses that end up causing the heart to grow to the left of center, with its left side receiving oxygenated blood via the pulmonary vein from the lungs and pumping it out through the aorta to the rest of the body, and its right half moving deoxygenated blood from the body into the lungs. In the process, the left and right halves of the future heart muscle are built of different kinds of cells.

Tracking down the effects of certain mutations that cause the determination of heart asymmetry to be random or reversed led to the discovery of the activin–sonic hedgehog signaling cascade. Eventually the masterful body of work that developed from that excellent approach, led in large part by Professor Michael Levin, hit a logical wall. If indeed all of this begins with the asymmetry of activin synthesis [or the synthesis of any one or more of those other candidate asymmetric asymmetry-signal molecules — it absolutely does not matter which one/s], then how is the asymmetry of activin synthesis established?

The gene that codes for the activin signal peptide molecule is not breaking the symmetry; it is not creating an asymmetry from which the asymmetries of everything thereafter can be determined. Activin reliably shows up to inhibit the synthesis of the sonic hedgehog protein on the right side of the primitive streak, thereby demonstrating and enforcing a very definite asymmetry that was absolutely necessarily already there, an asymmetric developmental algorithm already in place and in progress.

It Is Not about "Breaking Symmetry" — There Is No Symmetry that Needs to Be Broken

It is a considerable jump from there back to the serotonin transporter asymmetries of the first few cell divisions, and we know precious little so far about what happens in between, how the definition of a left-right difference is transmitted, but we do indeed know that there is a pathway connecting those first few cell divisions with heart looping and the rest of the grossly visible asymmetries. We do not know much about what it looks like, but it has to be there.

How is it that every cascade of left-right-asymmetry-determining molecules and reactions that we have discovered (including this earliest known subsystem that is already in place after only two or three cell divisions) *begins with* an asymmetry that is already reliably established?

How does the activin gene "know" [very reliably, in fact, and surely by way of information from a signal molecule product of some other cell] where to get turned on to make its product to inhibit the synthesis of the sonic hedgehog gene product on the side to the right of the node of the primitive streak (without which fundamental embryogenic asymmetries will be randomized)?

How do the serotonin transporter genes "know," very reliably, to synthesize their products earliest and most strongly in the right ventral blastomere, perhaps only two, at most only three, cell divisions into embryogenesis?

How can symmetry be "broken" so early? How can those serotonin transporter gene sequences "know" which of the first two or three or four cells is, or must become, the right ventral blastomere, in which they should first be expressed? Or, put another way: How does the right ventral blastomere "know" it is the cell in which those genes should first be expressed? In order for that "understanding" to be fixed, at least one — maybe two or all three — of the first three cell divisions had to be left-right asymmetric. It is already certain that those divisions are asymmetric; it is not so easily understood in which dimension/s they are asymmetric.

DNA Is the Reference Point for All Embryogenic Asymmetries

I believe, and hereby propose to you and all of developmental genetics, that the only possible answer lies in the inherent asymmetry of the DNA.

Molecules of DNA are reliably, non-randomly asymmetric …
in every cell since the beginning
of all cellular life that depends on the replication of DNA.

That asymmetry is of the molecular essence of DNA. The molecular structure of DNA just *is* asymmetric. It would not, could not, be DNA without being asymmetric. The understanding of that asymmetry was a fundamental part of what Watson and Crick got their Nobel Prize for — showing us how the DNA molecule is composed of two strands with opposite polarity wound around each other in opposite directions. As long as there is DNA in a cell, there is available a standard reference point for any asymmetry that any process in the cell may need to establish or to work from.

Except for several millions of quite small stretches of palindromic sequences that read the same in either direction for short distances (and which readily form hairpins and other secondary structures, which may serve as signal-recognition sites), the two strands of every natural molecule of DNA have different structures, different base-pair sequences and different functions.

Amar Klar, the one colleague of mine who spends the most time thinking about this part of the story, in informal discussions and a few of his papers, calls the two strands "Watson" and "Crick," not because he knows which is which, but to indicate that he knows they are different, because the cells he works with — mostly yeasts — have made it very clear to him that *they* "*know.*"

Every organism ever appropriately tested has essential functional mechanisms that depend on "knowing" which strand of the DNA is which. The two strands of every naturally occurring molecule of DNA are different, in their relative fractions of A=T adenine:thymine vs C≡G cytosine:guanine base pairs, in density of protein-coding versus noncoding DNA sequences, in preferential codon usage within the protein-coding sequences, in density of promoter and enhancer sequences that cause variations in the functionality of DNA sequences for coding, in density and distribution of replication-leading versus lagging strands — for a few of the better-known examples.

Human DNA replication starts simultaneously at multiple replication origins scattered around the genome. From each of those origins replication proceeds in both directions; it is bidirectional. The DNA has to be unwound and the intertwined single strands separated, to open up a "bubble"-like structure with "replication forks" at each end, within which the replication enzyme complex will work.

DNA replication can only add new bases to form the new copy in one chemical direction, so it copies only one strand (the *leading strand*) in that direction.

The other strand (the *lagging strand* in the first direction) is the leading strand for synthesis in the other direction.

The directional asymmetry of DNA comes from the fact that each "*nucleotide*" (base plus sugar plus phosphate) molecule in the chain has two different sides, and they are bonded together in the same direction all the way along each of the strands — and in the opposite direction on the opposite strand.

The difference can be considered to be on the deoxyribose sugar part of the nucleotide molecule. That sugar has five carbon atoms. In the number 1' ("one-prime") position there is attached an adenine, or a thymine, or a cytosine or a guanine — whichever nucleic acid base belongs in this position in the chain. In its proper position in the long chain of the DNA, the "backbone" of the long molecule is–sugar–phosphate–sugar–phosphate–, etc. Each sugar is attached to a phosphate on each side; one on the number 3' (three-prime) position and one on the number 5' (five-prime). We call those atoms in the sugar molecule one-prime, three-prime and five-prime (1', 3', and 5') to distinguish them from the 1, 3, and 5 positions on the ring of the base. During the biochemical synthesis of a molecule of DNA, each new nucleotide coming in to the chain adds itself on to the exposed 3' –OH end, binding by the phosphate on its 5' sugar ring position.

As the replication enzyme complex moves along the strand it is following, every 180 to 200 bases the complex turns around inside of the replication fork on its end of the bubble and goes back along the lagging strand, in the same *molecular* direction (but in the opposite physical direction because the nucleotide monomers in that other chain are all lined up going the other way). The new strand copied along the leading strand is continuous, and the new strand copied "backward" along the lagging strand is made in short pieces each about 180–200 bases long which must be bonded together to form the continuous final product copy of the lagging strand.

The leading and lagging strands are different, and every cell we have ever properly tested knows which is which and has ways in which it can and must use the difference.

As replication proceeds, the "old" parent copy of the leading strand is paired with a "new" daughter copy of the lagging strand, and the "old" parent copy of the lagging strand is paired with a "new" daughter copy of the leading strand. When replication is complete, and there are two daughter copies of each DNA molecule that entered this round of replication, then each of those daughter molecules includes one "old" strand from the parent double-stranded molecule, and one "new" strand that has just been synthesized.

So, the old Crick has a new Watson, the old Watson has a new Crick, and every organism properly tested to date has shown that it has functions for the proper execution of which it must properly use whatever that system of specifications may comprise.

Besides "knowing" "Watson" and "Crick," every organism appropriately tested has demonstrated essential functional mechanisms that depend on "knowing" which strand of the DNA is "old" and which is "new"; that is, which is "mother" strand and which is "daughter" strand from its most recent replication.

"Watson" and "Crick" differ in sequence, while "old" and "new" are epigenetic distinctions not representing sequence differences. The whole story of epigenetic functions and variations has only begun to be told, but we know enough to be certain that it is fundamental to the control of variations in development. Epigenetic variation shows some likelihood of being the best reason that monozygotic twins are not always concordant for certain genetic variations (especially in variations that are asymmetry-related — such as handedness and the fusion malformations and homosexuality and schizophrenia).

Perhaps in addition to, perhaps in parallel with, those Watson–Crick, new-old, and mother-daughter distinctions, in every portion of the double-stranded DNA molecule that functions as a gene, one strand is the "sense" strand, from which messenger RNA might be transcribed, from which in turn a protein amino acid sequence might be translated. The other strand is "antisense." Sense and antisense strands are different, and every cell ever properly tested "knows how" to recognize and use the difference. We are just beginning to discover a whole system of functions for bits of antisense RNA. Short pieces of antisense RNA can bind to the corresponding "sense" sequence of a messenger RNA molecule, making that bit double-stranded. Double-stranded RNA is not usable for translation into protein amino acid sequences and is exquisitely vulnerable to rapid degradation. These short RNA bits are therefore primarily effectors of negative controls on cell functions, and there is some recent evidence that they may sometimes have positive control functions.

Also working off the inherent molecular asymmetry of the DNA, there is a whole multilayered system of epigenetic control of expression of DNA gene sequences, which is also heritably strand specific and heritably asymmetric in at least some of its functions and manifestations.

Epigenetic controls on genetic expression are mediated in part by RNA molecules and protein molecules binding to the DNA, in part by chemical modifications of the DNA bases, of the RNAs and DNA-binding proteins. Many of the specific instances are known to involve both, interacting. We are also certain that different parts of this system operate in different parts of the cell differentiation system and that we do not yet know the whole story.

These variations are causes and consequences of asymmetric cell divisions in embryogenesis. They are "heritable" from a given cell to its daughter cells and are strong candidates for the mechanisms of causing cell-function differentiations and passing them down through the differentiated descendant cell line. They can be seen as a system of switches the changing sequence of which constitutes a combinatorial code representing the sequence of all cell divisions in the history of each given lineage, thus defining a virtually limitless variation of cell function specifications.

> If we suppose that there are 20,000 "protein-coding genes" in the human genome, and if each had only two states — all-the-way "on" and all-the-way "off," — the genome could generate $2^{20,000}$ different functional configurations — different sets/systems of proteins — present-or-absent in 20,000 separate choices, more than the number of objects in the Universe. That is actually a tiny fraction of the variability available because — for the simplest of many reasons — most genes can in fact occupy many conditions intermediate between full-on and full-off.

· The specific case of DNA base modifications such as methylation may be the easiest illustration. DNA methylation is the addition of $-CH_3$ methyl groups to certain positions on the DNA base molecules — most often, or at least best known, in the number five position on the ring structures of cytosine molecules in long $\cdots$ CGCGCGCG $\cdots$ runs of cytosine alternating with guanine, and/or of $\cdots$ GNCGNC $\cdots$ triplets, where N represents any of the four bases. The most important part is about the methylation of Cs next to Gs.

Replication does not remove methylation from the strands of a DNA molecule about to be replicated, nor in the process of replication. After replication, there are enzymes whose function is to recognize DNA which is methylated on the "old" strand only, and to fix the "new" strand to match. The daughter strands end up with the same set of modifications that were present on the mother strands — by definition a heritable change in the modification status and expression of a DNA sequence, but not in the sequence itself. A change in, or passed down by, this process defines an "epimutation."

This system of epigenetic processes is subject to variation as a means of controlling cell function and differentiation. There are times in the history of any given cell lineage when the ongoing functions of cells of the lineage in question must change. Epigenetic controls may be removed from some genes and/or added to others. The asymmetric differentiation divisions (where one daughter cell retains stem cell properties and the other goes into a differentiated state) would seem the most likely circumstance under which to enact such rearrangements.

Genome Imprinting

A recent paper shows that as many as 5% of our genes are expressed from only one of the maternal and paternal copies. The assortment appears to have some considerable degree of variation from one line of cells to another, and it appears to be stably transferred to daughter cells. This is genomic imprinting, or something very much like it, but far more extensive than we had seen or understood before.

In cells where the product of a particular gene is not needed, the DNA of that gene is reversibly bound up into an inaccessible complex with histone proteins in the nucleus. Before transcription and translation can activate that function by making that product, the DNA has to be made available to the enzymes of the transcription complex and the signal molecules that direct them to their targets. I have proposed that the sequence of epigenetic asymmetries — turning one system of genes on and perhaps another one off in consecutive cell divisions — may enact or enforce a combinatorial code that defines the specification of functions for each highly differentiated cell type. This could act as a sequence of binary switches, or as a bank of rheostats, primarily by way of changing the availabilities of specific sequences of DNA for expression by transcription of messenger RNA and translation of corresponding protein.

Bear in mind that there is a time when an asymmetric cell division determines the future of the newly differentiated one of the two daughter cells of that division, and there is another time, later, when the differentiated cell is far enough along in its newly differentiated functional role that we can see what it has become, and that those events may be separated by some considerable distance in developmental space and time. Such differentiations are, at their origins, molecular in scale, while our understandings of the results of cellular differentiations are in general much coarser, on a microscopic scale or even visible without magnification.

All of this taken together fits into, and contributes to, the hypothesis that all natural, spontaneous human twinning occurs by way of recruiting cells to establish two sets of body symmetry axes from within a single contiguous mass of cells, in which the establishment of left-right asymmetric cell functions has been under way since the first cell division of embryogenesis — which occurred just a few cell divisions before the definition of the structural axes.

It is through these mechanisms that both monozygotic and dizygotic twinning processes generate the same variety of anomalies in left-right asymmetries of embryogenesis and further development. Many, and quite possibly most, of the embryos resulting from the dizygotic process will have some cells in their bodies from each of the dizygotic embryonic genotypes. The majority of these chimeric individuals will have two normal cell lines of the same sex. They will therefore be undetectable to ordinary observation, and only with considerable technical difficulty will they be detectable even with the best of modern biomedical investigative methodologies.

I have been asked: If this chimerism business is so frequent and is reported to be even more likely in ART conceptions, then why have we never seen any sign of this in all the hundreds of IVFs we've done here? And why has it never been reported from all the tens of thousands of IVFs that have been done around the world?

This simply has to be the case for exactly the same reason that we never see chimerism in people walking around with two cell lines minding their own business — we do not see it because it does not look different from most of the others in any way that we have learned how to recognize. What should you look for, when you go back to your IVF lab with the question of chimerism in mind?

The fact that this question is asked in this way tells us just exactly what chimeric early embryogenesis must look like. It must look like every other (single-zygote) embryogenesis at the same stage — *or* ... maybe ... one cell division ahead.

There is no reason to expect the cell mass of the dizygotic chimeric embryo to be different in size or shape. It is the same cell mass of a single egg cell, perhaps increased slightly by the mass of the "retained second polar body" and the mass of the second sperm, either or both of which constitutes a small fraction of the "baseline" mass of the egg cell. And I have no data, but I seriously doubt that the size of human oöcytes is sufficiently constant that those increments would be noticeable.

There is no reason to imagine two separate and independent embryos having come together, so that the mass should be doubled to the sum of two egg-cell masses. The bulk of the structure comes from the substance of one egg cell and remains bounded by the same one single zona pellucida of that one oöcyte until hatching. There does need to be an extra sperm, but the incorporated sperm head is a small fraction of the volume of the oöcyte before and after sperm penetration, so this seems unlikely to make possible a reliable distinction based on size or mass.

A Prediction of This Hypothesis

The only difference I believe we might expect to see directly is the absence of the second polar body, because the haploid chromosome set that "should" have gone to form the second polar body had to stay inside to become the second maternal pronucleus needed to enter syngamy with the second sperm pronucleus to form the second zygote nucleus. However, reviewing photomicrographs of a number of such structures has left me with the impression that they do not in general show good resolution of the polar bodies, first and/or second. At best, it will not be easy to see.

Proposed test

I would like very much to have access to a large number of IVF "embryos" that have stopped dividing in cleavage stages, at two or three or four cells. I do not want to interfere in any way with any living ones; only the ones that have stopped. I want to separate the cells of a bunch of those and genotype them. In some fraction of them, I predict that we would find two different genotypes, and there is every reason to doubt that the fraction would be very small.

Another Prediction of This Hypothesis

I am convinced that most aneuploid "mosaic" individuals are in fact chimeras. Because they have been karyotyped (only) for the possibility of diagnosing a chromosome anomaly and have been found to have both normal and abnormal cell lines, then the original clinical question has been answered, and the investigation is considered complete. With very rare exception, they have not been genotyped for the possibility that they are chimeric.

Proposed test

I would like very much to have access to cell lines stored from those "mosaic" individuals, for genotyping. Isolated samples of the two cell lines would be ideal but not necessary wherever there are two samples with different mixtures of the two cell types.

I have been asked: If we should think that most chromosome-anomaly mosaics might be really chimeras instead, and spontaneous embryonic chimeras are dizygotic twins, why have we not found half of all "mosaics" to be mixed-sex?

(Discovered and documented) chimeras are not half mixed-sex either, but a sizeable fraction are, because hermaphroditism (visible as ambiguous genitalia on one of the twins) has been one of our best ways to discover them — along with double blood types, which is often missed in blood banks, especially in small admixtures.

I believe that the reason for this set of observations has to be the greater speed of male embryogenesis. In every mammal appropriately tested to date, female embryogenesis is slowed by the presence of a paternal imprint on one of the X-chromosomes. According to at least one published report, nearly all mixed-sex experimental chimeric mice at delivery have shown up with a normal, functional male phenotype.

Embryogenesis is faster for human males as well, and this may well bias the phenotypic outcome of human mixed-sex chimeras in ways similar to those of their mouse counterparts. Something about the paternal X-imprint varies over human racial groups, as evidenced by race-group variation in sex ratios. The frequency of opposite-sex twin pairs varies over race groups in strong positive correlation with the fraction female at birth — greatest in births from African ancestry, least in births of East Asian ancestry, with mothers of white European ancestry intermediate.

The Logic of DNA Genotyping and Forensic Implications of Chimerism

"... DNA doesn't lie!"

Okay ... I'll buy that.
But, you must understand, sometimes DNA tells
truths that most people in the DNA labs don't seem to understand.

It is surely happening with some frequency that fathers are being excused from parental responsibility because the child in question grew from an embryo initiated by a sperm cell from a chimeric second cell line.

It is surely happening with some frequency that people who have committed crimes are being excused from criminal responsibility because DNA from samples at the crime scene does not match exactly with DNA taken from the suspect, because the two sample sources represent the two different cell lines of a chimeric criminal suspect.

These errors should be stopped. It is possible, and not particularly difficult, to recognize and reliably document the relationship between the cell lines of a spontaneous embryonic chimera.

Testing for genetic identity or other genetic relationship between biological samples is conceptually not a very complicated thing. If DNA from that blood spot at yesterday's crime scene is mine, it will in general be expected to match for every marker in a DNA sample taken from my body today for comparison. It will not be expected to match DNA taken from anyone else, with the possible exception of a monozygotic twin if I have one.

Currently, it seems the most common sampling method is swabbing the inside of the cheek with a sterile long-handled cotton swab to collect cells shed from the epithelial layers inside of the cheek. This is not a large sample, but it can supply an amount of DNA that is adequate for today's technology. It is true, as you may have seen on television, that a cigarette butt or a licked stamp, or the rim of a vessel from which the suspect drank, may have enough such epithelial cells on it to provide enough DNA for a genotype — but it also may not. It is certainly preferable to have an amount more like what the cheek swab brings.

From the extracted DNA, we need to examine closely only 12 to 16 DNA sequences (that seems to be the standard in USA labs; fewer marker loci are routinely tested in some other places). This is a tiny fraction of the total number of markers that we are able to sample in this manner. The marker loci in this particular set have been chosen for their high degree of variation in the population, for the resulting probability that I have two different versions of each, of which both are different from the two that you have with a high probability, and the consequent certainty that my whole pattern of all the markers can be depended on to be different from your profile.

PCR — The Polymerase Chain Reaction

We can chemically synthesize primer DNA sequences which will bind quite specifically on either side of each of those target sequences in the chromosomal DNA strands extracted from the samples. Those flanking DNA sequence primers give the polymerase chain reaction (PCR) places to start, to produce many exact copies of the target sequences and only the target sequences.

> A polymerase is an enzyme that makes polymers (long chains of subunit molecules, which in this DNA context are combinations of a five-carbon sugar named deoxyribose and a phosphoric acid molecule with each of the famous DNA bases, adenine, guanine, cytosine and thymine). The backbone of the DNA molecule is sugar-phosphate-sugar-phosphate, and so on, and the bases attach to the sugar molecules on the side opposite from the phosphates.

The sample DNA is melted apart to single "template" strands. The primer binds exactly to a specific sequence that we know is next to the marker sequence we want to test. That makes a short double-strand starting point on the single strand. The polymerase extends the double-strand chain begun by the primer, matching each base as it goes to its proper partner on the single strand of melted sample DNA, all the way across the sequence of the marker locus.

There is another primer and polymerase working on the other strand, coming from the opposite direction. After a suitable period of time, we stop the extension of the copy chains by melting the strands apart. We repeat the cycle, and the copy strands we just made have now doubled the number of copies of each target sequence in the sample template DNA. Each copy strand is now copied as a template strand by the other primer in the other direction. In each cycle, the number of copies of each target sequence is doubled. In 30 cycles, we have amplified the sample about a billion times — probably enough to see. We can then run the product through any of several analyses which will sort the product pieces by size.

The target sequences typically used for these analyses all include a section in which a simple DNA sequence (most are three to five-base pairs in length) is repeated a variable number of times. The variation among people in the number of repeats in each of these segments is the main thing about these markers that makes them useful for these purposes.

The particular sequences that we have chosen for these uses have no known genetic function and seem therefore to be quite free to vary by mutation. For example, a fairly typical situation for any given one of these marker loci would be that the population might contain seven different alleles (different versions of that particular DNA sequence), varying in frequency something like: 25%, 21%, 16%, 14%, 12%, 8%, 3%, and 1%.

Many of these sequences exist in the population in even more versions than that. The allele-frequency numbers have been determined and updated frequently from large random samples of individuals from various subpopulations. The rarer any given allele, the fewer people who carry that allele, the more that marker will narrow down the list of people from whom a particular tissue sample might have come.

The markers that are used have been chosen from a large and well-documented history such that the random probability of finding more than one person in the population with any given combination of all the markers in the test set is vanishingly small, negligible.

No two of the marker loci we use are on the same chromosome, let alone close enough together for genetic linkage such that any of them could distort the probability distribution for any other one of them. The variations in each locus are scattered over the population each quite independently of every other locus in the set, so the proper mathematical treatment is to multiply the separate probabilities of each marker found, to find the joint probability of any particular set.

We might imagine a particular sample with one 25% and one 20% version at each and every one of these marker gene pairs. That is unlikely? Of course it is, and that's the point. Frequencies that high represent rather common allelic versions. With alleles this common at every locus, that would generate a profile about as common, and therefore weak for purposes of identification, as a profile can be. And yet, even that kind of pairing of common alleles at each locus would have a probability of being present in another person chosen randomly from the population of only 5% *per locus*. Five percent per locus, at 10 loci, is 0.05 to the tenth power — that is a very small number, $\sim 9.8 \times 10^{-14}$ or 0.0000000000000976.

It is also entirely possible (but also not at all likely) that an individual might have only rare alleles. For example, if each sequence pair had one marker with a population frequency of 1% and one with a frequency of 2%, that would give each pairing a probability of 0.02% or 1/5000.

The likelihood for matching 16 markers for that first hypothetical profile, with the commonest alleles at every locus, is 5% to the 16th power! That is, 1.52588×10^{-21}, or one chance in 6.5536×10^{20}. We should expect to find another one exactly like it only once in searching 655,360,000,000,000,000,000 more genotypes! That number is far larger than the number of humans who have ever lived. With only common alleles, this would be one of the "least unique" profiles that could be arranged.

If, on the other hand, the genotype were made up entirely of rare alleles, all at frequencies like 1% and 2%, as in the second example, then 0.02% to the 16th power gives one chance in a similar number with *40 more zeroes*.

Even with the most common alleles at every locus, the odds are overwhelmingly against there being any other human with the same configuration of DNA sequences. Rare alleles give the analysis even more power. It is of the nature of the system that most profiles will include some of the more common alleles and some of the rarer ones.

The calculations vary in detail according to the type of comparison being made, according to the genetic relationship being tested in any specific comparison. If you want to establish that a particular sample came from my body, then you must compare it with a sample taken from my body, and you will expect to find the same pair of alleles at every locus. The probability for matching each allele is one, 1.0, 100%, if it *is* mine, and the random population frequency is its probability under the hypothesis that it belongs instead to any unrelated individual randomly chosen from the population.

Even if every allele in a given DNA profile is a common one, the product of the individual allele probabilities will still be a very small number. A genotype generated by the full 16-marker set is unique among all humans who have ever lived and for much more than a foreseeable number of generations to come.

If the issue is parental responsibility, instead of criminal, then analysis of the child's genotype must be expected to show one of each pair of each parent's alleles at each locus tested. If that woman over there is the mother, and I am the father, the child should have — at each locus tested — one allele matching one of hers and the other matching one of mine. Here the likelihood ratio sets up as one-half (one when the parent's two alleles at that locus are the same) versus random population probability, *for each locus*. We cannot expect likelihood ratios as high as those we would find in comparing for unique identity, but they will still be overwhelming.

When, at any given locus, neither of the child's alleles matches either of the parent's alleles at that locus, this is considered by the most commonly used forensic laboratory software to indicate exclusion. However, remember that these loci have been chosen because of their high variability, which arises more or less exclusively through random mutation. One such failure-to-match result might conceivably be due to a new mutation and need not be considered exclusionary, but the presence of two or more such results is generally considered to be asking too much of Chance.

Forensic Complications from Chimerism

It should be obvious that chimerism can complicate these situations. If I am chimeric, I may have left a sample from one of my two cell lines in blood or semen at yesterday's crime scene, which appears not to match to me because the sample from the cheek swab you took today represents my other cell line.

The genetic understanding necessary for forensic resolution of such a situation is not really a very complicated idea: If I am chimeric as a result of a spontaneous mixing of embryonic cells, then my two cell lines are genetically representative of dizygotic twins — they are sibling cell lines. This knowledge, together with the understanding that chimerism is actually rather ordinary, will be all you need to sort out such problems.

> If two samples do not present exactly matching genetic profiles, but are consistent with having been taken from siblings, then *exclusion has not in fact occurred*. Chimerism must be considered, and only a living sibling showing a perfect match and no alibi has any chance of exonerating the original suspect. Only a living dizygotic twin of the suspect might provide a perfect match to the reference DNA sample and yet be innocent.

The presence of this situation should be fairly obvious on inspection of the raw data of the two DNA profiles. A chimeric second cell line will have a few loci showing no match, which is generally considered proof of exclusion. But it will also have at least one allele matching at about three-quarters of the tested loci, including matching for both alleles at about one-quarter of the loci tested. This is glaringly not what would be expected in a sample from any unrelated individual. Sorting out the truth is not only possible but conceptually simple, straightforward and quite powerful.

Remember what the genetic relationships will look like: If I am spontaneously chimeric, my second cell line is genetically a sibling (a full sibling unless we also share a chimeric parent or we have different fathers). Spontaneous chimeras are dizygotic twins. If blood or sperm cells I left at a crime scene come from a second cell line that differs from the cells in the cheek swab the police took as a test sample, then, most likely, there will be a different kind of match than has been looked for before.

If the crime scene sample and my cheek swab sample have the same genotype, all alleles are expected to match. If I am chimeric, but both of your samples represent the same one of my two cell lines, there is no problem — you have a match! This is entirely possible and will in fact usually be the case. The distribution of the second cell line in the body of a chimeric individual is random and patchy: its presence or absence in one sample need not mean anything with regard to the possibility of its presence in a sample from a different tissue, or even from the same tissue at a different time.

If, however, the crime scene sample and my cheek swab sample are genetically distinct because they come from dizygotic twin sibling cell lines in my chimeric body, then the DNA profiles will compare like those of siblings, which will prove to be a great deal different from the pattern to be expected between two randomly chosen unrelated individuals.

With each of my siblings, including my dizygotic twin if I have one, I may be expected to share *neither* allele at about one-fourth of the loci tested, to share *both* alleles at approximately one-fourth of the loci tested, and to share *only one* allele at about the remaining one-half of the loci tested. Each of those fractions is a probability — three or seven out of 16 loci instead of four or eight does not provide an exclusion.

Overall, there is an average expectation that one half of all alleles will be shared, just as between parent and child, but in a distinctively different pattern. It should be obvious. You will simply not see half of all the alleles in common between two unrelated randomly chosen individuals, especially not in that pattern of $\frac{1}{4}(2)$, $\frac{1}{2}(1)$, $\frac{1}{4}(0)$. No other relationship can be expected to show that pattern, and certainly no unrelated individual.

In our experience to date, comparison of a sibling profile against the probability of that same pattern of matching in any individual chosen randomly from the population typically results in a likelihood ratio value on the order of a few billion to one. This is substantially smaller than the same calculation with respect to an exact me-to-mine match between samples from genetically identical sources, but it will almost certainly be still compelling, still well beyond reasonable doubt.

If the question is not about a criminal concern but instead about paternity, the same considerations apply, but in a different pattern. Parent-child is also a first-degree relationship, sharing half of all DNA gene sequences identical-by-descent. The pattern is different and much simpler than the sibling pattern. By way of the segregation in gametogenesis, each of my parents passed to me one of the two alleles at each of their loci. I may be expected to have one of my two alleles at each locus matching one of my mother's two alleles at that locus, and the other one matching one of my father's two alleles. Each of my children may be expected to share with me one allele at each locus, and one with his or her mother.

Suppose a question of parentage arises, and I find myself needing to confirm or deny that I am the biological parent of a particular child, or the child of a particular parent. If everything is quite ordinary, the child will have one allele at each locus representing each parent. If, however, one parent is chimeric, results may show up rather differently: there will be one allele at each locus shared with the "ordinary," single-genotype parent, but only about half that many matches with the sample from the chimeric parent.

The sample from the chimeric parent will appear to be that of an aunt or uncle, a second-degree relationship, sharing about one-quarter of all alleles. The probability of such a match being found with any individual randomly chosen from the population will not be as distinctively small as that for a sibling comparison, but will still be telling.

It may help to collect other samples from the suspected individual, in the hope of finding the second cell line in another tissue. Because chimerism is patchy and, as far as we know now, randomly distributed among the tissues of the body, and because relative proportions of the two cell lines are known to change over time, a finding of or failure to find chimerism in one sample tells us nothing for certain about the possibility of finding it in any other tissue, or even in the same tissue at any other time. Blood and sperm cells, in particular, are produced in waves from permanent stem cell lines, and may be generated from one stem line now and from the other one at another time.

It certainly can help our understanding of the situation to test samples from other relatives. The parents and siblings of the parent in question will have the same (avuncular, second degree) relationship with the child as the chimeric second cell line does. Including their results in the likelihood ratio calculation (the whole family vs random unrelated individuals) certainly can generate a compelling conclusion.

Realizing that chimerism is actually rather ordinary, much too ordinary to be routinely dismissed out of hand, will be essential to progress.

Summary

A teacher's abiding duty to the individual student is to become unnecessary.
The honest teacher of science must be uncertain or wrong at times
if only because the apparent truth is incomplete and partly false.
Missing or poorly fitted pieces properly identified as such,
and apparent truths retested, provide toeholds
upon which the willing student may climb past the teacher.
The student not taught to question is not taught well.
The student taught not to question is damaged.

My answer to a request for my philosophy of teaching, a long time ago.

Every human life, past, present or yet to come, has its "beginning" at one same place and time — back when and where all Life on Earth had its beginning. Becoming human from that beginning took a very long time. It did not happen in an instant.

Some smart people believe that Life on Earth might even have been alive already when it came to Earth from somewhere else. Some other smart people have said that Darwin's Theory of Natural Selection is not much help with the very earliest stages of biogenesis — the first emergence of any form of life from non-living components on the planet. On the contrary, among the early conglomerations, the survival and increase in number of only those that were able to survive and increase in number is in fact clearly a process of natural selection and makes all the sense in the world.

Every life that is human is human entirely because of arising from human parents, and alive entirely because of having parents who could pass on their lives by way of living gametes.

The specifically human situation might be said to have begun when and where any pair of parents, both of whom could be called "human," first produced a viable, fertile offspring. That, however, would require that there must have been a "*first* human" and a second one — a very ordinary, normal thought, but one that will get us nowhere. The evolution of human-ness, the business of becoming human, has always been, as it is now, a gradual thing, with no end in sight in either direction.

It makes much more sense to consider these as horizons rather than events. Like "conception" as a beginning for a given individual, we may have a distinct impression that we understand it clearly at some considerable distance (at or near the limits of resolution of whatever sort of vision we are using), but at close range all certainty disappears and it has no instantaneous definition. Deal with it.

The genetic and epigenetic definition of "human" as species is demonstrably still undergoing constant and rapid change. We are not at or beyond any sort of endpoint. We are not the same species we were even a few years ago, or last night at bedtime for that matter.

(Only) if both of your parents are human, you are too; no way out of it — and there is no "beginning" of your being human anywhere in sight.

The definition of the human individual, of each or of any human individual, also constantly changes. You and I are not the same people we were yesterday, or even an hour ago, nor will we be the same tomorrow as we are today.

Human-ness is not a fixed state of being and is not subject to any rational definition confined to any fixed perspective.

Human-ness is constantly becoming.

There is no cellular developmental event, no point in developmental time, no *instant* before which an individual, personal human life does not exist and after which it does.

A definition in terms of developmental "potential" is uselessly abstract and imaginary because potential (the noun) is forever potential (the adjective), and unknowable and indemonstrable in present time by any reasonable definition thereof.

Twins provide a wealth of opportunity to explore the particulars of human embryonic development — because they did it differently. In just the same ways that we study how mutant organisms function differently from the "wild type" or "normal" organisms, I began with the prospect of learning about the consequences of the indisputably odd embryogenesis of monozygotic twins.

The folklore that has passed for a biological science of twinning had me considering dizygotic twins at the beginning as the comparison group, experimental controls. They were supposed to be exactly the same as singletons in their embryogenesis but could serve as a control for any developmental effects of the accident of having shared a pregnancy with a twin.

However, the data had me surrounded and kept telling me ... dizygotic twins are developmentally at least as different from singletons as monozygotic twins are, and in the same ways, all having to do with asymmetries established in the first few cell divisions of embryogenesis.

The classical idea that dizygotic twinning arises from two separate and independent ovulations, and proceeds from that point forward in development exactly as normal singleton development, is quite simple, but simply quite false. The necessary predictions of that hypothesis are all without the support of direct evidence, and the predictions of that hypothesis that have been tested are contradicted by sound evidence. The process cannot be working that way.

Naturally conceived human dizygotic twins
do not arise from double ovulation.
The truth is not that simple.

Simplicity is indeed a hallmark of truth, but not a guarantee:

Not all simple things are true.

Not all true things are simple.

The cellular mechanism of the dizygotic twinning event is the same as the monozygotic twinning mechanism, a recruitment of cells from among a single contiguous mass of cells, in/by which to establish two sets of embryonic axes as a double armature for twin embryogenesis and fetal development.

As a result of this origin, dizygotic twins, whether born alive together or (much more often) born alone, are often chimeric, either or both carrying in their bodies some cells of the other twin's genotype. This is an entirely expectable consequence of having recruited cells to define two separate bodies from within a single contiguous mass of cells composed of cells of two genotypes. Sorting them perfectly would be miraculous.

Spontaneous human chimeras are not fantasies — people are born every minute of every day with two cell lines, two kinds of cells in their bodies, cell lines that are genetically different in many ways — not just in the number of copies of one chromosome, but carrying different versions of the genes on many chromosomes. There are too many differences to imagine they got that way by having a great many mutations early in the embryo, but they surely had to get that way before they were born.

One pregnancy with two genetically different embryos is, by definition, dizygotic twins. Nothing but a pair of dizygotic twins answers to that description. So, dizygotic twins, sometime between formation of the zygote or zygotes and delivery or death before delivery, somehow exchange cells, so that one or both of them now has his or her "own" cells plus some cells that came from the other twin. The mixture may be anything from a handful of cells genetically different from most of the other cells in the body to — at least in theory — an equal 50:50 mixture.

For decades, the cover story has been that the placentas of chimeric dizygotic twins must have fused and placental blood vessels must have grown connections between them. So, the twins must have traded some cells by way of connections between their placental circulations, including some stem cells so that the mixing would last through life and not just go away in days or weeks with the normal turnover of blood cells.

We know that theory does not work, because smart and careful investigators like Professor Benirschke have looked at a great many twin placentas, searching for just such connections. A very, very few such connections have been found — several orders of magnitude (powers of 10) less often than chimerism has been found in other, direct studies — and only by people who were not looking for them — only by accident.

There are also some dizygotic twins who are born inside a single set of gestational membranes. That means they were together, inside, when the outer layer of the cells of a morula embryo differentiated into the trophoblast cells that would become the chorion (the heavier outer layer of the membranes). That normally happens before hatching, while the morula is still inside the zona pellucida ... and they can only be inside a single zona pellucida if all of this is happening within the substance of a *single* egg cell.

When chimeric dizygotic twins have separate chorions (which we have long assumed but have no evidence to prove is the way they do it most of the time), that means they were together and became twins even before the trophoblast differentiated and even more surely inside the same zona pellucida. This means they had to come from the substance of a single secondary oöcyte (egg cell).

The secondary oöcyte normally contains, at ovulation, DNA for one maternal pronucleus and one second polar body. This mass of DNA is equivalent to, and adequate for forming, two maternal pronuclei, and it is apparently easily enough arranged to do that instead of discarding one of them by pinching out the second polar body. The situation then calls for two sperm, to bring in two paternal pronuclei and two sets of centrioles. Although this is considered abnormal from a perspective such as this one, it is also nevertheless a very common occurrence, not at all a limiting requirement, as evidenced by the frequency of triploids (having three copies of each chromosome instead of the normal diploid two) among the products of spontaneous abortions.

In the majority of chimeric individuals, both cell lines are normal and of the same sex. Therefore, the majority of chimeric individuals are normal people whose chimerism will be discovered, if at all, only by accident, almost always only by extensive DNA genotyping done for reasons quite unrelated to their individual phenotypes.

Chimerism can hardly be less frequent than 10% to 15% of the population. It is rather more likely to be higher than that because that 10%–15% is the conservatively estimated frequency of twins among all live births, when including those born alone as sole survivors from twin conceptions (much more frequent than liveborn pairs).

This conservatively estimated frequency of chimerism has serious implications for forensic considerations of parenthood and of criminal identity. Everything necessary for dealing properly with the fact of frequent chimerism is in place in the forensic system, except for the acknowledged understanding that chimerism is in fact there to be dealt with. While chimerism has a definite aura of mystery, it nevertheless has a straightforward and well-defined biology, which generates well-defined genetic patterns in the standard forensic DNA profile and in parentage testing.

Double ovulation as the hypothetical cellular origin of dizygotic twinning does not work. No one has ever offered any direct evidence for it. There is, however, ample, clear and consistent evidence against it, and I have presented here a plausible alternative mechanism.

The hypothesis presented here, that the dizygotic twinning mechanism is the same as that of monozygotic twinning, does work. This hypothesis explains the observations summarized here, that dizygotic and monozygotic twins have the same system of distortions of their embryogenesis.

Bibliography

Abeliovich D, Leiberman JR, Teuerstein I, Levy J (1984) Prenatal sex diagnosis: Testosterone and FSH levels in mid-trimester amniotic fluids. *Prenat Diagn* **4**(5), 347–353.

Anderson PW (1972) More is different: Broken symmetry and the hierarchical structure of science. *Science* **177**(4047), 393–396.

Aoki R, Honma Y, Yada Y, Momoi MY, Iwamoto S (2006) Blood chimerism in monochorionic twins conceived by induced ovulation: Case report. *Hum Reprod* **21**(3), 735–737.

Aston KI, Peterson CM, Carrell DT (2008) Monozygotic twinning associated with assisted reproductive technologies: A review. *Reproduction* **136**, 377–386.

Be C, Velásquez P, Youlton R (1997) Spontaneous abortion: Cytogenetic study of 609 cases. *Rev Med Chil* **125**(3), 317–322.

Bergson H (1911) *L'Evolution Creatrice* [Creative Evolution]. Henry Holt & Co. Translation by A Mitchell [online at http://web.archive.org/web/20060516195812/ http://spartan.ac.brocku.ca/~lward/Bergson/Bergson_1911a/Bergson_1911_toc. html].

Bering JM (2006) The folk psychology of souls. *Behav Brain Sci* **29**, 453–498.

Bieber FR, Nance WE, Morton CC, Brown JA, Redwine FO *et al.* (1981) Genetic studies of an acardiac monster: Evidence of polar body twinning in man. *Science* **213**, 775–777.

Boklage CE (1976) Embryonic determination of brain programming asymmetry: A neglected element in twin-study genetics of human mental development. *Acta Genet Med Gemellol* **25**, 244–248.

—— (1977a) Schizophrenia, brain asymmetry development, and twinning: A cellular relationship with etiologic and possibly prognostic implications. *Biol Psychiat* **12**(1), 19–35.

—— (1977b) Embryonic determination of brain programming asymmetry: A caution about the use of data on twins in genetic studies of human mental development. *Ann NY Acad Sci* **299**, 306–308.

—— (1978) On cellular mechanisms for heritably transmitting structural information. *Behav Brain Sci* **1**, 282–286.

—— (1980) The sinistral blastocyst: An embryonic perspective on the development of brain-function asymmetries. In Herron J (ed.), *Neuropsychology of Left-Handedness*. Academic Press, New York. [Also in Japanese translation.]

—— (1981a) On the distribution of nonrighthandedness among twins and their families. *Acta Genet Med Gemellol* **30**, 167–187.

—— (1981b) On the timing of monozygotic twinning events. In Gedda L, Parisi P, Nance WE (eds.), *Twin Research 3, Part A: Twin Biology and Multiple Pregnancy*. Alan R Liss, New York, pp. 155, 165. [Also can be found as *Prog Clin Biol Res* 69A, 155–165.]

—— (1984a) Differences in protocols of craniofacial development related to twinship and zygosity. *J Craniofac Genet Devel Biol* **4**, 151–169.

—— (1984b) Twinning, handedness and the biology of symmetry. In Geschwind N, Galaburda A (eds.), *Cerebral Dominance: The Biological Foundations*. Harvard University Press, Cambridge, Massachusetts.

—— (1984c) Differences in protocols of craniofacial development related to twinship and zygosity. *J Craniofac Genet Devel Biol* **4**, 151–169.

—— (1984d) On the inheritance of directional asymmetry (sidedness) in the starry flounder *Platichthys stellatus*: Additional analyses of Policansky's data. *Behav Brain Sci* **7**, 725–730.

—— (1985) Interactions between opposite-sex dizygotic fetuses, and the assumptions of Weinberg difference method epidemiology. *Am J Hum Genet* **37**(3), 591–605.

—— (1987a) The unmapped methodological territory between one gene and many comprises some intriguing environments. *Behav Brain Sci* **10**(1), 18–19.

—— (1987b) Twinning, nonrighthandedness and fusion malformations: Evidence for heritable causal elements held in common. Invited Editorial Essay. *Am J Med Genet* **28**, 67–84.

—— (1987c) Developmental differences between singletons and twins in distributions of dental diameter asymmetries. *Am J Phys Anthro* **74**(3), 319–332.

—— (1987d) Race, zygosity, and mortality among twins: Interaction of myth and method. *Acta Genet Med Gemellol* **36**, 275–288.

—— (1987e) The organization of the oöcyte and embryogenesis in twinning and fusion malformations. *Acta Genet Med Gemellol* **36**, 421–431.

—— (1990) The survival probability of human conceptions from fertilization to term. *Int J Fertil* **35**(2), 75–94.

—— (1992) Method and meaning in the analysis of developmental asymmetries. In Lukacs JR (ed.), *Culture, Ecology and Dental Anthropology*. Kamla-Raj Enterprises, Delhi. [Can also be found as *J Hum Ecol* **3**(13), 147–156.]

—— (1995) The frequency and survival probability of natural twin conceptions. In Keith LG, Papiernik E, Keith DM, Luke B (eds.), *Multiple Pregnancy: Epidemiology, Gestation and Perinatal Outcome*. Parthenon, New York, pp. 41–50.

—— (1996) Effects of a behavioral rhythm on conception probability and pregnancy outcome. *Hum Reprod* **11**(10), 2276–2284.

—— (1997) Scientific correspondence: Sex ratio unaffected by parental age gap. *Nature* **390**, 243.

—— (2005a) The epigenetic environment: Secondary sex ratio depends on differential survival in embryogenesis. *Hum Reprod* **20**(3), 583–587.

—— (2005b) Biology of human twinning: A needed change of perspective. In Blickstein I, Keith LG (eds.), *Multiple Pregnancy: Epidemiology, Gestation & Perinatal Outcome*. Taylor & Francis, Parthenon Publishers, New York, Chap. 36, pp. 255–264.

—— (2006a) Embryogenesis of chimeras, twins and anterior midline asymmetries. *Hum Reprod* **21**(3), 579–591.

—— (2006b) Embryogenesis of chimeras, twins and anterior midline symmetries. *Hum Reprod* **21**(3), 579–591. [Republished in May 2006. *Hum Reprod* (*Indian Ed*) **2**(6), 267–279.]

—— (2006c) Reply to Golubovsky: Mosaics/chimeras and twinning in the current reproductive genetics perspective. *Hum Reprod* **21**(9), 2461–2462.

Boklage CE, Elston RC, Potter RH (1979) Cellular origins of functional asymmetries: Evidence from schizophrenia, handedness, fetal membranes and teeth in twins. In Gruzelier JH, Flor-Henry P (eds.), *Hemispheric Asymmetries of Function in Psychopathology*. Elsevier-North Holland, London.

Boklage CE, Kirby CF Jr, Zincone LH (1992) Annual and sub-annual rhythms in human conception rates: I. Effective correction and use of public records LMP dates. *Int J Fertil* **37**(2), 74–81.

Boklage CE, Zincone LH, Kirby CF Jr (1992) Annual and sub-annual rhythms in human conception rates: II. Time-series analyses show annual and weekday but no monthly rhythms in LMP dates. *Hum Reprod* **7**(7), 899–905.

Brajenović-Milić B, Petrović O, Krasević M, Ristić S, Kapović M (1998) Chromosomal anomalies in abnormal human pregnancies. *Fetal Diagn Ther* **13**(3), 187–191.

Braude P, Bolton V, Moore S (1988) Human gene expression first occurs between the four- and eight-cell stages of preimplantation development. *Nature* **332**(6163), 459–461.

Chang HJ, Lee JR, Jee BC, Suh CS, Kim SH (2009) Impact of blastocyst transfer on offspring sex ratio and the monozygotic twinning rate: A systematic review and meta-analysis. *Fertil Steril* **91**(6), 2381–2390. [doi:10.1016/j.fertnstert.2008.03.066 Epub ahead of print.]

Cohen-Bendahan CC, Buitelaar JK, van Goozen SHM, Cohen-Kettenis PT (2004a) Prenatal exposure to testosterone and functional cerebral lateralization: A study in same-sex and opposite-sex twin girls. *Psychoneuroendocrinology* **29**(7), 911–916.

Cohen-Bendahan CC, Buitelaar JK, van Goozen SH, Orlebeke JF, Cohen-Kettenis PT (2005a) Is there an effect of prenatal testosterone on aggression and other behavioral traits? A study comparing same-sex and opposite-sex twin girls. *Horm Behav* **47**(2), 230–237.

Cohen-Bendahan CC, van Goozen SH, Buitelaar JK, Cohen-Kettenis PT (2005b) Maternal serum steroid levels are unrelated to fetal sex: A study in twin pregnancies. *Twin Res Hum Genet* **8**(2), 173–177.

Côté F, Fligny C, Bayard E, Launay JM, Gershon MD, Mallet J, Vodjdani G (2007) Maternal serotonin is crucial for murine embryonic development. *Proc Natl Acad Sci USA* **104**(1), 329–334.

Culbert KM, Breedlove SM, Burt SA, Klump KL (2008) Prenatal hormone exposure and risk for eating disorders: A comparison of opposite-sex and same-sex twins. *Arch Gen Psychiat* **65**(3), 329–336.

Dassule HR, Lewis P, Bei M, Maas R, McMahon AP (2000) Sonic hedgehog regulates growth and morphogenesis of the tooth. *Development* **127**(22), 4775–4785.

de Chardin Teilhard P, Humain LP (1955) *The Phenomenon of Man* (1959) Harper Perennial, New York.

de Waal FB (2008) The thief in the mirror. *PLoS Biology* **6**(8), e201.

Derom C, Vlietinck RF, Derom R, Boklage CE, Thiery M, Van den Berghe H (1991) Genotyping macerated stillborn fetuses. *Am J Obstet Gynec* **164**(3), 797–800.

Derom C, Thiery E, Vlietinck R, Loos R, Derom R (1996) Handedness in twins according to zygosity and chorion type: A preliminary report. *Behav Genet* **26**(4), 407–408.

Derom C, Jawaheer D, Chen WV, McBride KL, Xiao X, Amos C, Gregersen PK, Vlietinck R (2006) Genome-wide linkage scan for spontaneous DZ twinning. *Eur J Hum Genet* **14**(1), 117–122.

Dobson AT, Raja R, Abeyta MJ, Taylor T, Shen S, Haqq C, Pera RA (2004) The unique transcriptome through day 3 of human preimplantation development. *Hum Mol Genet* **13**(14), 1461–1470.

Dubé F, Amireault P (2007) Local serotonergic signaling in mammalian follicles, oöcytes and early embryos. *Life Sci* **81**(25–26), 1627–1637.

Dumoulin JC, Derhaag JG, Bras M, Van Montfoort AP, Kester AD *et al.* (2005) Growth rate of human preimplantation embryos is sex dependent after ICSI but not after IVF. *Hum Reprod* **20**(2), 484–491.

Ehrman BD (2008) *God's Problem: How the Bible Fails to Answer Our Most Important Question — Why We Suffer*. Harper Collins Publishers, New York.

Ekelund CK, Skibsted L, Søgaard K, Main KM, Dziegiel MH, Schwartz M, Moeller N, Roos L, Tabor A (2008) Dizygotic monochorionic twin pregnancy conceived following intracytoplasmic sperm injection treatment and complicated by twin-twin transfusion syndrome and blood chimerism. *Ultrasound Obstet Gynecol* **32**(6), 832–834.

Elston RC, Boklage CE (1978) An examination of fundamental assumptions of the twin method. In Nance WE, Allen G, Parisi P (eds.), *Twin Research: Part A — Psychology and Methodology*. Alan R Liss, New York, pp. 189–199. [Also can be found as *Prog Clin Biol Res* **24A**, 189–199.]

Fairbanks DJ (2007) *Relics of Eden: The Powerful Evidence of Evolution in Human DNA.* Prometheus Books, Amherst, New York.

Fernandes M, Hébert JM (2008) The ups and downs of holoprosencephaly: Dorsal versus ventral patterning forces. *Clin Genet* **73**(5), 413–423.

Forget-Dubois N, Pérusse D, Turecki G, Girard A, Billette JM, Rouleau G, Boivin M, Malo J, Tremblay RE (2003) Diagnosing zygosity in infant twins: Physical similarity, genotyping, and chorionicity. *Twin Res Hum Genet* **6**(6), 479–485.

Frankfurter D, Trimarchi J, Hackett R, Meng L, Keefe D (2004) Monozygotic pregnancies from transfers of zona-free blastocysts. *Fertil Steril* **82**(2), 483–485.

Fukumoto T, Kema IP, Levin M (2005) Serotonin signaling is a very early step in patterning of the left-right axis in chick and frog embryos. *Curr Biol* **15**(9), 794–803.

Galis F (1999) Why do almost all mammals have seven cervical vertebrae? Developmental constraints, Hox genes, and cancer. *J Exp Zool (Mol Dev Evol)* **285**, 19–26.

Gardner DK, Lane M (2005) *Ex vivo* early embryo development and effects on gene expression and imprinting. *Reprod Fertil Dev* **17**(3), 361–370.

Geng X, Speirs C, Lagutin O, Inbal A, Liu W, Solnica-Krezel L, Jeong Y, Epstein DJ, Oliver G (2008) Haploinsufficiency of Six3 fails to activate sonic hedgehog expression in the ventral forebrain and causes holoprosencephaly. *Dev Cell* **15**(2), 236–247.

Glinianaia SV, Magnus P, Harris JR, Tambs K (1998) Is there a consequence for fetal growth of having an unlike-sexed cohabitant in utero? *Int J Epidemiol* **27**(4), 657–659.

Goldman RD, Blumrosen E, Blickstein I (2003) The influence of a male twin on birthweight of its female cotwin: A population-based study. *Twin Res* **6**(3), 173–176.

Golubovsky MD (2003) Postzygotic diploidization of triploids as a source of unusual cases of mosaicism, chimerism and twinning. *Hum Reprod* **18**(2), 236–242.

Guilherme R, Drunat S, Delezoide A-L, Oury J-F, Luton D (2009) Zygosity and chorionicity in triplet pregnancies: New data. *Hum Reprod* **24**(1), 100–105. [doi: 10.1093/humrep/den364, PMID: 18945712 Advance Access Epub.]

Hamatani T, Ko MSh, Yamada M, Kuji N, Mizusawa Y, Shoji M, Hada T, Asada H, Maruyama T, Yoshimura Y (2006) Global gene expression profiling of preimplantation embryos. *Hum Cell* **19**(3), 98–117.

Harlap S, Shahar S, Baras M (1985) Overripe ova and twinning. *Am J Hum Genet* **37**, 1206–1215.

Hoekstra C, Zhao ZZ, Lambalk CB, Willemsen G, Martin NG, Boomsma DI, Montgomery GW (2008) Dizygotic twinning. *Hum Reprod Update* **14**(1), 37–47.

Hofstadter D (1979) *Godel, Escher, Bach: The Eternal Golden Braid*. Basic Books, New York. [20th Anniversary Edition (1999). Random House, New York.]

Hofstadter D (2007) *I Am a Strange Loop*. Basic Books, New York.

Iselius L, Lambert B, Lindsten J, Tippett P, Gavin J *et al.* (1979) Unusual XX/XY chimerism. *Ann Hum Genet* **43**(2), 89–96.

James WH (1971) Excess of like sexed pairs of dizygotic twins. *Nature* **232**(5308), 277–278.

James WH (1979) Is Weinberg's differential rule valid? *Acta Genet Med Gemellol* (*Roma*) **28**(1), 69–71.

Kauffman S (1996) *At Home in the Universe: The Search for the Laws of Self-Organization and Complexity.* Oxford University Press, New York.

Kauffman S (2008) *Reinventing the Sacred: A New View of Science, Reason, and Religion.* Basic Books, New York.

Kim SC, Jo DS, Jang KY, Cho SC (2007) Extremely rare case of cephalothoracopagus characterized by differences of external genitalia. *Prenat Diagn* **27**(12), 1151–1153.

Kläning U, Pedersen CB, Mortensen PB, Kyvik KO, Skytthe A (2002) A possible association between the genetic predisposition for dizygotic twinning and schizophrenia. *Schizophr Res* **58**(1), 31–35.

Klar AJ (2004) An epigenetic hypothesis for human brain laterality, handedness and psychosis development. *Cold Spring Harb Symp Quant Biol* **69**, 499–506.

Klar AJ (2007) Lessons learned from studies of fission yeast mating-type switching and silencing. *Annu Rev Genet* **41**, 213–236.

Koopmans M, Kremer Hovinga ICL, Baelde HJ, Fernandes RJ, de Heer E, Bruijn JA, Bajema IM (2005) Chimerism in kidneys, livers and hearts of normal women: Implications for transplantation studies. *Am J Transplant* **5**, 11495–11502.

Lanza R, Berman B (2009) *Biocentrism: How Life and Consciousness Are the Keys to Understanding the True Nature of the Universe*. Ben Bella, Dallas, TX.

Levin M (2005) Left-right asymmetry in embryonic development: A comprehensive review. *Mech Dev* **122**(1), 325. [Erratum in: *Mech Dev* **122**(4), 621.]

Lummaa V, Pettay JE, Russell AF (2007) Male twins reduce fitness of female co-twins in humans. *Proc Natl Acad Sci USA* **104**(26), 10915–10920.

Maher ER, Afnan M, Barratt CL (2003) Epigenetic risks related to assisted reproductive technologies: Epigenetics, imprinting, ART and icebergs? *Hum Reprod* **18**(12), 2508–2511.

Maher ER (2005) Imprinting and assisted reproductive technology. *Hum Mol Genet* **14** (Spec No 1), R133–138.

Mahtani MM, Willard HF (1988) A primary genetic map of the pericentromeric region of the human X chromosome. *Genomics* **2**(4), 294–301.

Marzullo G (2006) Seasonal conceptions in neural tube defects, schizophrenia and hemispheric laterality: Evidence implicating differential embryonic survival. *Birth Defects Res* (*Part A*) (*abstracts*) **73**, 343.

Marzullo G, Boldage CE (2009) The epigenetic environment II: Excess of spring conceptions in schizophrenia, nonrighthandedness, and neural tube defects, and the general-population annual conception-rate rhythm. Manuscript in preparation.

Marzullo G, Fraser FC (2005) Similar rhythms of seasonal conceptions in neural tube defects and schizophrenia: A hypothesis of oxidant stress and the photoperiod. *Birth Defects Res (Part A)* **73**, 1–5.

Marzullo G, Fraser FC (2008) The conception season and cerebral asymmetry among American baseball players: Implications for the seasonal birth effect in schizophrenia. *Psychiatry Research* **167**(3), 287–293.

Miura K, Niikawa N (2005) Do monochorionic dizygotic twins increase after pregnancy by assisted reproductive technology? *J Hum Genet* **50**(1), 1–6.

Mortimer G (1987) Zygosity and placental structure in monochorionic twins. *Acta Genet Med Gemellol (Roma)* **36**(3), 417–420.

Nylander PP, Osunkoya BO (1970) Unusual monochorionic placentation with heterosexual twins. *Obstet Gynecol* **36**, 621–625.

Paoloni-Giacobino A, Chaillet JR (2004) Genomic imprinting and assisted reproduction. *Reprod Health* **1**(1), 6–12.

Pierucci O, Zuchowski C (1973) Non-random segregation of DNA strands in Escherichia coli B-r. *J Mol Biol* **180**(3), 477–503.

Pinker S (2003) *The Blank Slate: The Modern Denial of Human Nature.* Penguin, New York.

Prior H, Schwarz A, Güntürkün O (2008) Mirror-induced behavior in the magpie (*Pica pica*): Evidence of self-recognition. *PLoS Biology* **6**(8), e202.

Quintero RA, Mueller OT, Martinez JM, Arroyo J, Gilbert-Barness E *et al.* (2003) Twin-twin transfusion syndrome in a dizygotic monochorionic-diamniotic twin pregnancy. *J Matern Fetal Neonatal Med* **14**(4), 279–281.

Red-Horse K, Yan Zhou, Genbacev O, Prakobphol A, Foulk R, McMaster M, Fisher SJ (2004) Trophoblast differentiation during embryo implantation and formation of the maternal-fetal interface. *J Clin Invest* **114**(6), 744–754.

Ridley M (2000) *Genome: Autobiography of a Species in 23 Chapters.* Harper Collins, New York.

Ridley M (2004) *The Agile Gene* (previously published as *Nature via Nurture*). Harper Perennial, New York.

Rydhstroem H, Heraib F (2001) Gestational duration, and fetal and infant mortality for twins vs singletons. *Twin Res* **4**(4), 227–231.

Sathananthan H, Menezes J, Gunasheela S (2003) Mechanics of human blastocyst hatching *in vitro. Reprod Biomed Online* **7**(2), 228–234.

Schrödinger E (1935, 1936) Discussion of probability relations between separated systems. *Proc Camb Philos Soc* **31** (1935), 555–563; **32** (1936), 446–451.

Schultz RM (2002) The molecular foundations of the maternal to zygotic transition in the preimplantation embryo. *Hum Reprod Update* **8**(4), 323–331.

Shermer M (2004) *The Science of Good & Evil: Why People Cheat, Gossip, Care, Share, and Follow the Golden Rule.* Times Books, Henry Holt & Co., New York.

Shiota K, Yamada S (2005) Assisted reproductive technologies and birth defects. *Congenit Anom (Kyoto)* **45**(2), 39–43.

Souter VL, Kapur RP, Nyholt DR, Skogerboe K, Myerson D, Ton CC, Opheim KE, Easterling TR, Shields LE, Montgomery GW, Glass IA (2003) A report of dizygous monochorionic twins. *N Engl J Med* **349**(2), 154–158.

Souter VL, Parisi MA, Nyholt DR, Kapur RP, Henders AK, Opheim KE, Gunther DF, Mitchell ME, Glass IA, Montgomery GW (2007) A case of true hermaphroditism reveals an unusual mechanism of twinning. *Hum Genet* **121**(2), 179–185.

Spemann H (1938) *Embryonic Development and Induction.* Yale University Press, New Haven.

Spemann H, Mangold H (1924) Induction of embryonic primordia by implantation of organizers from a different species. In Willier BH, Oppenheimer JM (eds.), *Foundations of Experimental Embryology.* Hafner, New York, pp. 144–184.

Takahashi S, Kawashima N, Sakamoto K, Nakata A, Kameda T, Sugiyama T, Katsube K, Suda H (2007) Differentiation of an ameloblast-lineage cell line (ALC) is induced by sonic hedgehog signaling. *Biochem Biophys Res Commun* **353**(2), 405–411.

Tarkowski AK (1998) Mouse chimeras revisited: Recollection and reflections. *Int J Dev Biol* **42**, 903–908.

Tarlatzis BC, Qublan HS, Sanopoulou T, Zepiridis L, Grimbizis G, Bontis J (2002) Increase in the monozygotic twinning rate after intracytoplasmic sperm injection and blastocyst stage embryo transfer. *Fertil Steril* **77**, 196–198.

Thijssen JM (1987) Twins as monsters: Albertus Magnus's theory of the generation of twins and its philosophical context. *Bull Hist Med* **61**, 237–246.

van Anders SM, Vernon PA, Wilbur CJ (2006) Finger-length ratios show evidence of prenatal hormone-transfer between opposite-sex twins. *Horm Behav* **49**(3), 315–319.

van de Beek C, Thijssen JH, Cohen-Kettenis PT, van Goozen SH, Buitelaar JK (2004b) Relationships between sex hormones assessed in amniotic fluid, and maternal and umbilical cord serum: What is the best source of information to investigate the effects of fetal hormonal exposure? *Horm Behav* **46**(5), 663–669.

van Dijk BA, Boomsma DI, de Man AJ (1996) Blood group chimerism in human multiple births is not rare. *Am J Med Genet* **61**(3), 264–268.

Vernadsky VI (1926) *The Biosphere* [in Russian]. English translation (1998) by DB Langmuir. Copernicus, New York. [Also Vernadsky VI, McMenamin MAS, Margulis L, Ceruti M (1988) *The Biosphere: Complete Annotated Edition.* Kindle Edition.]

Vietor HE, Hamel BC, van Bree SP, van der Meer EM, Smeets DF, Otten BJ, Holl RA, Claas FH (2000) Immunological tolerance in an HLA non-identical chimeric twin. *Hum Immunol* **61**(3), 190–192.

Voracek M, Dressler SG (2007) Digit ratio (2D:4D) in twins: Heritability estimates and evidence for a masculinized trait expression in women from opposite-sex pairs. *Psychol Rep* **100**(1), 115–126.

Waye JS, Willard HF (1986) Structure, organization, and sequence of alpha satellite DNA from human chromosome 17: Evidence for evolution by unequal crossing-over and an ancestral pentamer repeat shared with the human X chromosome. *Mol Cell Biol* **6**(9), 3156–3165.

Weeks DE, Nygaard TG, Neystat M, Harby LD, Wilhelmsen KC (1995) A high-resolution genetic linkage map of the pericentromeric region of the human X chromosome. *Genomics* **26**(1), 39–46.

Williams CA, Wallace MR, Drury KC, Kipersztok S, Edwards RK, Williams RS, Haller MJ, Schatz DA, Silverstein JH, Gray BA, Zori RT (2004) Blood lymphocyte chimerism associated with IVF and monochorionic dizygous twinning: Case report. *Hum Reprod* **19**(12), 2816–2821.

Wołczyński S, Kulikowski M, Szamatowicz M (1993) Triploidy as a cause of failure in human reproduction. *Ginekol Pol* **64**(3), 154–160.

Wu JS, Myers S, Carson N, Kidd JR, Anderson L, Castiglione CM, Hoyle LS, Lichter JB, Sukhatme VP, Simpson NE (1990) A refined linkage map for DNA markers around the pericentromeric region of chromosome 10. *Genomics* **8**(3), 461–468.

Yamagishi C, Yamagishi H, Maeda J, Tsuchihashi T, Ivey K, Hu T, Srivastava D (2006) Sonic hedgehog is essential for first pharyngeal arch development. *Pediatr Res* **59**(3), 349–354.

Yang MJ, Tzeng CH, Tseng JY, Huang CY (2006) Determination of twin zygosity using a commercially available STR analysis of 15 unlinked loci and the gender-determining marker amelogenin — A preliminary report. *Hum Reprod* **21**(8), 2175–2179.

Yoon G, Beischel LS, Johnson JP, Jones MC (2005) Dizygotic twin pregnancy conceived with assisted reproductive technology associated with chromosomal anomaly, imprinting disorder, and monochorionic placentation. *J Pediatr* **146**(4), 565–567.

Glossary

"When I use a word," Humpty Dumpty said, in rather a scornful tone,
"it means just what I choose it to mean — neither more nor less."

Lewis Carroll

You should find nothing really idiosyncratic here, but many of these words have other definitions that are irrelevant here and are therefore not included.

abdomen: most posterior of the three major segments of the generalized animal embryo (head, thorax and abdomen)

ABO blood group: the most important of the transfusion/transplant antigens of human tissues; first major blood cell surface antigen system discovered, by Landsteiner in 1901; based on variation of specific sugars in a side chain on cell surface proteins

abortion, spontaneous: also known as *miscarriage*; the spontaneous failure of an ongoing recognized pregnancy in the first few weeks after maternal and clinical recognition of the pregnancy

abortion, therapeutic: medical termination of pregnancy for therapeutic reasons

acardia: absence of the heart; a severe embryogenic malformation representing failure to properly establish the posterior portion of the anterior–posterior axis in embryogenesis; usually missing all body structures posterior to the chest region; seen at birth only in monozygotic twins, where the malformed fetus can survive beyond very early embryogenesis only with the support of a (usually normal) pump twin to supply circulation.

acephalus: absence of the head; a severe malformation representing failure to properly establish the anterior portion of the anterior–posterior axis of embryogenesis; seen only in monozygotic twins, where the malformed fetus can survive beyond very early embryogenesis only with the support of a (usually normal) pump twin to supply circulation.

accretion: process in and by which the size of a thing increases by the continuing addition of smaller components

adenine: one of the four DNA bases; a purine, with a double aromatic ring (a six-member [4C, 2N] ring and a five-member [3C, 2N] ring, sharing one [2C] side) as the main body of the molecule

African: descended from the indigenous peoples of sub-Saharan Africa

African-American: American citizens of African ancestry

agglutination, agglutinating: clumping; here used in the context of blood cells clumping together and falling out of suspension because of being stuck together by the action of antibody molecules

aging gametes: egg or sperm cells going beyond their useful lifespan; 12 hours or so for egg cells, a few days for sperm cells after deposition in the female reproductive tract

Albertus Magnus, Albert of Cologne, Albert the Great: 1206–1280; a Dominican priest, scientist, theologian and philosopher; major advocate for peaceful coexistence of science and religion; substantially responsible for bringing the teachings and methods of Aristotle into medieval scholarship

algorithm: a method; a procedure or formula for solving a problem, usually in the form of a stepwise listing of calculation instructions

allele: a specific version of a given gene's DNA sequence

ambidextrous: able to use either hand with more or less similar grace and skill

amnion, *pl.* **amnia** (Gk) or **amnions:** the finer, inner layer of the fetal/gestational membrane ("bag of waters"); most twins have one each, even when inside a shared chorion

amniotic fluid: contents of the gestational membranes, the "water" of pregnancy in which the fetus is bathed

amphibian: in biology, an animal of the vertebrate group (class Amphibia) including frogs, toads, salamanders, newts and caecilians; typically hatch, as tadpoles, from eggs laid in water, then spend adult life on land, in water or both

ampulla (of the oviduct): wider section of the oviduct nearer the ovary than the uterus

anastomosis, *pl.* **anastomoses:** a connection, a joint, between hollow, tubular structures; in the present context, a joining between blood vessels of two placentas

androgen: from Greek roots meaning "making male"; a corticosteroid hormone of the type more characteristic of males; the best known androgen is testosterone

anencephaly: severe neural tube defect malformation in which the neural tube of the embryo does not close at the anterior (head) end, forming no cranium and no cerebrum; no head above a line running from the brow over the ears to the nape of the neck; survival after live birth is usually only a matter of hours

aneuploidy: an error in number or structure of chromosomes; errors in number include departures from disomy — having any number other than two copies of one or more individual chromosomes, and departures from diploidy — having any number other than two full sets of chromosomes; errors in structure include deletions, insertions, translocations and inversions, because of which one or more chromosomes has missing or extra DNA sequences, or sequences in wrong places resulting in errors of control of functions

aniridia: absence (rarely partial, usually bilateral) of the irides (plural of iris, the colored rings that surround the pupils and control the amount of light entering the eye chambers); often due to mutation in the gene PAX6, the human homolog of the gene known as eyeless in the fruit fly

annual seasonal rhythm: a rhythm varying with the seasons, with a period of one year; here in the context of an annual rhythm in human conception rate

anomaly, *pl.* **anomalies:** something wrong, sometimes only by being out of the ordinary; abnormality, peculiarity, deviation; human "chromosome anomalies," for example, are usually disorders of the number or structure of chromosomes, and thus represent large mutations

anterior (vs posterior): toward the head end (vs tailward)

anteroposterior: the direction, or axis, running from head to tail

antibodies, fluorescent: labeling agent for microscopic analyses, exquisitely specific as characteristic of antibodies, and capable of demonstrating the exact location of any cell component against which an antibody can be generated; antibodies can be made fluorescent by attaching any of a variety of fluorescent molecules, such as fluorescein and rhodamine

antidepressant: drug to counteract depression

antigen: may be any substance that elicits an antibody response, which causes cells of the immune system to make antibodies

antimere: the "same" structure from the other side of the body; left hand and right hand are antimeric to each other

antisymmetric: this one is more than a little tricky; most definitions or discussions of the concept are from mathematics; in a biological context, I am using it here to indicate: functionally, programmatically, consistently different between sides, in approximately equal and opposite ways

anxiolytic: drug to counteract anxiety

aorta: the largest artery; leaves the left ventricle to distribute oxygenated blood to the body

arachnid: a joint-legged (arthropod) invertebrate animal of the class Arachnida, which includes the spiders, scorpions, ticks, and mites (arachnids in general have eight legs, not six like insects)

Aristotle: 384–322 BCE; Greek philosopher, from the time when the term philosophy included everything worth knowing; major contributor to human cultural evolution by way of promoting the use of human reason as a means to know and understand

ART — artificial (or assisted) reproduction technology: umbrella term for all methods to achieve pregnancy by means other than natural combinations of ovulation, copulation, insemination, and fertilization

artificial chimerism: state of chimerism achieved by artificial means, such as tissue transplantation (which includes blood transfusion)

ascertain, ascertainment: to find and choose individuals as members of a sample — usually with some particular trait — to participate in a research effort

Asian: descended from, or otherwise pertaining to, the peoples of the eastern part of the continent of Asia — primarily Chinese, Japanese, Korean; for reasons not yet understood, distinct from the people of the Indian subcontinent and other South Asians

assumption: supposition; something taken for granted or posited as true; in this context, for the sake of simplifying an argument or other reasoning process, for example, the (disastrously false) assumption that boy-girl twins may be considered representative of all dizygotic twins

asymmetry: etymologically, the word means a departure from symmetry, up to the level of an absence of symmetry, but in most of its biological uses it is only through such departures that any impression of symmetry is to be perceived; in biological usage, the symmetry at issue is the (only approximate, never complete or exact) bilateral symmetry of animal organisms

asymmetry, brain function: the human brain depends deeply on differences in function between the left and right cerebral hemispheres; I have proposed that the most "human" of mental functions depend on reflection, whereby the individual mind/self considers, or reflects upon, itself or its behaviors

asymmetry, developmental: the installation, in embryogenesis and fetal development, of the cellular bases for the structural and functional asymmetries of the adult human organism

asymmetry, directional: a consistent tendency for a given structural or behavioral asymmetry to be in a specific direction; quantitatively, a non-zero mean left-right difference

asymmetry, fluctuating: a component of biological asymmetry that is not consistently directional, which is random as to direction; variation of left-right differences about their non-zero mean left-right difference

atrial septal defect: a defect (hole) in the wall between the atria (the paired, smaller, upper chambers) of the heart (the left atrium pumps oxygenated blood from the pulmonary vein, returning from the lungs, into the left ventricle for pumping out into the body; the right atrium pumps deoxygenated blood from the vena cava, returning from the body, into the right ventricle to be pumped into the lungs via the pulmonary artery)

Axenfeldt anomaly: a malformation of the anterior chamber of the eye, seldom isolated, rather typically one of a cluster of anomalies of the eye: occurs in several different syndromes of anomalous development, seems often due to a mutation in the PAX6 gene (the human homolog of the gene called eyeless in the fruit fly)

barb: the largest of the component substructures of a feather, attached directly to the main shaft of the quill

barbule: the third-level substructure of a feather; branches directly from the barb

basal body: also known as *basal granule* and *kinetosome*: cylindrical structure at the base of a cilium or flagellum, generally considered homologous to, and derived from, the centriole/s, by a mechanism that remains mysterious; basal body and centriole have spiral arrangement of nine skewed (neither parallel nor perpendicular to the circumference) triple-tube structures; cilia and flagella have nine double tubes, parallel to the circumference, surrounding another pair of tubes; much of the literature seems to represent the belief that these two kinds of structure are continuous, but details remain lacking

behavior: what a thing (living or otherwise) does

Belgium: a small nation (population ~10.5 million) of western Europe, bordered to the north by the North Sea and the Netherlands, to the east by Germany and the Duchy of Luxembourg, and to the south and west by France. Composed primarily of French-speaking Wallonia and Flemish-speaking Flanders, with a small ~1% contingent of German speakers in a region near the border with Germany; home of the East Flanders Prospective Twin Survey — since 1946, a population-based bank of data about twins and higher multiple births collected from birth

Benirschke, Kurt, MD: born May 26, 1924; reproductive pathologist who probably knows more about the placentas of mammals, and specifically *Human Placental Pathology* (the name of his textbook, with Peter Kaufmann), and placentation of human twins, than the rest of us put together, and most of what we know we learned from him

binomial distribution: describes the behavior of a variable representing the number of observations in a sample of a certain size with one of exactly two mutually exclusive outcomes, when: (i) the number of observations sampled represents a small fraction of the population sampled; (ii) each observation is independent, no case in the sample having any influence on the outcome of any other observation in the sample; and (iii) the probability of the specified outcome is constant at each observation. The use discussed here is the distribution of sex-pairing among dizygotic twin pairs as a function of sex fractions

binovular follicle: an ovarian follicle containing two egg cells

biocentrism: the theory that life creates the Universe instead of the other way around, consistent with the quantum-mechanical understanding that reality becomes fixed by observation; in this view, current theories of physical existence do not work and cannot be made to work until they fully account for life and consciousness

biogenesis: the emergence of the properties of life from a system composed entirely of non-living components

biometrics (*biometric*): the study of (*having to do with*) the relative sizes of body parts; biometrics is the science and technology of measuring and analyzing biological data; in some applications, biometrics refers to technologies that measure and analyze human body characteristics, such as fingerprints, eye retinas and irises, voice patterns, facial patterns and hand measurements, for authentication purposes

blastocoel: the fluid-filled chamber of the blastocyst

blastocyst, blastula: the human embryo at the stage when the inner cell mass (future embryo proper) has just differentiated from the trophoblast (future extraembryonic support structures, including the placenta and amnion to be developed from the polar trophoblast, and the chorion to be developed from the mural trophoblast)

blastomere/s: the individual cells of the blastocyst

blastopore: in the amphibian embryo, the site of invagination of cells from the blastocyst surface in the process of forming the gastrula embryo; analogous to the primitive streak in the human embryo

blood typing: genetic characterization of individuals according to their red blood cell surface antigens

boundary: definition encompassing a system or structure that distinguishes that system or structure from all that does not belong to it; prototype system boundary is the active membrane of a living cell

branchial arches: embryonic structures homologous to the gill arches of fish embryos; embryonic precursors to structures of the face, jaws, throat, and ears

buccolingual: from cheek to tongue, a dimension of measurement of tooth sizes, vs mesiodistal, the dimension of measurement of tooth sizes along/around the dental arch from the incisors to the molars

calcified: hardened by deposition of calcium

canonical correlation analysis: a multivariate statistical analysis method involving systems of structural variables in each of two groups which are most highly correlated between the groups; used here in analyses of developmental left-right asymmetries in the teeth

capacitation: a stage in sperm cell maturation; a system of physiological processes taking place in the female reproductive tract, in response to secretions encountered there, whereby sperm cells become capable of penetrating an egg cell

cardiac looping: the embryonic primordium of the heart is formed from the midline fusion of the cardiac tubes; as the heart tube loops, the cephalic (headward) end of the heart tube bends ventrally (toward the belly), caudally (toward the tail), and slightly to the right, then proceeds to further development of the heart

central incisors: the "two front teeth" and their counterparts in the lower jaw, bracketed by the lateral incisors, all eight being relatively wide, shallow and sharp, for cutting

centriole: cylindrical organelle found in most eukaryotic cells; typically about 400–500 nm long and 200 nm in diameter; walls usually formed of nine triplets of microtubules, surrounding a central pair; centrioles typically occur in pairs, oriented perpendicularly, being most of the mass of the **centrosome**; major organizing focus of the **cytoskeleton**, and thus of the spatial arrangement of all cellular substructures; involved in the building and functioning of the mitotic spindle; required for formation of flagella and cilia; involved in establishment of left-right asymmetries in embryogenesis

centromere: appears as a major constriction on fully compacted chromosomes, as in cells in the midst of cell division; attachment point for spindle fibers to pull the chromatids of freshly duplicated chromosomes apart into the daughter cell nuclei in cell division

centromere position, chromosomes grouped by:
  metacentric: describes chromosomes the centromeres of which are near the middle of the length of the chromosome
  submetacentric: describes chromosomes the centromeres of which are distinctly off-center of the length of the chromosome
  acrocentric: describes chromosomes the centromeres of which are near an end; in the human genome, these chromosomes (#s 13, 14, 15, 21, 22) have short arms with largely redundant sequences dedicated primarily to transcribing ribosomal RNA; they are composed of a narrow "stalk" and a "satellite" (so named because these parts were found always near to the rest of their chromosomes, but their attachments via the stalks were not visible in the microscopy of the time)
  telocentric: describes chromosomes the centromeres of which are at an end; the human has none of these; the mouse only has these

cephalic: having to do with the head

cerebrospinal fluid: made by, and secreted into the brain ventricles from, the choroid plexus; circulates inside the **meninges**; bathes and cushions the brain and spinal cord

cervical: having to do with the neck; probably the most familiar use of the word is in reference to the cervix (the neck) of the uterus

cetacean/s: the order **Cetacea** includes whales, dolphins and porpoises, totaling 90 species, all marine except for four species of freshwater dolphins. There are two suborders: the baleen whales (which feed by filtering large quantities of small prey from great mouthfuls of water, using the comb-like keratinaceous baleen) and the toothed whales, which includes the dolphins and porpoises, the sperm whale, and the orca. They are mammals (warm-blooded, breathe air with lungs, live born young are fed the mother's own milk, and they have — very little — hair). They range in size up to the blue whale, the largest animal ever to live on Earth. Their closest living relative among terrestrial mammals is the hippopotamus. Their tail fins are horizontal flukes, not vertical as are those of fish

chemotaxis: behavior of single-celled organisms orienting themselves, and moving, toward or away from substances sensed in the environment

Chile: long (north-south ~4300 km), narrow (average width less than 180 km) nation on the west coast of the South American continent, between the Pacific Ocean on its west and the ridge of the Andes mountain range on its east. Largest variation of latitude within its borders of any nation. Source of data best demonstrating the correlation of human conception rate with day length and temperature

chimera/s: in biology, individual organism/s with cells of two or more different genotypes in the same body

chitin: the substance composing the shells of arthropod animals (insects, arachnids, crustaceans); second most abundant biological substance in the world after cellulose, the primary skeletal material in plants; like cellulose, chitin is composed primarily of sugar polymers (polysaccharides)

chondrocytes: cells the function of which is to produce cartilage

chorion: the thicker outer layer of the gestational membranes, developed from the mural trophoblast cells of the blastocyst

chromatid: daughter chromosome, one of two identical copies of a single "old" chromosome, consisting of one "old" strand from the mother chromosome and one newly replicated strand; to be segregated into daughter cell nuclei in mitosis; it is only a "chromatid" as long as the sisters are attached at the centromere after replication — "chromosome" as soon as they separate

chromosome: one of a species-characteristic number of packages into which nuclear DNA is packed for moving in cell division; composed of chromatin = DNA + proteins, most of which is histone proteins

circular logic: flawed, false, pointless "logic" in which the conclusion/s restate/s the premise/s; *see tautology*

cleavage: process whereby the single-cell zygote, inside the zona pellucida, divides without growth into one or two dozen cells to become the morula

cleft chin: a generally harmless anomaly of the fusion of the mandibular processes, seldom resulting in more than a dimple or a shallow groove in the chin

cleft lip, CL/P: failure of the fusion of a maxillary process with the frontonasal process, resulting in a cleft in the upper lip; unilateral cases are ~80% left side, with ~5% sibling repeat risk; right side and bilateral cases represent greater multifactorial genetic loading and have higher repeat risk; sometimes involves the seam between the palatal shelf and the palatal contribution from the frontonasal process, as well as the lip

cleft palate, CP: cleft of the hard palate, resulting from failure of the palatal shelves to fuse

clinical pregnancy: clinically recognized/confirmed pregnancy; usually beyond six to eight weeks post-fertilization

Collaborative Perinatal Project: a multi-center project of the National Institute of Neurological and Communicative Disease and Stroke; followed some 53,000 children from first prenatal care visit during pregnancy, through their first year after birth

commensal: from Latin, "sharing the table"; describes relationships between organisms in which one organism derives food or other benefits from another organism without hurting or helping it, for example, commensal bacteria are part of the normal flora in the mouth

complex, complexity: in general usage, complexity is often used as descriptive of something with many parts in intricate arrangement; science involves numerous approaches to characterizing complexity, with dozens of definitions; often involved with "system" concepts, especially with the idea of *emergence* as a threshold in complexity beyond which a system has properties that are not among, and could not have been predicted from, the properties of the parts; it seems most people who recognize the word think they know what it means until they have reason to try to define it

conception: becoming pregnant — not an instantaneous event, but a developmental horizon, distinct only from a distance

concordant vs discordant: matching vs not-matching; in twin studies or pairs-of-relatives studies in genetic analysis, a concordant pair matches for the phenotype in question while the members of a discordant pair differ

conditional mutation: mutation generally induced for research purposes, which allows the gene to make a functional product under certain conditions, but not in other conditions

congenital cataract: cloudiness of the lens/es of the eye/s present at birth or shortly thereafter; if not promptly repaired, can result in permanent visual loss (by preventing the brain from learning how to interpret visual input and see); often associated with mutations in PAX6, the human homolog of the eyeless gene in the fruit fly

congenital heart defects, CHD: developmental anatomical malformations of the heart, including but not limited to: aortic stenosis, atrial septal defects, coarctation of the aorta, double outlet right ventricle, hypoplastic left heart, pulmonary atresia, pulmonary stenosis, transposition of the great vessels, truncus arteriosus, ventricular septal defects

consciousness: this is another of those things everybody understands until they set out to explain it; state of being awake, awareness, perception, that component or property of mind by which, or state of mind in which, we are awake and aware and actively perceiving our circumstances; its cellular basis remains certain, but mysterious in detail

consciousness of self: reflection of awareness and perception of itself, of the "person" from which it originates; considered by some to be the exclusive province of the human mind, an attribute generally understood to be held by no non-human organism

cornea: the clear outer layer covering the anterior chamber of the eye; changes in shape provide a major fraction of the eye's focusing power

corneal dystrophy: cloudy-to-opaque obstruction of the cornea; may arise from a variety of causes; including mutations in PAX6, the human homolog of the eyeless gene in the fruit fly

corpus luteum: (Latin: yellow body); after ovulation, if conception ensues and progresses to implantation, when the trophoblast begins to secrete chorionic gonadotropin, the remains of the ovulatory follicle become the corpus luteum, which will function as a gland secreting progesterone to support the continuation of the pregnancy

correlation: a statistical, mathematical relationship of shared variance between two or more randomly varying traits or measures, such that variation in one trait or measurement tends to follow that in another, or to move in opposite directions in negative correlation

cortex: (Latin: bark, rind); the outer layer of an organ or part, as the egg cell cortex; vs **medulla**, the inner part

corvid/s: *Corvidae*; the family of birds including rooks, ravens, crows, magpies, jays and jackdaws; mostly omnivorous and highly social; "more intelligent than mammals of similar weight"; have been taught to count and to vocalize human words; have been observed to make and use simple tools and devise other schemes to achieve objectives

craniofacial: of or pertaining to the cranium and the face; bones of the craniofacies include: ethmoid, frontal, two inferior nasal conchae, two lacrimal, mandible, two maxillae, two mastoid, two nasal, occipital, two parietal, two palatine bones, sphenoid, two temporal, vomer, two zygomatic bones; most of those listed as single are in fact developed by fusion of bilateral halves; craniofacial and brain development are deeply inter-related

crossover: physical exchange between DNA sequences of different copies of chromosomes during meiosis; genetic consequence of crossing over is recombination

crown: in dentistry, the occlusive/occlusal surface of the tooth

crustacean: arthropod invertebrate organism; over 50,000 species including the crabs, shrimps, lobsters, krill, copepods and barnacles, as well as a few terrestrial species; must moult the exoskeleton to grow

cumulus oöphorus: "egg-bearing cloud"; a cloud of support cells surrounding the egg cell on its release from the ovarian follicle in ovulation

cusp: one of the "points" on the crown of a tooth

cyclops malformation: due to the failure of the frontonasal process to progress down through the midface, to separate the eye-field into parts for two eyes and to properly form the nose; often results in a proboscis with a single "nostril" on the "forehead" above the single eye

cytogenetic/s: genetics at the microscopic, chromosomal level

cytoplasm: contents of a cell outside of its nucleus; once seen as liquid, it is so full of cytoskeleton and organellar structures, such as endoplasmic reticulum and vesicles of many sorts, as to more closely fit the label "gel"

cytosine: one of the nucleic acid bases; a pyrimidine, the main structure of which is a six-member ring, 4C, 2N

cytoskeleton: primarily proteinaceous (includes actins in filaments, tubulins in microtubules, intermediate filaments, microfilaments, vimentins, lamins, keratins, dyneins, kinesins, and others) scaffolding or skeleton contained within the cytoplasm of all cells; maintains cell shape, enables cellular motion via its substructures the flagella and cilia; plays important roles in the movement of vesicles and organelles within the cell, and provides crucial framework for cell division

daylight time, daylight saving time, summer time: in most modern societies gradually since the 1950s, an official advancing of clock time by one hour, with the result of moving the customary work period earlier in the day, lengthening the time between the end of the workday and sunset

deciduous dentition: the first set of teeth, the "baby" teeth, "milk" teeth, that will be replaced by the secondary "adult" dentition

degrees of freedom: number of dimensions of the parameter space; number of values or dimensions in a system under analysis that are free to vary

dental diameters: measures of tooth size and shape; herein the maximum measurements across the crown of a tooth in buccolingual (cheek-to-tongue across the dental arch) and mesiodistal (incisor-to-molar along the dental arch) directions

dentin: with enamel, cementum, and pulp, one of the four major components of teeth; typically the largest fraction of the mass of the tooth; calcified with hydroxylapatite; a living, cellular tissue, which may continue cellular response throughout life

detritus: non-living particulate organic matter; usually composed primarily of bodies of dead organisms or fragments thereof, or fecal material thereof

development: unfolding; here in the context of unfolding the plan in the zygote for the embryo and fetus, and eventually the adult body

dextrocardia: "rightward heart"; major midline fusion malformation, in which heart-looping has gone the wrong way; may be relatively benign if the great vessels are also inverted, very dangerous if not; the cardiac component of situs inversus

dictyotene: a prolonged state intermediate in prophase of Meiosis I in oögenesis, in which the primary oöcyte is suspended from about the 16th week of fetal life until discharged from the ovary after puberty. The chromosomes are slightly relaxed from their level of fullest condensation, allowing transcription of the DNA for some of the functions of preparing the egg cell for eventual ovulation

dihydrotestosterone, DHT: androgenic hormone derived from testosterone by chemical reduction by the enzyme 5-alpha-reductase; metabolite responsible for most actions of testosterone in androgen-sensitive tissues

diphthong: a combination of two or more consecutive vowels, in which the individual vowel sounds are not separately pronounced, but merged in a different vowel sound characteristic of that particular combination and the word in which it appears, e.g., ae, ai, oo, ou, au, eu, ei, ie, and so on

diploid: having two copies of the organism's genetic information, usually in the form of two copies of each chromosome; this is the state of most eukaryotic organisms

diprosopus: "two face"; a major midline a/symmetry malformation in which the structures of the head and face are substantially duplicated; may be complete, resulting in two heads, or incomplete, with one head having two more or less complete faces; often associated with mutations in sonic hedgehog (SHH)

discriminant function analysis: a multivariate statistical analysis method, based on that linear combination of the variables at hand which maximizes differences among groups in the sample; closely related to multiple regression on the grouping variable

dispermy/dispermia: a condition of double fertilization, which apparently most often results in an abnormal triploid embryo (given a single maternal pronuclear contribution), but which may also initiate chimeric development if two maternal pronuclear contributions are available

dispermic: of chimeras, cell lines of which show different paternal genetic contributions, derived from two fertilizations

dizygotic: of twins, derived from two zygotes

Doctor Universalis: "universal teacher"; "teacher of everything"; an honorific "nickname" for Albertus Magnus, Albert the Great, *q.v.*

doctrine, dogma: body of belief relating to morality and faith, established by authority; an authoritative principle, considered to be unarguably true

dorsal vs ventral: toward the back vs toward the belly

ectoderm: "outer skin, or layer"; embryonic layer of cells which will generate structures on the "outside" of the body; generally precursor to skin and nervous systems

ejaculate/ejaculation: *v.* to release sperm cells in seminal fluid; *n.* the substance of such a release

embryology: the scientific study of prenatal development

emergence, emergent: a fundamental concept in evolution and developmental biology, in systems theory, and in philosophy; refers in rough generality to ways in which the patterns and properties of complex entities arise from the multiplicity of simpler interactions among the components of the complex entity in question; that property or phenomenon whereby it comes to be the case that ("how come") the whole is greater than the sum of its parts; an emergent whole has functions and properties arising from interactions among its parts that are not present in, and are not predictable from, the properties of any or all of its parts ("irreducible")

enamel knot, enamel organ: growth center of a tooth bud

endoderm: ventral layer of the trilaminar stage of embryogenesis, primordial to the gut and its derivatives

environment/al: having to do with the physical, social and cultural conditions or circumstances surrounding an organism, extraneous to the organism itself, which influence the development and survival of the organism

entanglement: from quantum mechanics (Schrödinger, 1935); a relationship between paired particles that have arisen from a shared prior quantum state in such a way as to generate a state of correlation between the properties (position, momentum, spin, charge, etc. — all properties) of the paired particles at any given instant in their further existence (see http://plato.stanford.edu/entries/qt-entangle)

epistemology: the study of knowledge, of knowing, of the ways and means of knowing, of the nature of knowledge

epithelium, epithelial: concerning tissue/s composed of layers of cells that line the external and internal surfaces of the body, and which form many of the glands and the developmental primordia of many body structures, particularly derivatives of the endoderm and ectoderm

Equator: circumference of the Earth equidistant between North and South Poles

equinox: first day of spring or autumn, when day length is 12 hours worldwide; usually near to March 21 and September 21, respectively

estrogen: a steroid reproductive hormone produced primarily by the ovary, responsible for promoting development and maintenance of female sexual characteristics and functions

evidence: clear indication, example or proof; observation or experience bearing on, supporting or negating, a hypothesis under consideration; required in a scientific context to be empirical, to be documented as appropriate to the particular field of inquiry at issue, and to be straightforwardly subject to repetition or refutation

exon: part of the sequence of the primary messenger RNA of a functional gene which will ordinarily be included in the mature message by splicing; as opposed to the **intron**s, which are "intervening sequences" that will be spliced out of the primary message and not be a part of the mature message to be translated into protein

extremophile, extremophilic: (concerning) organism/s tolerating or requiring extreme physical conditions for life (acidity, alkalinity, temperature, pressure, aridity, salinity, etc.), largely microbial, but including some multicellular worms, insects, crustaceans and tardigrades

facultative: contingent; optional; elective, discretionary, not required or compulsory; capable of behaving or functioning differently under differing conditions

Fairchild, Lydia: phenotypically normal, chimeric American woman charged with welfare fraud and threatened with loss of her children when her DNA genotype was interpreted as being incompatible with her being the mother of her children

fallacy: flaw in an argument that renders the argument invalid

fate map: a representation of the developmental progress of each cell in the body of an organism from the zygote to the adult

fecundability: the ability to become pregnant; not the same as fertility, which requires the ability to bear children, but only part of it

fertility: the ability to have live offspring

fertilization: the process of interaction between sperm and egg cell, ending with syngamy when paternal and maternal pronuclei succeed in merging to produce a functional zygote

fetus: prenatal form of the organism during the gestational period between completion of embryogenesis and delivery

fibroblasts: connective tissue stem cells; most common cells in animal connective tissues; make and maintain extracellular matrix; providing structural framework for many tissues, and play critical role in wound healing

Finland: small (population five million) democratic nation of northern Europe, bounded by Sweden, Russia, Norway, and the Gulf of Finland; least densely populated nation in Europe, northernmost (60–67 degrees north latitude) except for a small part of Norway; 51 days of constant polar night in extreme north

flagellum, *pl.* **flagella:** three biologically distinct kinds of motility organelles — the eukaryotic (homologous with cilium, cilia), bacterial and archaebacterial; eukaryotic version has basal bodies homologous with centrioles (which are cell-division organelles apparently fundamental to the structure and function of the mitotic spindle)

folate, folic acid: vitamin (B9) particularly important for reactions involving methyl transfer, including modification of DNA by methylation; proper levels in the body of women of reproductive age, beginning before conception and continued through pregnancy, can lower the frequency of neural tube defect malformations by as much as 70%; frequency of congenital heat defects is also reduced

follicle (ovarian follicle, ovulatory follicle, Graafian follicle): cystic structure near the surface of the mature ovary, in which an oöcyte matures and ripens, surrounded by the zona pellucida, the cumulus oöphorus and the granulosa cell layer, in preparation for ovulation by the rupture of the follicle

follicular phase of menstrual cycle: that portion of the menstrual cycle (from onset to ovulation) during which the egg cell is undergoing final preparations for ovulation

forensics: the science and technology used to investigate and establish facts in courts of law

foveal hypoplasia: one of a group of anomalies of eye development which can be caused by mutations in PAX6, the human homolog of the eyeless gene in the fruit fly

fractal: having to do with a class of geometric constructions with simple, recursive equational definitions, which can be subdivided into parts, each of which is (at least approximately) a reduced-size copy of the whole, a property called self-similarity; typical of the shapes of clouds, coastlines, lightning and snowflakes, and of the structures of many tissues, especially branched tissues such as lung and blood vessels

fraternal twin, twinning: common misnomer for dizygotic twin, twinning

freemartin: sterile heifer calf from a male-female bovine twin pair, believed due to exchange of hormones prenatally through placental anastomoses

frontonasal process: component tissue of the developing face; from forehead to the philtrum of the upper lip; splits the eye field, generates most of the middle of the nose, joins with lateral nasal processes forming the nostrils and with the maxillary processes forming the rest of the upper jaw and lip

fusion malformations: also called midline malformations; developmental anomalies of structures formed in the embryonic midline by the fusion and remodeling of bilateral half structures; neural tube defects, congenital heart defects and orofacial clefts are the best known

Gaia Hypothesis: proposes that the planet Earth is and functions as a single living organism/system able to maintain conditions optimal to life by way of the indigenous biosphere acting as a homeostatic feedback system

Galton, Sir Francis: 16 February 1822–17 January 1911; founder of the scientific study of behavioral genetics; generally credited with the concept of the Genetic Twin Study; coiner of the "Nature vs Nurture" concept

Galtonian Twin Study: a form and method of genetic analysis after the concepts of Galton, in which within-pair relationships in monozygotic twin pairs are compared with those of dizygotic twin pairs as an estimator of the fraction of variation in a given trait that is genetic in origin

gamete aging: time beyond maturity of the egg and sperm cells

ganglion, *pl.* **ganglia:** an aggregation of nerve cell bodies and dendritic structures — often interconnected with other ganglia; the dorsal root ganglia pass sensory information from distal sensory neurons to spinal neurons

Gardner, Richard: Oxford University zoology professor; developmental biologist, leader in the use of chimeric and transgenic mice, more recently deeply involved in stem cell research

gastrula, gastrulation: the stage of embryogenesis following the blastocyst or blastula stage; passage from the bilaminar stage to the trilaminar stage, via the establishment of the third embryonic layer of cells; occurs in human embryogenesis when cells from the primitive streak ingress between the ectoderm and endoderm layers, driving anteriorly toward the prochordal plate and laterally, generating the notochord and other mesodermal structures

Gell-Mann, Murray: (born September 15, 1929); American physicist; won 1969 Nobel Prize for developing a system for classifying subatomic particles; coined the term "quark"; introduced quantum chromodynamics and the weak interaction; now at the Santa Fe Institute, contributing to development of complexity theory among other interests

gene products: proteins and RNA molecules produced from codes in DNA base pair sequences

genome: the entire system of genetic material of a particular organism; the whole set of genetic information in each individual member of the species; the human genome comprises 23 pairs of the DNA-plus-protein packages called chromosomes, totaling about 3.2 billion base pairs, a bit over a meter in length when unpacked

genome scan: a method of testing for evidence of genetic linkage with a set of markers covering the whole genome at intervals small enough to make it unlikely that any linkage relationship can be missed, for example, it takes about 400 markers to cover the human genome at an average interval of 10 map units, so that one or two will necessarily be within five map units of any trait-determining gene we might study

genotype: genetic constitution of an individual, usually with reference to a particular trait

germ cells: cells from which gametes may develop; oögonia, spermatogonia

germ cell line: vs soma, somatic cells, somatic cell line; those cells of the body from which gametes may develop vs cells without reproductive capacities

gestation: the carrying of a developing embryo or fetus in the uterus of the female of a viviparous species; may be multiple

gill arches: structures in vertebrate embryos, six in number, also known as branchial arches and pharyngeal arches; grow into gills in fish; primordial to many of the muscles, bones, nerves and arteries of the human face and neck, especially several of the cranial nerves and the bones of the ear, the carotid, aorta, pulmonary artery and ductus arteriosus, the parathyroid glands

gills: respiratory structures in fish, through which oxygen is absorbed from water moving over the gills

gonosome/s: another, less common, name for the sex chromosomes, X and Y in the human

guanine: one of the nucleotide bases; a purine, major element of its molecular structure being a nine-member double ring — a six-member [4C, 2N] ring sharing one 2C side with a five-member [3C, 2N] ring

Haldane, John Burden Sanderson ("Jack"): 1892–1964; British polymath; master of multiple languages; wrote in history, politics and fiction (Daedalus, predicting IVF); made important contributions to chemistry, biology, mathematics and particularly to statistical aspects of population genetics

hamuli (hooklets): fourth-order structural component of a flight feather

handedness, hand preference: preference for one hand or the other for most motor activities; a clear majority of the human population is fairly strictly righthanded; the minority, nonrighthanded population includes primarily ambidextrous individuals; the brain-function asymmetry organization of the nonrighthanders is more flexible

Harlap, Susan, MD: Israeli reproductive epidemiologist; demonstrated increase of twinning and aneuploidy in conceptions with overripe oöcytes

hatching: on about the fifth day after successful syngamy, the embryo sheds the zona pellucida, in preparation for its embedment into the wall of the uterus; this usually coincides closely with the advancement of the morula stage embryo to the blastocyst stage, with the inner cell mass being differentiated from the trophoblast

heritability: a concept from statistical genetics; the fraction of the measurable variation in a particular trait in a particular population which can be explained by correlation with Mendelian genetic relatedness

hermaphrodite: a **true hermaphrodite** is an individual whose body has both male and female reproductive tissues, most particularly gonadal tissues; a

pseudohermaphrodite generally has only ambiguous genitalia; many true hermaphrodites are found to be chimeric, with both male and female cell lines

hermaphroditism: state of being hermaphroditic; being a hermaphrodite

heterokaryotic: "different nuclei"; found in twins with different karyotypes, as in otherwise apparently monozygotic twins who differ in sex by way of one of them having lost the Y chromosome and continued development as a Turner syndrome female

heteroplasmy: a mixture of genetically distinct mitochondrial DNA molecules among the cells of an individual

hippocampus, *pl.* **hippocampi:** a brain substructure lying under the medial temporal lobes on each side of the brain; critical for formation of new memories, particularly vulnerable to most forms of potentially-damaging brain insult, and one of the earliest components to decline in dementia

Hispanic: American with ancestry in south or central America, or the Caribbean; at this time, the majority are from Mexico; the term "Hispanic" derives from the fact that most grew up in Spanish-speaking nations; Spanish ancestry is not in fact common and is usually minor when present

histone/s: family of basic proteins (as opposed to acidic, carrying an excess of positive charges) with primary functions in binding to DNA in chromatin; tightness of binding is a control variable — accessibility of the DNA is essential for replication and transcription; acetylation, methylation, phosphorylation and several other modifications of the histone proteins are involved in the regulation of tightness of histone binding

holoblastic: pertaining to the cleavage stage of embryogenesis, each cell division goes all the way through, resulting in two roughly equal-sized daughters; as opposed to **meroblastic** cleavage, in which very asymmetric cell divisions pinch off small parts of the whole

holoprosencephaly: major midline a/symmetry malformation in which the embryonic forebrain (the prosencephalon) does not properly develop into a two-sided structure; severity of brain malformation roughly inversely proportional to degree of separation

alobar: complete absence of separation of the prosencephalon into left and right lobes

semilobar: partial separation

lobar: near normal separation

The holoprosencephaly sequence includes the cyclops malformation, single central incisor, gapped central incisors, medial cleft lip, and diprosopus, often involving mutations of sonic hedgehog. Sirenomelia may be a related caudal manifestation

homosexual: adjective, pertaining to, or having, a sexual orientation, or attraction, to persons of the same sex; use as a noun is rudely pejorative

homosexuality: the state of having a sexual orientation or attraction to others of one's same sex; the behavioral expression of such an attraction in sexual activity between individuals of the same sex; a normal feature of cooperative social interaction behaviors in hundreds of species, but considered by some, on religious grounds, to be perverse and evil among humans (www.seedmagazine. com/content/article/the_gay_animal_kingdom)

Human Genome Project: international cooperative scientific effort to learn the entire DNA base-pair sequence structure of the human complement of DNA, toward improved understanding of its functions http://en.wikipedia.org/wiki/ Human_genome

hydrophilic: "water-loving"; physical property of molecules that can form hydrogen bonds with water molecules, leading to stable solution in water and other polar solvents

hydrophobic: "water-fearing"; physical property of molecules that are repelled by/from water; preferring neutral and nonpolar solvents; hydrophobic molecules in water tend strongly to cluster together to minimize their total contact with water

hypoplastic left heart: congenital heart defect; a malformation in which the whole left side of the heart (including the aorta, aortic valve, left ventricle and mitral valve) is underdeveloped; invariably lethal without prompt treatment

identical twin/s: there is no such thing; for the closest thing we have, see "monozygotic"

ignorance: not knowing; state of lacking knowledge, information or awareness

implantation: the attachment of the blastocyst to, and its penetration of, the epithelial surface of the lumen of the uterus and, in humans, its embedment in the endometrium, normally occurring on the sixth or seventh day after fertilization

imprinting, genomic: parent-of-origin-specific gene expression, whereby the expression of a gene DNA sequence in development depends on which parent provided which copy of the chromosome in question. A major component of the system of epigenetic functions involved in control of gene expression, "genome imprinting" is in general set during gametogenesis, and plays a suite of critical roles in embryogenesis and fetal development. The primary underlying physical mechanism appears to be variation in availability of DNA for transcription, probably modulated by variations in chromatin structure. The "setting" of these expression "switches" is heritable, passing down through progeny of that particular DNA strand by resetting the modification on daughter strands during or after every replication of that strand. Discovered in experiments with swapping pronuclei in fertilized mouse egg cells — only with a proper set of one paternal and one maternal pronucleus can embryogenesis proceed to a proper functional result, caused by the differential modification of the DNA in the gamete

independence, statistical: in a condition of statistical independence, the outcome of any one event or process in the system under analysis has no effect on, and offers no predictive information about, the outcome of any other such event or process

Independent Assortment: second of Mendel's Principles; the alleles at any one genetic locus segregate in meiosis independently of the alleles at any other locus; true of pairs of loci on different chromosomes or on the same chromosome at such a distance that the loci are not genetically "linked"; exceptions to independent assortment are all situations of "genetic linkage", where the loci are close enough together on the same chromosome that they will not reliably be separated by recombination and will therefore tend to segregate together at least part of the time

induction, embryonic: in embryonic induction, cells of one type trigger changes of the form or function in other cells or cell types; subject of Hans Spemann's 1935 Nobel Prize for Medicine or Physiology

infant: the child in the first year after birth

iniencephaly: a rare major neural tube defect; malformation in structures of the cervical spine resulting in extreme backward bending of the head, which is often disproportionately large; the neck is usually absent, with skin of the face and scalp attached to the skin of the chest and back; usually associated with other anomalies such as anencephaly, hydrocephalus, cyclopia, absence of the mandible, clefting of the lip and palate, congenital heart defects — all of the structures depending on the cervical and posterior cranial neural crest cells are disrupted by the absence of those cells

insemination: the deposition of semen into the female reproductive tract, normally making sperm cells available for fertilization of the egg cell

intron: an untranslated "intervening sequence" in a primary messenger RNA; will be removed by splicing during preparation of the mature messenger RNA

iris, *pl.* **irides:** the colored ring around the pupil of the eye; normally functions by dilation and contraction to control the amount of light entering the eye; absent or defective in **aniridia**

James, William H. (Bill): veteran researcher in relationships between the circumstances and the outcomes of fertilization and conception — particularly twinning, sex fraction at birth, and environmental factors which may influence those variables

Keegan, Karen: a teacher in Massachusetts; found to be chimeric after genotyping for purposes of a kidney transplant returned results indicating that she is not the mother of two of her three sons; no evidence in her appearance or history indicated her chimerism; a second cell line compatible with being the mother of the two disputed sons was found in some old surgical samples — representing her unborn dizygotic twin sister

kinesthesia/kinesthetic: a sense mediated by receptors located in muscles, tendons, and joints and stimulated by bodily movements and tensions; *also*: sensory experience derived from this sense; also known as **proprioception/ proprioceptive**

labile: constantly undergoing or likely to undergo change; unstable

Landsteiner, Karl: 1868–1943; Austrian anatomical pathologist, physiologist and immunologist; won 1930 Nobel Prize for discovery of the ABO blood groups and elucidation of the physiology thereof; also discovered the Rhesus blood group and a number of others

largemouth bass: *Micropterus salmoides*; perciform fish of the sunfish family; a favorite of sportfishers because of its aggressive feeding and dramatic, and often successful, efforts to escape the hook; top of the North American freshwater food chain, with the consequence that many stocks are presently carrying excessive accumulations of mercury in their tissues

Le Douarin, Nicole: 1930–; French developmental biologist; currently director of the Embryology Institute at the CNRS (Centre National de la Recherche Scientifique, French counterpart of US NSF-NRC); distinguished career making and studying quail-chick chimeric embryos, focusing on developmental functions of neural crest cells, especially in the nervous, blood and immune systems

Lewontin, Richard: 1929–; American leader in development of statistical population genetics and evolutionary theory; brought methods of molecular biology into evolutionary considerations; also noted for social commentary in several directions, especially concerning his argument against there being any genetic basis (in allele frequency distributions alone) for the social construct "race"

likelihood: in statistical genetic analysis usage, the joint probability of an entire outcome (such as a whole pedigree structure), which varies as a function of varying hypotheses about the underlying mechanism

likelihood ratio: the ratio of likelihoods under different hypotheses, the maximum value of which (the maximum likelihood) can be taken to indicate the hypothesis which best fits the observations

lipid bilayer: a structure characteristic of cell and organelle membranes, forming the main framework of membrane structure from two layers of phospholipid molecules, with the hydrophilic phosphate groups on the outside facing the aqueous environment and the hydrophobic lipid tails on the inside away from the water

lizard: scaly reptile of the suborder Sauria, differing from snakes by having four legs, movable eyelids and external ear openings; differing from salamanders by being reptiles instead of amphibians

locus, *pl.* **loci:** Latin, place; the position of a gene in a linkage map or on a chromosome; the DNA sequences in a particular place in the genome of interest because of one or more genes located there

longitude: distance, in degrees, east or west of the Prime Meridian through Greenwich, England

lumbar: pertaining to the loin region of the vertebrate skeleton; posterior to the rib cage and anterior of the pelvis

luteal phase: that part of the non-conceptive menstrual cycle after ovulation

lysosome: a subcellular organelle of eukaryotic cells, containing acid-hydrolase digestive enzymes, the function of which is the destruction and elimination of waste materials

macrophage/s: from Greek "big eater"; large white blood cells derived from monocytes, acting in nonspecific defense characteristic of innate immunity and in cell-mediated immunity's specific defense; operate by engulfing and digesting cellular debris and pathogens, and by presenting antigens from destroyed pathogens to lymphocytes and other immune system cells to stimulate the specific immune responses of those cells to the pathogen

malformation: birth defect; congenital anomaly; specific structural defect of an organ or any part of the body due to abnormal or defective developmental processes; a developmental error resulting in faulty development of a body structure

mandible: the movable lower jaw

mandibular process: either of a pair of embryogenic structures arising from the first branchial arch and driving toward fusion in the midline to form the primordial structure of the mandible

mantra: Sanskrit word, literally meaning "mind protection," from roots *man* "to think" and *trai* "to protect (the mind from intrusive mundane conceptions)"; sacred syllable or word, or set of syllables or words through the repetition of, and reflection upon, which one attains perfection or a desired state of mind; a verse, phrase or syllable believed to have particular religious significance and power

marker, genetic: a DNA sequence known to be variable in the population due to mutation and polymorphism, used as a point of reference in genetic analyses, particularly in mapping or in forensic genetic identification; there are several thousands of such sequences scattered throughout the genome and known to be available for such analyses

Matheny, Adam: psychologist and behavioral geneticist; longtime director of the Louisville Twin Study

maxilla: the upper jaw

maxillary process: either of two derivatives of the first branchial arch, which grow toward the midline to fuse with the frontonasal process, forming the basic structure of the maxilla

mechanochemical: of or relating to conversion of chemical energy into mechanical work; of or pertaining to molecular structure and/or catalytic capabilities of the living cell

meiosis: the "reduction division"; a sequence of two highly specialized cell divisions required for the generation of haploid gametes from diploid germ line cells; results in egg and sperm cells having single copies of each chromosome with which to reconstitute the diploid nucleus of the zygote

meninges (*pl.*)**:** (proper singular is **meninx**)**:** three layers of protective membranes surrounding the brain and spinal cord; outermost is dura mater, the toughest of the three; innermost is pia mater, most delicate of the three; in between is the arachnoid; inflamed in meningitis

menses: the menstrual cycle, often used particularly for the menstrual period

menstrual cycle: cyclical, hormone-driven process of preparing the uterus for the possibility of implantation and conception by building up the endometrium, and sloughing that lining after the opportunity for conception has passed without result

menstrual cycle length: the names "menses," "menstrual," "menstruation" derive from the similarity of cycle length to the length of the lunar cycle; commonly erroneously understood to average 28 days, the average is 29.5 days

mesenchyme: has been called "embryonic connective tissue," but connective tissues of various sorts occupy a relatively small part of the list of tissue types derived from mesenchyme; tissue derived primarily from the mesoderm, the middle layer of the trilaminar stage of the embryo; contains fibroblasts and collagen bundles and is in fact the origin of many forms of connective tissue, but also of many other cell types — especially from neural crest mesenchyme

mesiodistal: from the middle to the periphery, as from sternum to fingertips; of dental diameters, measured across the crown of the tooth along a line from the midline (between the central incisors) along the dental arch, to the rearmost surface of the molars; perpendicular to the buccolingual dimension

mesoderm: the middle layer of the trilaminar stage of embryogenesis; precursor primarily to muscle and bone

messenger RNA: single-strand RNA, transcribed from DNA, spliced into final appropriate sequence by removal of noncoding sections and joining of its coding sections into a continuous mature message, which may then be translated into amino acid sequence of protein gene products

meta-analysis: a statistical process in which the results of multiple prior statistical analyses can be combined to generate a more robust result than was achieved by any of the results previously obtained

microarray: molecular genetics research device consisting usually of a glass slide spotted with microscopic dots of specific DNA gene sequences in a precisely known arrangement. When incubated with RNA extracted from living cells, the messenger RNA present in the extract (varying according to what genes were being expressed in the cells at the time of the extraction) binds to the specific sequences on the microarray representing the gene from which the message was transcribed. Quantification of the binding on each of the spots can tell us which genes were being read. Comparison with extracts from cells in a different physiological or developmental state can tell us what genes change their expression between the different states

MTOC, microtubule organizing center: in eukaryotic cells, generally a fuzzy, morphologically somewhat indistinct organelle from which microtubules appear to emerge; involved in organizing microtubules and other cytoskeletal elements involved in form and function of flagella, cilia and the spindle which separates the replicated daughter chromosomes during cell division; in human cells, essentially congruent with the centrosome, comprising the centrioles and the pericentriolar "bodies" and other pericentriolar material

midline: sagittal (longitudinal) plane through the center of the body, dividing the body into approximately equal left and right halves

mikveh (also transliterated from Hebrew as **mikve, mikvah**): ritual bathing pool in which a person immerses as part of conversion or ritual purification. This pool and its water are precisely prescribed by Jewish law. Immersion is the common core component of every traditional Jewish conversion process. Throughout Jewish history, unmarried women have immersed in the mikveh prior to their wedding; married women immerse at the end of seven days of stainless purity from the end of each monthly menstrual cycle, in preparation for the resumption of family relations in the most fertile days of their cycles

Mintz, Beatrice: (born January 24, 1921); a pioneer in the use of chimeric mice for studies of developmental biology and genetics

miracle: any occurrence which cannot be explained from existing knowledge or technology, and which therefore might be imagined or superstitiously considered to be supernatural in origin

mirror-imaging: commonly identified with discordance for handedness between co-twins, and attributed to disturbances of asymmetry development in embryogenesis by the "splitting" that is imagined to have initiated the twinning event; has some basis in reality that is not at all represented in this concept as stated here — twins of both zygosities equally have left and right side patterns in their craniofacial development which lack the usual left-right differences observed in singletons

mirror twins: a pair of twins in whom "mirror-imaging" is thought to have occurred

miscarriage: spontaneous abnormally early end of a recognized pregnancy

mitochondria, *sg.* **mitochondrion:** organelle/s in which most of the cell's energy-yielding metabolism takes place; has its own DNA in a small circular "chromosome" with genes not present in the nucleus; biochemically more like bacteria than the rest of the cell

mitosis: "ordinary" cell division process underlying growth by increase in cell number, or turnover in tissues

molar pregnancy: false pregnancy in which the "conceptus" is a mole

mole, hydatidiform mole: a "conceptus" constituting an anomalous growth of trophoblast tissue, implanted in the endometrium and proliferating, without the embryogenic inner cell mass of a normal embryo; may be "complete mole," in which the abnormal tissue contains only paternal genetic contributions, or "incomplete" or "partial," in which genetic material in the mole represents both parents; origin of complete mole from "empty egg" with endoreduplication of the sperm's genetic contribution, is a fable, based on the apparent absence of a maternal genetic contribution when the growth is analyzed after delivery; in fact a sizable fraction (up to half) of complete moles have contributions from two different sperm cells; partial moles are usually triploid, requiring either dispermic fertilization or "retention of the second polar body" or anomalous duplication of the maternal pronucleus. Causes of the condition are even less well understood than the mechanisms by which they are generated

molecule: the smallest indivisible unit of a chemical compound substance composed of more than one chemical element or of more than one atom of a single chemical element

mollusk (also **mollusc**): member of the invertebrate phylum Mollusca; includes bivalves (two-shelled) such as clams and oysters, univalves (single-shelled) or gastropods such as snails and abalones, and the cephalopod octopus, squid and cuttlefish species; the cephalopods are the most neurologically advanced and sophisticated of all invertebrates — their eyes rival ours in functionality

monoembryonic: derived from a single embryo; often used inappropriately as synonym for monozygotic

monozygotic: derived from a single zygote

monster: according to Aristotle "a mistake of purpose in nature" (including twins); later, an animal with a malformation or birth defect; into English from Latin *monstrum*: "that which is shown, omen, portent, sign"

morula: Latin, berry; the embryonic stage resulting from cleavage of the zygote; a solid ball of up to perhaps as many as 32 apparently undifferentiated cells, prior to the differentiation of the outer layer of those cells into the trophoblast separating from the inner cell mass

mosaic: in genetics, the monozygotic counterpart of a chimera; an individual with two or more genotypically different cell lines derived from a single original zygote by way of a mutation early in embryogenesis; this is exclusively a cytogenetic diagnosis, assigned to individuals with mixtures of normal and cytogenetically abnormal cells, the most common of whom are partially trisomic

mosaic embryo: a type of embryo in which specific cells have specific futures, where future developments may be specifically disrupted if a specific precursor cell is damaged or killed: as opposed to **regulative embryos**, which can "regulate" and rearrange development to compensate for a missing cell or cell line

Mosaic Law: Law of Moses; guiding rules of Judaism

motor nerve: a nerve the function of which is to cause movement

multifactorial: a major model of genetic transmission, postulated to involve "many genes" "of small, equal and additive" effect, plus "environmental" contributions; generally considered the primary mechanism for inheritance of continuous or quantitative traits, such as height, weight, blood pressure and IQ; also considered the primary mechanism of behavioral inheritance, for explaining variations that are clearly genetically familial, but which cannot be adequately modeled by any single-gene mechanism or any mechanism involving a small number of interacting genes

multiple regression: a multivariate statistical technique for testing the relative impacts of multiple independent predictor variables on a dependent outcome variable

multivariate: of statistics, involving multiple variables

mutation: any heritable change in a DNA sequence

mystery: that which is, for the time being, unknown, incomprehensible to reason and rational understanding; closely related to miracle

NINCDS, National Institute of Neurological and Communicative Disease and Stroke: a member institute of the US National Institutes of Health

NOMOTC, National Organization of Mothers of Twins Clubs: in the USA, an umbrella organization of support groups for parents of twins and higher multiples; has numerous counterpart organizations in other countries

nematode: roundworm; the most numerous of multicellular animals; mostly microscopic or nearly so; unsegmented worms with spindle-shaped bodies; prime subjects for studies of developmental genetics

neotenous: characterized by neoteny

neoteny: in development, a process that carries into the mature organism features formerly (in evolution or in development) seen only in the immature organism; may be considered a process of advancing sexual maturation into states of development previously characteristic of the immature; in the human, invoked with respect to the larger adult brain-to-body-weight ratio formerly seen only in the juvenile; thought to have driven evolution of a relatively larger birth canal and delayed fusion of cranial sutures allowing benign and reversible deformation of the head during delivery, perhaps among other features which allow delivery of larger-brained infants

neural crest: a population of stem cells long thought to be induced by specific mesodermal cells, in the neuroectoderm at both lateral edges of the neural plate, although recent evidence suggests that they may be specified much earlier and independently of both mesoderm and neuroectoderm. When the edges of the neural plate fold and roll to the dorsal midline and fuse to form the neural tube, neural crest cells migrate from the crests of the neural folds for great distances along specific paths, and differentiate into a fabulous array of cell types. Derivatives include at least most of the cells of the peripheral nervous system, the pigment-producing cells, endocrine cells, hair follicles and the cartilage and bone of the head, face and neck — particularly including the bones of the ears. Crucial to proper development of the heart, the brain, and the teeth. Neural crest stem cells found in adult hair follicles have been found able to differentiate into neurons, nerve supporting cells, cartilage/bone cells, smooth muscle cells, and pigment cells

neural tube: embryonic precursor to the central nervous system, formed from the neural plate by fusion of its lateral edges in the midline

neural tube defect, NTD: any of several midline/fusion anomalies in the formation of the neural tube caused by failure of complete or proper closure of the neural plate rolling up to form the neural tube in the midline

neurotransmitter: molecule which carries nerve impulses from one nerve cell to another across synapses; acetylcholine, epinephrine and norepinephrine, serotonin, histamine, gamma-amino butyric acid, dopamine and glutamine are among the better known

neurula: embryonic stage when the neural tube forms

neurulation: formation of the neural tube, formation of the neurula stage of embryogenesis

Newman, HH: zoologist; student of the biology and physiology of human twins and twinning, and of the use of twins for genetic analysis, most active in the second and third decades of the 20th century

newt: salamandrid amphibian, with characteristic vertebrate body plan; lays eggs singly in water, vs clumps for frogs and strings for toads; popular subject for embryological studies because of the relative ease of observation

Nigeria, Federal Republic of: West African nation bordered by Benin to the west, Chad and Cameroon in the east, Niger in the north, and the Gulf of Guinea in the south; most populous nation on the African continent at ~140 million; human habitation since at least 9000 BCE; contains subpopulations with the world's highest frequency of twin births

nonrighthanded, nonrighthandedness: lefthanded or ambidextrous; not having a strictly right-sided motor hand preference; a minority of the human population, roughly 10%, whose brains are less strictly functionally lateralized

nucleic acid: first discovered and named in the form of the acidic contents of the nuclei of living cells; a polymeric compound of nitrogenous bases (adenine, guanine, cytosine, thymine and uracil), five-carbon sugar molecules (ribose or 2-deoxyribose), and phosphate groups. Two major forms, DNA and RNA; D- with deoxyribose, R- with ribose; DNA uses thymine, RNA uses uracil in place of thymine; DNA exists mostly in a helical arrangement of antiparallel double strands. DNA is primary molecular repository for transmission of genetic information between generations in most eukaryotes; RNAs of various kinds function primarily in various mechanisms required for the expression of the information in DNA

nucleoside: compound of nitrogenous nucleic acid base and sugar ribose or deoxyribose; adenosine and deoxyadenosine, guanosine and deoxyguanosine, cytidine and deoxycytidine, thymidine (usually with deoxyribose), uridine (usually with ribose)

nucleotide: phosphorylated nucleoside;
    monophosphate: __MP; AMP, dAMP, GMP, dGMP, etc.
    diphosphate: __DP
    triphosphates: __TP

nucleus, *pl.* **nuclei:** the largest of the intracellular organelles; the intracellular repository of genetic information in nucleic acids; the site of synthesis of most RNAs

odontoblast: cell, member of cranial neural crest mesenchyme population; on outer surface of dental pulp; creates dentin

ommatidium, *pl.* **ommatidia:** individual component unit of arthropod compound eyes

oöcyte: the egg cell

oöcytoplasm: the cytoplasm of the oöcyte

oögenesis: development of the egg cell from its precursor oögonium, complete at ovulation

oöplasm: oöcytoplasm

organelle: often membrane-bounded substructures, with specific subsystem functions within cells; the major organelles and cellular structures include: nucleus, nucleolus, ribosomes, several sorts of vesicles, the endoplasmic reticulum (smooth and rough), Golgi apparatus, cytoskeleton (including centrioles and spindle apparatus in dividing cells), mitochondria, several sorts of vacuoles, lysosomes

Organizer, Spemann's: a patch of cells that forms on the dorsal lip of the blastopore in amphibian embryos, homologous to cells at Hansen's node of the avian and mammalian primitive streak; these cells migrate under the ectoderm to form the mesoderm/mesenchyme inducing the neural plate and numerous other specific cell-type differentiations

orthodox, orthodoxy: from Greek "right thought" or "right teaching"; typically used to refer to correct worship or correct doctrinal observance in religious or intellectual activity shared by an organization or movement, usually as determined and promulgated by a governing body

osteoblast: cell of neural crest mesenchyme or mesodermal origin which functions to form bone

ovary, *pl.* ovaries: egg-producing reproductive organ of the female; the female gonad; also produces numerous hormones, including: estrogens, progesterone, follicle stimulating hormone, luteinizing hormone, and testosterone

ovulate, ovulation: release of the mature oöcyte from the ovary

palate: roof of the mouth in vertebrate animals, separates the oral cavity from the nasal cavity, consists of the anterior bony hard palate and the posterior soft palate

palatal shelves: embryonic precursors of the palate, flat structures arising from the maxillary processes, grow up and over the tongue, reaching toward fusion in the midline — anteriorly with the primary palate from the frontonasal process and posteriorly with each other

palpebral fissure: the openings in the skin for the eyes, between the eyelids; normally about 10 mm vertically, 30 mm horizontally; may be reduced in size in fetal alcohol syndrome

papilla: several uses in anatomy; generally a nipple-like structure; herein, the dental papilla is a collection of neural crest mesenchymal odontoblasts in the developing tooth lying below the enamel organ; gives rise to dentin and pulp

paradox (paradoxical): (pertaining to) an apparently true statement that leads to a contradiction or other counterintuitive outcome; OR, an apparent contradiction that is not in fact a true contradiction; the statements at issue may not truly imply the supposed contradiction, the counterintuitive result is not in fact a contradiction, or the premises cannot all be true at once

parity: the number of children a woman has delivered

pedigree: a diagram representing a family structure; particularly, in genetic analysis or genetic counseling, representing the segregation of some trait/s in the family

penetrance: a finagle factor in genetic analysis, used to explain apparent departures from normal Mendelian segregation of a trait in a pedigree. If and when one or more individuals, in direct line of descent, having both affected ancestor/s and affected descendant/s and therefore obligate transmitter/s of the trait and the gene causing it, do not express the gene and might thus be supposed not to possess the gene — that is a case of incomplete penetrance. "Penetrance" is a fraction, with denominator equal to the number of people in the pedigree who must have the gene — because they are in direct line of descent — and with numerator equal to the number of those people who express the trait. Penetrance may vary by sex, by age, by parity, etc., such that penetrance may be (more or) less complete in males, or in younger mothers, or in mothers of fewer children

peptide: a polymer of amino acid monomers; an oligopeptide if only composed of a few amino acids; a polypeptide if composed of many amino acids; a protein molecule

periodontal: surrounding a tooth; concerning tissues and structures surrounding a tooth

Peters' anomaly: eye malformation resulting in opacity of the central cornea associated with lens-cornea adhesion; one of numerous expressions of various PAX6 mutations

pharyngeal arches: also called **"branchial arches"**; embryogenic substructures on each side of the neck, including gill arches in fish embryos; interacting extensively with cranial and postcranial neural crest mesenchyme as precursors to numerous cartilage, bone, muscle and nerve structures of the head, face, jaws, ears, throat, neck; including maxilla, mandible, pharynx, larynx, trachea; trigeminal, facial, glossopharyngeal and vagus nerves; carotid, subclavian, aortic and pulmonary arteries

phenome: the system of all phenotypes of an organism; construction as in genotype/genome

phenotype: the appearance of an individual organism, generally with respect to a specific trait

pheromone: any of a large and diverse group of species-specific volatile hormones, with a fascinating variety of chemical structures over all the species that use them and all the things they are used for. They are released into the environment from various parts or surfaces of the body, to be perceived by chemical senses (not including the conscious sense of smell in the human), by means of which many organisms communicate about the state of affairs between them and to trigger specific behavioral responses appropriate to the circumstances — alarm, ovulation, copulation, availability and location of food, etc. Human pheromones identified to date are apparently secreted primarily by modified sweat glands or sebaceous glands in the axillae — the armpits and the groin

philtrum: variable-depth V-shaped groove in the upper lip, from the nasal sill and columella to the vermilion border of the lip; formed by fusion of the most ventral extent of the frontonasal process with the maxillary process on each side

phylogeny: the history of organismal lineages as they change through time; often represented as a tree – "The Tree of Life"

pilonidal cyst: a relatively minor to minimal neural tube defect of the posterior neuropore; generally discovered as an abscess in the cleft between the buttocks

pineal gland: named for its pine-cone shape, pea-sized endocrine gland hanging between the brain hemispheres, in the roof of the third ventricle, beneath the dorsal end of the corpus callosum; declared by Rene Descartes to be the single point of connection between the mortal body and the immortal soul; influenced by a great variety of neurotransmitters; receives neural inputs from multiple sources, apparently including the superior cervical ganglia, the pterygopalatine ganglion, and the trigeminal ganglion; synthesizes and secretes serotonin, norepinephrine, and melatonin; synthesizes melatonin from serotonin in a rhythmic pattern generally inverse to the daylight cycle; generally contains thyrotropin releasing hormone, luteinizing hormone release hormone, and somatostatin release inhibiting factor; may also have paracrine functions involving parts of the pineal body regulating activities of other parts

pituitary gland: *"the Master Gland," "conductor of the endocrine orchestra"*; secretes hormones controlling several of the other endocrine glands including at least the thyroid, the gonads, and the adrenals; controls the menstrual cycle with follicle stimulating hormone, leutinizing hormone; also contributes vasopressin, oxytocin, antidiuretic hormone, thyrotropin releasing hormone, follicle-stimulating hormone, adrenocorticotropic hormone, luteinizing hormone, melanocyte-stimulating hormone, prolactin and growth hormone; of double embryonic origin — cells from embryonic gut and embryonic brain move together (your tongue can probably feel a small pit in your hard palate in the middle of the roof of your mouth, called "Rathke's Pouch" — origin of the gut cells that migrated to meet cells coming from the floor of the neural tube, to form the pea-sized pituitary hanging on a stalk from the hypothalamus)

placenta: temporary vascular organ joining mother and fetus; built from maternal and fetal components, induced by invasion of the endometrium by the polar trophoblast of the blastocyst embryo

placental anastomosis: fusion of blood vessels between twin placentas; routine in monochorionic twin pregnancies; extremely rare in dichorionic pregnancies of either zygosity

placental fusion: coalescence of twin placental masses into a single mass; happens in about half of all dichorionic twin pregnancies, regardless of zygosity

placentation: formation of a placenta in the uterus; the type or structure of a placenta, according to the number of chorionic membranes attached, or the point of insertion of the umbilical cord

plankton/planktonic: incalculably numerous drifting organisms moving only with currents in the pelagic zone of oceans or bodies of fresh water

polyamine: any of a variety of basic (as opposed to acidic) organic compounds, such as putrescine, spermine or spermidine, characterized by multiple amino groups; important in the extreme condensation of DNA in the sperm head

polygenic: multilocal; of a trait — determined by multiple genetic inputs

polymer: a large molecule composed of multiple (monomer) subunits, most commonly in linear arrays; biological examples include nucleic acid polymers of nucleotides and protein polymers of amino acids; silk and spider webs are proteins, most natural plant fibers are cellulose — polymers of glucose; synthetic fibers and plastics such as nylon, polyesters, polypropylene, polyethylene, etc., can be made from many different simple monomers that will react to form long chains

polymerase: an enzyme which catalyzes a polymerization reaction, in which monomeric units are covalently bound into long polymeric chains; here, particularly the enzymes which assemble DNA and RNA molecules, DNA polymerases and RNA polymerases respectively

polymerase chain reaction, PCR: a procedure for amplifying specific nucleic acid sequences, particularly of DNA. Useful for many analytic procedures. Requires primer sequences matching specific base sequences on either side of that part of the whole DNA sequence which is to be amplified. Also requires template DNA molecules, heat-stable DNA polymerase enzyme, and supplies of nucleotide triphosphates. Double-strand DNA to be amplified is melted to single "template" strands in high temperature; temperature is reduced to allow primers to bind specifically to matching sequences on the templates and for polymerase to extend the primer sequences by polymerizing onto the end of the primer the sequence of the strand complementary to the template strand. This makes for short regions that are double stranded, to the ends of which the DNA polymerase may add single nucleotides, elongating the primer and generating the strand complementary to the template sequence. After a reasonable interval for polymerization, the newly "replicated" double strands are separated by melting at high temperature and the process repeated, doubling the concentration of the target sequence in each repetition of the cycle — each 10 cycles amplifies about a thousand-fold

postdoctoral fellow: journeyman scholar in training/mentorship beyond the doctoral degree; commonly involves moving from the research area of the doctorate into another area which may occupy the new scholar's attention for an independent career

posterior: behind, toward the tail; opposite of anterior

potential: possibility of doing or becoming something that is not a matter of present fact, imagined or otherwise projected from present appearances and recollection of previously observed entities to which the one in question appears to be similar

pregnant, pregnancy: carrying a fetus

preimplantation genetic diagnosis: extreme version of prenatal diagnosis, performed in the course of *in vitro* fertilization and embryo transfer, by removal of one cell from a morula embryo for genetic analysis prior to transfer to the uterus for implantation

prenatal diagnosis: genetic testing of a fetus

primer: a short (usually 15 to 20 bases) nucleic acid sequence, specifically complementary to a sequence flanking a target sequence for polymerase chain reaction, used to bind to template DNA to provide an initiation point for polymerization of the strand complementary to the template

primitive streak: a key structure in embryogenesis, which forms in the posterior midline of the ectoderm or epiblast layer of the bilaminar stage of embryogenesis; and which generates, from its "node"; the cells of the intraembryonic mesoderm; the mammalian counterpart of Spemann's organizer in amphibian embryos

primordium, *pl.* **primordia:** cell or (more usually) group of cells from which a particular embryonic structure may be expected to arise; the earliest recognizable stage of the development of an organ

primordial: having to do with, or having the nature of, a primordium

principal components analysis: a multivariate statistical technique of the multiple regression family, which extracts statistically independent (non-overlapping and noncorrelated) linear combinations of the data variables, representing successively smaller fractions of the multivariate variance of the dataset

probability: the chance of a particular event occurring, expressed as a unitless number p in the range $0 \leq p \leq 1$; an impossible event has probability zero; an inevitable event has probability one

prochordal plate, prechordal plate: midline patch of endodermal cells in contact with ectoderm near the anterior end of the long axis of the bilaminar stage embryo; anterior target of the migration of the notochordal process within the mesoderm as the mesoderm is formed from the primitive streak between the ectoderm and endoderm; the plate itself will remain two cell layers thick, becoming the pharyngeal membrane when the head folds ventrally toward the belly and tail

product, gene product: may be an RNA transcript of the gene DNA sequence in question, used as such; may be a protein translated from a messenger RNA; numerous RNA products are structural (the family of transfer RNAs, the ribosomal RNAs) or regulatory

profile, DNA: most commonly mentioned in forensics contexts; a highly-specific individual pattern of specific alleles at various tested loci; output of genotyping

DNA extracted from a sample in evidence, for purposes of identifying the individual source of that sample

pronucleus, *pl.* **pronuclei:** either of two (maternal and paternal) haploid nuclei in the fertilized oöcyte; these will meld in the process of syngamy to form the diploid nucleus of the zygote, which will soon take over the guidance of embryogenesis

prosencephalon: most anterior of the three segments of the early embryonic brain, divides into the telencephalon (precursor to the cerebrum) and the diencephalon (precursor to all components of the thalamus and the pretectum); failure of the prosencephalon to divide properly into left and right half structures produces the defects of holoprosencephaly

protamine: highly basic protein that replaces histones in the condensation of DNA for packaging in sperm; allows for substantially greater condensation of DNA (e.g., for packaging in a sperm head) than is possible with histones

proteasome: an organelle of intracellular digestion, specialized for degrading endogenous proteins that are damaged, misfolded, beyond their "useful lives" or otherwise extraneous, to recycle their amino acids for new protein synthesis

prototype: original, typical form or example of a thing; the basis or standard for other instances of the same

pulmonary artery: major artery carrying deoxygenated blood from the right side of the heart to the lungs to be re-oxygenated

pulp (of a tooth): live soft tissue in the center of a tooth

purine: organic nitrogenous base; nine-member double aromatic ring structure (pyrimidine plus imidazole rings sharing a two-carbon side) of five carbon and four nitrogen atoms; includes adenine and guanine

pyrimidine: organic nitrogenous base; six-member aromatic ring structure of four carbon and two nitrogen atoms; includes cytosine, thymine, and uracil

quail: small chicken-like bird, used to profound effect in quail-chick chimeric embryo experiments originated by Nicole Le Douarin, on the migration and functions of neural crest cells in the development of nervous and immune systems

quantum entanglement: see *entanglement*

quickening: the onset of recognizable fetal movement, usually in the 14th to 16th week of gestation; a traditional criterion for maternal recognition of pregnancy, and thus for "conception"

RNA: see *ribonucleic acid*

RNA polymerase: see *polymerase*

race: a taxonomic concept in population genetics that, largely for want of understanding of the differences between its valid biological meaning and various egregious social abuses thereof, has become burdened with intensely difficult and noisy political connotations. In population genetics, race is a sound probabilistic concept representing differences among subpopulations in sets of genetic variations that are more or less coherent within, and different among, the subpopulations in question. Never absolute, never without some degree of overlap among the subgroups, never without the variation among the subgroups being imagined as a relatively minor portion of all variation in the population as a whole, such distinctions usually arise as consequences of geographical separation and the consequent random genetic drift (especially if the geographical separation began with a "population bottleneck" — a relatively small subset of the parent population). When selective pressures differ among the various environments occupied by the various subpopulations, the level of differentiation among such subgroups may proceed to speciation. In the human population, there are at least three major such groups (major continental subpopulations) which differ quite sufficiently to qualify for this definition. A history of social abuses predicated upon severe distortions of the meanings of the real differences among these groups, and misapplication thereof for justification of social wrongs, has left an evil stink around the whole concept. It is a simple fact, as a consequence of the nature of the underlying biology, that any given individual may be a member of a particular racial group according to an overwhelming majority of the statistical differences and not score among the others within that group with respect to any one given variable. Proper medical application of observed differential liability to various diseases and pharmacogenetic differences in drug responses may be delayed while we work out our concerns for the fact that group membership is a statistical shortcut which takes nothing away from the fact that each human individual is unique

recombination: the genetic result of breakage-and-reunion (crossing-over) exchanges between maternal and paternal homologs of each chromosome during meiosis; rearrangement of parental combinations of alleles at linked loci

recombinogenic: given to attracting or inducing recombination events

red herring: from fox hunting lore; a smoked herring (coppery red in color), dragged across the trail of the running fox could be depended upon to make dogs lose the scent trail; a fallacy in which irrelevant argument is used to distract, to divert attention from the real issue at hand; an argument in which premises are introduced which are logically irrelevant to the conclusion

reflection: self-contemplation, meditation, rumination; recursive mental operation, in which the mind/self considers its own behavior

reflex ovulation: ovulation in response to stimuli associated with coitus or insemination

regulative embryo: vs **mosaic**; embryo in which a missing or damaged component can be repaired or replaced by remaining cells

rhythm method: a method of avoiding or encouraging the initiation of pregnancy by appropriate timing of intercourse

ribonuclease (RNase): any of a large family of enzymes which cut RNA molecules

ribonucleic acid, RNA: polymeric nucleic acid composed of ribose-phosphate backbone and the nucleic acid bases guanine, cytosine, adenine and uracil; may have been the first nucleic acid to evolve; usually single stranded, but readily forms many different sorts of secondary structure which sometimes include double-stranded sections; performs many functions in the cell: messenger RNA is transcribed from the DNA, and is read by the ribosomes which are composed of both RNA and proteins. Transfer RNA is the adapter molecule (in a specific form for each different amino acid — codon combination) which carries each amino acid into the ribosomal translation complex to match with the proper three-base "codon" sequence on the messenger RNA. Some RNA molecules have catalytic properties, like enzyme proteins; these are called "ribozymes." RNA has —OH hydroxyl groups on both the 2' and 3' carbons of the ribose sugars in its chain. It is more easily hydrolyzed (chemically broken) than DNA, which is more stable because it lacks the 2' hydroxyl

salamander: an amphibian with legs and tail

schizophrenia: severe, chronic disabling psychotic thought disorder, characterized by various combinations of anhedonia, hallucinations, delusions, bizarre thought patterns, personality disintegration, distortion in the perception of reality, and psychomotor retardation

segment, segmentation, segment formation: a stage of animal embryogenesis occurring after anteroposterior and dorsoventral axis definitions, in which the length of the embryo is divided by various processes of cell signaling and differentiation into components will have different developmental futures

segregation — Mendel's "first law": the two copies of any given gene DNA sequence in any individual segregate among the offspring of that individual; the process whereby that occurs

SSRI, selective serotonin reuptake inhibitor: any of a class of drugs used primarily as antidepressants and anxiolytics, the chemical action of which is concentrated on prolonging the presence and activity of serotonin as neurotransmitter in serotonergic synapses by inhibiting its reuptake after release into the synaptic cleft

self: essence of the individual, reflective awareness of one's own identity, contents of one's own system boundaries; in immunology, that which the immune system recognizes as requiring protection, not to be attacked as foreign

self-referential: reflexive; referring to the entity doing the referring; in general having to do with logical structures

self-similar: a property of fractal structures, such that the structure is basically the same at any level of magnification

sensory: vs **motor**; of nerves, functioning to receive sensations or stimuli and transmit them to the spine or brain

septal defects: subgroup of congenital heart defects; midline fusion malformation in which the dividing walls (septa) between atrial or ventricular chambers of the heart are not fully or properly formed, leaving a hole between the chambers which disrupts the proper path of blood through the heart and lungs

serotonergic: class of neurons with serotonin as their dominant neurotransmitter

serotonin: 5-hydroxy-tryptamine, a hormone and (primarily inhibitory) neurotransmitter (communicates between nerve cells), synthesized from the amino acid tryptophan in serotonergic neurons in the central nervous system and in certain cells of the gastrointestinal tract of animals including humans; major reservoir in the body is in blood platelets; also found in mushrooms, fruits and vegetables; "happiness hormone," thought to be deficient in brains of depressed individuals; also may be involved in some forms or cases of anxiety; recently found involved in a/symmetry determinations very early in embryogenesis

sesquizygotic: hypothetical "type" of twins supposed to have developed from 1.5 zygotes, genetically intermediate between monozygotic and dizygotic, imagined to occur by virtue of sharing the same maternal half-genome with different paternal contributions; the first case ever observed was recently reported

sex ratio at birth: number of boy babies per girl baby, normally about 1.05 for white European babies, higher for babies of East Asian (including Hispanic) ancestry, lower for babies of African ancestry

sib, sibling: person with the same parents; brother, sister; half-sibs have only one parent in common

simultaneous, simultaneity: occurring at the same time

SNP, single nucleotide polymorphism: a genetic variation (polymorphism) in a species population, in which the difference among members of the population concerns a single base pair

singleton: individual product of a single pregnancy

sirenomelia: from Greek, "mermaid limbs"; "mermaid syndrome"; a rare major midline/fusion malformation in which the legs develop fused, not properly separated by pelvic structures, associated with mutations of the SHH, sonic hedgehog gene

situs inversus: inverted position of thoracic and abdominal organs; heart, lungs, liver, spleen may be on the opposite side of the body to their usual positions; relatively benign when complete and concordant, but quite dangerous and disruptive when the position of the heart is discordant with the positions of the great vessels

situs solitus: "normal" — usual, customary, majority — position of thoracic and abdominal organs

solstice: extremes of the annual seasonal variation of day length, summer solstice around June 21 being the longest day of the year in the Northern Hemisphere and shortest in the Southern Hemisphere; winter solstice vice versa around December 21; points in the Earth's orbit around the Sun when the tilt of its rotational axis is along a radius of its orbit

sonic hedgehog, SHH: one of three proteins in the mammalian *hedgehog* protein family, the others being desert hedgehog (DHH) and Indian hedgehog (IHH). SHH is the best studied element of the hedgehog signaling pathway. It plays a key role in regulating vertebrate organogenesis, such as in the growth of digits on limbs, bilaterally asymmetric organization of the forebrain (prosencephalon) and organization of the embryogenic building of the teeth

Spemann, Hans: (27 June 1869 to 9 September 1941); Austrian developmental biologist; won the 1935 Nobel Prize for elucidating embryonic induction, whereby one embryogenic cell type induces changes in form and function of other cell (type)s; identified the "Organizer" functions of a small patch of cells in the amphibian blastula embryo which move to the interior, differentiating and causing the differentiation of multiple other cell types

spermatocyte: a precursor of the sperm cell; primary spermatocyte is daughter of the spermatogonium, and divides to produce two secondary spermatocytes, which will divide to yield four spermatids

spermatogonium: a sperm stem cell, divides to produce a stem cell and the primary spermatocyte

spina bifida: a neural tube defect malformation in which the lateral edges of the neural plate fail to meet in the midline and close properly to complete formation of the neural tube; varies in severity according to which layers of the structure remain open; most common in the lumbar (abdominal) region, but also occurs in a thoracic (chest) form

spindle (fiber): fibrous microtubular protein structure, composed of tubulin, responsible for guiding and performing the proper segregation of duplicated chromosomes to daughter cells in cell division; some run pole-to-pole and elongate during division to increase the distance between the poles of the dividing cell; some run from centromeres to poles and shorten during division to pull the chromosomes apart and into the daughter cell spaces

spontaneous: not in response to any stimulus or other manipulation

spontaneous abortion: spontaneous failure and loss of a recognized pregnancy

spontaneous otoacoustic emissions, SOAE: sounds spontaneously generated by the cochlea (the spiral sensory organs of the inner ear)

splicing, RNA: in most eukaryotic genomes, most functional gene DNA sequences, and the messenger RNAs transcribed from those DNA sequences, are interrupted by sequences that are not parts of the code for the gene's protein product. The parts that will be expressed are exons, the unexpressed intervening sequences are introns. In general, the introns must be removed from the messenger RNA sequence before it can be translated into the amino acid sequence of the protein product; this is called splicing. A given gene DNA sequence may be used to code for multiple different final products by alternative splicing, wherein splices are performed at varying sets of splice sites, resulting in different mature messages to be translated

stenosis: a narrowing, generally of a tubular structure such as a blood vessel or the spinal canal

stupidity: inability to learn, for whatever reason; deliberate, therefore culpable, ignorance

subpopulation: population within a larger population, usually sharing one or more specific traits to distinguish its members from the members of other subpopulations

subsegment: segment within a larger segment

subsequence: sequence within a longer sequence

subsystem: a system which is a component of a more complex system

superstition: belief, generally arising from ignorance or fear, not based on evidence, reason or knowledge, in the significance and value of a particular thing, happening, ritual, behavior, or the like, maintained by ignorance of the laws of nature or by faith in magic or chance; particularly, an irrational belief that future events are influenced by specific behaviors, with no evidence for a causal relationship; originally used pejoratively as label for religious ritual behaviors above and beyond apparent straightforward understandings of New Testament practices; expanded to include all sorts of religious practices and beliefs that seemed unfounded or primitive in light of modern knowledge

supersystem: a system composed of lesser systems, composed of subsystems

synapse: specialized joint between nerve cells, axon of one transmitting to dendrite or cell body of the other, across which nerve impulses are transmitted by neurotransmitters released from vesicles inside the synaptic knob of the presynaptic (sending) cell, to travel across the synaptic cleft, to excite or inhibit the excitation of the postsynaptic (receiving) cell. The neurotransmitter molecules are taken up from the synaptic cleft by receptors on the postsynaptic cell membrane

synapsis: a joining, here in the context of the joining of all four strands of each duplicated chromosome in the zygotene through diplotene stages of the prophase of the first division of meiosis in gametogenesis. There is a point-to-point alignment, end-to-end, which is only recently seen as beginning from multiple crossover points along the length of the chromatid strands. An obligate event in meiosis of most organisms, without which gametogenesis fails

synaptinemal complex: a system of proteins that forms among the DNA strands of each chromosome in synapsis, overall like a double ladder in structure, with lateral elements as rungs reaching between the central element and the condensed duplicated chromosome structures

syngamy: the process of forming the nucleus of the zygote from the melding of the paternal and maternal pronuclei

system: a complex set of interacting or interdependent entities, real or abstract, forming an integrated whole, with properties or functions not present in, or predictable from, the properties of its component parts

tadpole: fishlike larval form of many amphibians

Tarkowski, Andrzej: (born 1933); a pioneer in the use of experimental chimeras in studies of developmental biology and genetcs

tautology: a logical structure in which the conclusion restates the premise/s; a statement that is true entirely and only by virtue of saying the same thing twice

template: in general, a pattern; in molecular genetics and particularly PCR, usually a single strand of DNA to be copied

teratogen: from Greek, "monster maker"; any chemical substance or physical impact which can distort a developmental process to cause birth defect or malformation

teratogenesis: the process of inducing a distorted developmental process through chemical or physical means

teratogenesis, behavioral: teratogenesis resulting in altered behavioral development

testes, testicles: the male reproductive organs/glands, site of spermatogenesis, primary source of testosterone in the male body

testosterone: an endocrine corticosteroid hormone, the most prominent of the androgens in human male physiology, derived from cholesterol, made in and secreted from the testicles

tetragametic: form of chimerism acknowledged to have arisen from four gametes, i.e., from two zygotes

tetraploid: carrying four copies of each chromosome, human tetraploids very rarely survive to term birth

theodicy: (from Greek *theos*, "god"; *dikē*, "justice"), effort to explain why or how an all-good, all-powerful, and all-knowing God permits the existence of evil in the world

third kind (third type) **of twin or twinning:** twins arising from a single egg cell and two sperm, popularly imagined to be intermediate between monozygotic and dizygotic in within-pair resemblance, by virtue of having the same maternal genetic contribution. This was an imaginary construction designed to explain twins who were declared fraternal by the delivering physician (usually because of dichorionicity alone), but who were considered to resemble each other more than should be expected of dizygotic twins

thoracic: pertaining to the thorax

thorax: the chest or rib cage portion of the vertebrate body

thymine: one of the nucleic acid bases; a pyrimidine, the main structure of which is a six-membered [4C, 2N] ring

thymus: a major organ of the immune system; single, bilobed but generally considered unpaired, and apparently derived in embryogenesis from a single midline primordium; located behind the upper end of the sternum; primary function is to host the maturation of major subsets of the white blood cells

thyroid: an endocrine organ, unpaired, but formed in embryogenesis by midline fusion of bilateral components

Tierra del Fuego: Spanish, "Land of Fire"; archipelago at the southern tip of the South American continent, governed in parts by Argentina and Chile, the southernmost inhabited land mass, separated from the South American mainland by the Strait of Magellan, the southern point of which is Cape Horn

tinnitus: "ringing" in the ears

tooth sizes: see *dental diameters*

topology: the branch of mathematics dealing with the structures of surfaces

torus: a three-dimensional ring structure with a single hole, the simplest version of which is a "donut," or any shape that can be simply transformed to that structure without changing the number of holes (this would include a teacup or pitcher with one ear, as long as the hole inside the handle is the only space, which is not a part of the solid, but which is surrounded by the solid)

transcription: the process of reading information from DNA base pair sequences into messenger RNA, which can serve as template for protein synthesis in "translation"

transcription factors: small molecules, often small proteins, which bind to DNA or to (complexes of) other DNA-binding proteins to mark DNA sequences for transcription, and modulate transcription rates by varying the speed and activity level of RNA polymerases binding to the DNA to transcribe the information into messenger RNA

transgenic: of an organism, having genetic material of another organism artificially inserted in its genome

translation: process using messenger RNA as template, and transfer RNAs specific to each amino acid — codon combination to select each proper amino acid in turn, to be assembled into polypeptide protein chains, which may serve either structural or catalytic ("enzyme") functions in the cell

transmission, genetic: the passing of genetic information between generations, between parent and child

transposition of the great vessels, TGV: misalignment of the aorta attached to the right ventricle instead of the left and the pulmonary artery attached to the left ventricle instead of the right

triploid: having three copies of each chromosome, 69 chromosomes total in the human

triploidy: state of being triploid; incompatible with survival after birth, seldom survives to birth

trisomy: state of having a third copy of one chromosome; most cases of Down syndrome are caused by trisomy of chromosome 21, for example

trizygotic: of triplets, arising from three zygotes

ubiquitin: small (76 amino acids) protein, best-known function of which is labeling proteins for degradation by proteasomes; also regulates stability and intracellular location of many proteins

ultrasound: an imaging technology commonly applied to examination of fetal condition and the progress of pregnancy

ungulate: the hoofed animals, including the artiodactyls (even-toed ungulates — cow, sheep, goat, deer), and the perissodactyls (the odd-toed ungulates — horses, tapir, rhinoceros)

unilateral: vs **bilateral**; confined to one side, as for example most cleft lips involve only one side, and that most often the left

union-intersection test: a procedure of multivariate analysis appropriate for testing the equality of multivariate means or covariance matrices when the tested samples are correlated

Universe: it seems the word was originally meant to include All of Everything in existence; all matter and energy, including the solar system around our star Sol, all the other stars in our galaxy The Milky Way, all the other stars in all the other galaxies that exist, and all the contents of intergalactic space, considered as a whole; there is new cogent reason to suppose that our Universe, as originally envisioned, may in fact be one among an infinity of such entities

uracil: one of the nucleic acid bases; takes the place of thymine in RNA molecules; a pyrimidine; pairs with adenine

vanishing twins: most twin pregnancies discovered in the first trimester of pregnancy do not result in birth of a live pair of twins

variance, phenotypic: variation among the members of a sample from a population in individual development with respect to a particular phenotypic trait

ventricular septal defect: a hole in the septum between the ventricles of the heart

vertebra, *pl.* **vertebrae:** bone/s of the neck, spinal column and tail (when present)

viscera (visceral): (having to do with) the internal organs

vitamin D: a vitamin involved primarily in the biochemistry of calcium; a corticosteroid chemically related to the androgens and estrogens and involved in regulation of reproductive function; biosynthesis from cholesterol like the sex steroids, but requires ultraviolet light on the skin

wavelength (of light): electromagnetic parameter of light related to color (red and infrared, longer wavelength; violet and ultraviolet, shorter) and inversely related to frequency and energy level of light

Weinberg, Wilhelm: 1862–1937; German general practitioner and obstetrician, whose academic work in population genetics contributed, independently of British mathematician G. H. Hardy, to the Hardy–Weinberg formula describing equilibrium allele frequencies in populations; he also contributed the fatally flawed Weinberg Difference Method for estimating the fractions of a sample of twins belonging to each zygosity group

zona pellucida: a tough, elastic, non-cellular layer surrounding the oöcyte; the "eggshell." Zona pellucida structure is species-specific and can only be penetrated by species-appropriate sperm

zygote: a single cell formed from the substance of the egg cell and the sperm; first cell of embryogenesis. The nucleus of the zygote cell is formed by the integration of a paternal pronucleus (unwrapped and reconstructed inside the egg cell from the head of a sperm cell) with the maternal pronucleus in the egg cell. The zygote nucleus is formed by the fusion, in syngamy, of maternal and paternal pronuclei. In the human, its cell-division apparatus is derived from centrioles contributed by the sperm

Index